GEOPOLITICAL TRADITIONS

Condemned as an intellectual poison by the late American geographer Richard Hartshorne, geopolitics has confounded its critics. Today it remains a popular and important intellectual field despite the persistent allegations that geopolitics helped to legitimate Hitler's policies of spatial expansionism and the domination of place. Using insights from critical geopolitics and cultural history, the contributors focus on how geopolitics has been created, negotiated and contested within a variety of intellectual and popular contexts.

Geopolitical Traditions argues that geopolitics has to take responsibility for the past while at the same time reconceptualizing geopolitics in a manner which accounts for the dramatic changes in the late twentieth century. The book is divided into three sections: first 'Rethinking Geopolitical Histories' concentrates on how geopolitical conversations between European scholars and the wider world unfolded; second 'Geopolitics, Nation and Spirituality' considers how geopolitical writings have been strongly influenced by religious iconography and doctrine, with examples drawn from Catholicism, Judaism and Hinduism; and third 'Reclaiming and Refocusing Geopolitics' contemplates how geopolitics has been reformulated in the post-war period with illustrations from France and the United States.

Geopolitical Traditions brings together scholars working in a variety of disciplines and locations in order to explore a hundred years of geopolitical thought.

Klaus Dodds is Lecturer in Human Geography at Royal Holloway, University of London, and **David Atkinson** is Lecturer in Human Geography at the University of Hull.

CRITICAL GEOGRAPHIES
Edited by
Tracey Skelton, *Lecturer in International Studies,*
Nottingham Trent University,
and **Gill Valentine**, *Professor of Geography,*
The University of Sheffield.

This series offers cutting-edge research organized into four themes of concepts, scale, transformations and work. It is aimed at upper-level undergraduates, research students and academics, and will facilitate inter-disciplinary engagement between geography and other social sciences. It provides a forum for the innovative and vibrant debates which span the broad spectrum of this discipline.

GEOPOLITICAL TRADITIONS

A century of geopolitical thought

*Edited by Klaus Dodds
and David Atkinson*

London and New York

First published 2000
by Routledge
2 Park Square, Milton Park, Abingdon, Oxon, OX14 4RN

Simultaneously published in the USA and Canada
by Routledge
270 Madison Ave, New York NY 10016

Routledge is an imprint of the Taylor & Francis Group

Transferred to Digital Printing 2009

Typeset in Perpetua by
Curran Publishing Services Ltd

British Library Cataloguing in Publication Data
A catalogue record for this book is available from the British
Library

Library of Congress Cataloging in Publication Data
Geopolitical traditions: a century of geopolitical thought
Klaus Dodds and David Atkinson
408 pp. 15.6 x 23.4 cm
Includes bibliographical references and index
1. Geopolitics. I. Dodds, Klaus. II. Atkinson, David, 1969–
JC319.G483 2000
327.1'09'04 21--dc21   99-044347

ISBN 0-415-17248-9 (hbk)
ISBN 0-415-17249-7 (pbk)

Publisher's Note
The publisher has gone to great lengths to ensure the
quality of this reprint but points out that some
imperfections in the original may be apparent.

CONTENTS

CONTENTS

FIGURES

TABLES

CONTRIBUTORS

David Atkinson teaches Geography at the University of Hull. His research
interests revolve around the histories of geographical knowledge, and the historical, cultural and political geographies of modern Italy and Italian Africa.

Sanjay Chaturvedi is Reader in Political Science at the Centre for the Study of
Geopolitics at the Panjab University, Chandigarh. He is author of *The Dawning
of Antarctica* (Segment Books 1990) and *The Polar Regions: A Political Geography*
(John Wiley 1996) and was a Leverhulme Research Fellow at the Scott Polar
Research Institute, University of Cambridge between 1992 and 1995.

Paul Claval was Professor of Geography at the Sorbonne. He taught there for the
last twenty five years and specialized in the history of geographical thought
which led him to explore the various components of human geography:
social, economic, urban, regional, cultural and political. His latest book *La
Géographie française depuis 1870* (French Geography Since 1870) was published
by Nathan in 1998.

Klaus Dodds is Lecturer in Geography at Royal Holloway, University of London.
He is author of *Geopolitics in Antarctica: Views from the Southern Oceanic Rim* (John
Wiley 1997) and *Geopolitics in a Changing World* (Longman 2000). His research
interests include critical geopolitics and the international politics of Antarctica
and the Falklands/Malvinas. At present, he is involved in a Leverhulme Trust
funded project on the 'Falklands/Malvinas in a Changing World'.

Michael Heffernan is Professor of Geography at the University of Nottingham.
He has taught at the Universities of Cambridge, Loughborough and at UCLA.
In 1999–2000 he was Alexander von Humboldt Research Fellow at the Geo-
graphisches Institut in the Ruprecht-Karls-Universitat Heidelberg. His research
examines the politics of geographical knowledge, the role of geography in the
construction of political identities and the relationship between geography
and memory. His most recent book is *The Meaning of Europe: Geography and*

Geopolitics (Arnold 1998) and is currently working on a volume entitled *The Politics of Geography*.

Leslie W. Hepple is Senior Lecturer (and Director of the M.Sc. programme 'Society and Space') in Geography at the University of Bristol. His research interests include geopolitics, the cultural history of landscape and spatial econometrics. He was the author of the influential paper 'The revival of geopolitics' in *Political Geography Quarterly* in 1986.

Andrew Kirby is Professor of Social Sciences and Chair of the Social and Behavioural Sciences Department at Arizona State University West in Phoenix. He has also taught at the Universities of Arizona, Colorado and Reading, and has been a visiting scholar at both Stanford and Berkeley. He was reviews editor of *Political Geography Quarterly* from 1982–1992, and is currently editor for the international journal, *Cities*. His most recent books include the edited collection *Pentagon and the Cities* (Sage 1992) and *Power/Resistance* (Indiana University Press 1993).

Timothy W. Luke is Professor of Political Science at Virginia Tech and the author of numerous books and articles. His work engages the problematic of informationalization and how it has transformed social structures, political institutions, environmental politics, notions of art, practices of education and the nature of geopolitics. His latest books are *Eco-Critique* (University of Minnesota 1997) and *Capitalism, Democracy and Ecology: Departing from Marx* (University of Illinois Press 1998).

David Newman is Professor and Chairperson of the Department of Politics and Government at Ben Gurion University of the Negev. He is editor of the journal *Geopolitics* and a member of the Commission on the World Political Map of the International Geographic Union. He has published extensively on geographical and territorial aspects of the Arab–Israeli conflict. His most recent book is *The Dynamics of Territorial Change: A Political Geography of the Arab–Israeli Conflict* (Westview 1999).

Gearóid Ó Tuathail (Gerard Toal) is Associate Professor of Geography at Virginia Tech. His research interests range from the history of geopolitics to international political economy, the US foreign policy/mass media relationship to information technology and education. He is author of *Critical Geopolitics* (University of Minnesota Press 1996) and co-editor of *Critical Geopolitics* (Routledge 1998), *An Unruly World? Globalization, Governance and Geography* (Routledge 1998) and *The Geopolitics Reader* (Routledge 1998).

Joanne Sharp is Lecturer in Geography at the University of Glasgow. Her research interests include popular geopolitics and national identity, particularly in the context of twentieth-century US cultural politics. She has recently co-edited the feminist geography collection *Space/Gender/Knowledge* with Linda McDowell. The University of Minnesota Press is publishing her forthcoming book *Condensing Communism: the Reader's Digest and American Identity 1922–1994*.

James Derrick Sidaway is Lecturer in Geography at the University of Birmingham. His research interests are in 'East–West' and 'North–South' relations and his current research focuses on transformations in Southern Africa, Portuguese and Spanish geopolitical discourses and the sociology of geographical knowledge. He has recently completed a European Union funded project on the borderlands between Spain and Portugal.

Keiichi Takeuchi is Emeritus Professor at Hitotsubashi University and Professor of Geography at Komazawa University, Tokyo. He has written a number of books on the history and methodology of geography, and on the regional problems of Mediterranean countries, especially Italy. He is currently President of the Japan Association of Economic Geographers.

Peter J. Taylor is Professor of Geography and Associate Dean for Research in the Faculty of Social Science and Humanities at the University of Loughborough. He was editor of *Political Geography* (1982–1998) and an editor of the *Review of International Political Economy* (1992–1997). His previous books include *The Way the Modern World Works: World Hegemony to World Impasse* (John Wiley 1996) and *Modernities: A Geohistorical Perspective* (Polity 1999).

Nigel Thrift is Professor of Geography at the University of Bristol. His research interests include social theory, money and the international financial system, Third World socialism and time. His books include *Writing the Rural* (Paul Chapman 1994), *Spatial Formations* (Sage 1996), *Money/Space* (Routledge ·1996) and, as Co-Editor, *Mapping the Subject* (Routledge 1995).

PREFACE

Klaus Dodds and David Atkinson

Geopolitics has provoked personal passion and intellectual outbursts ever since the emergence of the term in the 1890s. From the earliest expressions of geopolitics at the start of this century, the individuals involved such as Rudolf Kjellén, Halford Mackinder, Karl Haushofer and Isaiah Bowman constantly sought to influence national and international politics with their theories about the contemporary world. In the mid-twentieth century, when the late American geographer Richard Hartshorne exclaimed in 1954 that geopolitics was 'an intellectual poison', he confirmed the widespread opinion that geopolitical reasoning was synonymous with Nazi spatial expansionism, and that the theories and approaches gathered under the label 'geopolitics' were little more than a bogus 'pseudo-science' whose political contamination brought shame upon academic geography (Hartshorne 1954). Moreover, such opinions proved to be remarkably resilient. Forty years later, the preliminary meeting of the International Geographical Union Commission on the World Political Map was jeopardized when the Soviet delegation protested against a proposal to include 'geopolitics' in the title of the new organization (Vitkovskiy 1981). Once more, geopolitics was condemned for its associations with totalitarianism and political extremism. Clearly, within some quarters of academic geography, the geopolitical was unacceptable.

Yet despite the frequent condemnation of the term, geographers in Europe and North America never altogether abandoned the intellectual terrain demarcated by geopolitics. As Leslie Hepple noted in his examination of Anglo-American geopolitics from 1945 onwards, some geographers continued to undertake geopolitical investigations under the guise of different studies, while others displayed considerable intellectual courage by refusing to abandon 'geopolitics' from their vocabularies (Hepple 1986). By contrast, during the 1970s, beyond the sensitivities of the academy, the 'security intellectuals' and foreign policy advisors of the United States enthusiastically rehabilitated the term. Ironically though, their understanding of geopolitical ideas was probably prompted by their own long-standing use of geopolitical concepts such as 'Heartland' and 'Rimland' to sustain and justify American Cold War strategies and interventions throughout the world for the previous two decades. It is only in recent years that geographers have begun to re-engage more rigorously with geopolitics; the interrogation of these Cold War geopolitical representations of the world, for example, has underpinned the emergence of what is called 'critical geopolitics' (Ó Tuathail 1996; Ó Tuathail and Dalby 1998).

This edited collection presents further evidence of the thorough revival of geopolitics in the mainstream academic arenas of the Euro-American world and beyond. It presents accounts of geopolitical thinking from different contexts throughout the century, and examines anew the histories and concerns of this problematic genre. However, it also seeks to disrupt the neat and progressive histories of academic geopolitics implicit within stories of emergence, decline and revival. The narrative outlined above is a simple caricature. To be blunt, the development of geopolitical conversations in Europe (let alone the wider world) do not support such a contention. The geopolitical experiences of Spain, Portugal and France, for example, provide powerful counter-evidence to the patterns of growth and decline in the Anglo-American world. Indeed, geopolitics has constituted a powerful set of travelling theories that have been negotiated in a variety of intellectual and geographical locales ranging from Jesuit anti-communism and British adventure fiction, to French left-wing circles and contemporary environmentalism. Elsewhere, the recasting of geopolitics in Argentina, India, Israel, and Japan provoked new articulations of these ideas, and further exchanges which challenge our received Euro-American understandings of the history of geopolitics. The ensuing chapters demonstrate that these controversial histories deserve careful consideration and elaboration rather than blanket condemnation or uncritical rehabilitation.

Through three main sections that rethink the histories of geopolitics, consider questions of nation and spirituality, and visit contemporary articulations of these notions, this collection has deliberately sought to demonstrate the tremendous diversity and richness of geopolitics and the range of sites where they are produced. At the same time, underlying these disparate geopolitical expressions are often faint but identifiable 'traditions' of geopolitical ideas. Although frequently amorphous and intermittent, these traditions have nevertheless informed and influenced the debates around geopolitics for much of the century, as 'geopoliticians' invoke and re-work these ideas in the light of prior geopolitical thought. These currents of thought have been connected not only by formal academic exchanges between German and Japanese scholars in the 1930s for example, but also by a tranche of ideas, concepts and languages concerning space, identity, politics and the physical environment. While recognizing the specific contexts of geopolitics, we also want to trace these enduring traditions through the century. In the late 1990s, questions of geopolitics continue to demand our attention as political leaders mobilize military forces in defence of the same European region that fascinated early geopoliticians such as Halford Mackinder one hundred years ago. While geopolitical questions continue to occupy the political agenda, geopolitical knowledges should occupy the attention of critical scholars.

This collection of essays had its origins in a conference session entitled 'Rethinking Geopolitics' that was held by the Institute of British Geographers at the University of Strathclyde in January 1996. Thanks are due to our co-convenor James Sidaway, the Political Geography Research Group of the Institute of British Geographers for its sponsorship, and to the participants and discussants. We are indebted to the following for their comments, advice and expertise in translation: Denis Cosgrove, David Demeritt, Editha Dodds (German), Felix Driver, Mike Heffernan, Andy Jonas, Sylvie Gray (French) and David Simon. We appreciate the support of our colleagues at the Departments of Geography of the University of Hull, University of Wales, Lampeter and Royal Holloway, University of London. We also owe a debt of gratitude to our editors at Routledge, Sarah Lloyd, Casey Mein and Sarah Carty for their patience and good humour. Finally, many thanks to our authors for their efforts, enthusiasm, and contributions to this project.

Bibliography

Hartshorne, R. (1954) 'Political geography', 211–14 in P. James and C. Jones (eds) *American Geography: Inventory and Prospect*, Syracuse: Syracuse University Press.

Hepple, L. W. (1986) 'The revival of geopolitics', *Political Geography Quarterly* vol. 5 supplement: 21–36.

Ó Tuathail, G. (1996) *Critical Geopolitics*, London: Routledge.

Ó Tuathail, G. and Dalby, S. (eds.) (1998) *Rethinking Geopolitics*, London: Routledge.

Vitovskiy, O. (1981) 'Political geography and geopolitics: a recurrence of American geopolitics (and a reply to A. K. Henrikson)', *Soviet Geography, Review and Translation* 22: 586–97.

1

INTRODUCTION

Geopolitical traditions: a century of geopolitical thought

David Atkinson and Klaus Dodds

Originally coined by the Swedish political scientist Rudolf Kjellén in 1899 (Holdar 1992), few terms in the modern history of geography have been as controversial and emotive as 'geopolitics'. From its obscure origins at the twilight of the nineteenth century through to the widespread and sometimes indiscriminate contemporary uses of the term, the label has frequently been the focus of intense and often acrimonious debate. At certain times, 'geopolitics' even attracted such enduring notoriety that political geography and even academic geography in general were tainted by what, in 1927, the geographer Carl Sauer felt able to call the 'wayward child of the geographic family' (Sauer 1927).

To many geographers since Sauer, geopolitics has remained an enigmatic, shadowy, contested and sometimes shameful category. Yet despite occasional outrage, periodic soul-searching, denial and notoriety, geopolitics has continued to lurk insistently within the discipline. Moreover, geographers and other scholars have returned again and again to this intellectual domain. For many writers – and not just those who might be identified as politically conservative and reactionary – geopolitics seem to offer the seductive promise of a privileged perspective upon current affairs and a unique insight into the political world.

This book therefore brings together reflections upon the enduring problematic of 'the geopolitical' throughout the twentieth century. It does so in the light of approaches enabled by critical histories of geography and critical geopolitics. It engages with some of the substantive expressions of geopolitics, but also deals with less well-known, non-anglophone experiences. It recognizes the roles of elite and popular, formal and practical geopolitics, while acknowledging that geopolitics are contingent and context-bound. There is no attempt to define or delimit the topic, but analysis revolves around those movements and individuals who have consciously used or appropriated the term 'geopolitics'. As such, the

1

book is a modest attempt to understand some of the natures of 'the geopolitical' through history, and the continuing use of geopolitical ideas, however fractious and incoherent this tradition may be. However, until relatively recently, a book with this theme might have been deemed unacceptable by some geographers. For much of the Cold War period in the Anglo-American world, geopolitics was shunned by many within academic geography. One result was that histories of geopolitics in the English speaking world were both partial and problematic. Therefore, these first pages discuss these partial histories and their subsequent revision.

Partial histories

Geopolitical thought emerged at the close of the nineteenth century as geographers and other thinkers sought to analyse, explain and understand the transformations and finite spaces of the *fin de siècle* world (Heffernan 1998, and this volume; Kearns 1983, 1993; Kern 1983; Smith 1999). Geopolitics is thus related to other elements of geography that underpinned western imperialism in the period. However, while the entangled nature of geography's links with imperialism has recently attracted sustained academic attention, by contrast, geopolitical thought has suffered notoriety for much of the post-war period. In Anglo-American geography for example, until recently, histories of the discipline reproduced an artificial dualism whereby geography's imperial heritage remained largely unspoken, but geopolitics was routinely shunned as a bastardized form of geography that had been conscripted to political service.

This notoriety is often traced to the hysterical reactions to Karl Haushofer and his development of German *Geopolitik* in early 1940s American journalism. Writing for the mass audiences of *Current History* and *The Reader's Digest* in 1941, for example, Frederick Sondern (1941a, 1941b) wrote of 'The Thousand Scientists behind Hitler'. He told a sensational tale of:

> the work of Major General Professor Dr. Karl Haushofer and his Geopolitical Institute in Munich, with its 1,000 scientists, technicians and spies. These men are almost unknown to the public, even in the Reich. But their ideas, their charts, maps, statistics, information and plans have dictated Hitler's moves from the very beginning.
>
> (Sondern 1941b: 45)

It was such breathless accounts, syndicated across the newspapers and magazines of early 1940s America, that prompted further interest in this secretive *Institut für Geopolitik* and its supposed domination over Hitler and Nazi Foreign Policy. *Geopolitik* cast a long shadow in the American popular imagination, and

2

these sinister themes even found their way into popular books and cinema (de Seversky 1942; Meilinger 1995; Ó Tuathail 1996).

At an official level, Roosevelt's administration was worried enough to commission academic studies of *Geopolitik* (Whittlesey, Hartshorne and Colby 1942). And although elements of this analysis qualified Sondern's hyperbole, self-styled 'experts' such as the Jesuit-academic Edmund Walsh (see Ó Tuathail, this volume) were still agreed that:

> The basic, incontestable truth is that Haushofer, directly in some instances indirectly in others, co-ordinated, integrated, and rationalised the whole field of comparative geography for the uses of the Führer . . . [geopolitics] became a dynamic driving rod in the mechanics of states craft. A huge personnel was mobilised by Haushofer to comb the earth for significant facts and geographic information.
>
> (Walsh 1944: 22)

Indeed, these ideas took such a firm hold in the United States that after the collapse of the Third Reich, proceedings began to indict Haushofer at the Nuremberg war trials. At least one of the prosecuting team thought that in the absence of Hitler, they had nevertheless found the true mastermind behind German expansionism (Jacobsen 1979).

It is thought that Haushofer would not have been tried at Nuremberg, although his 1946 suicide pre-empted any indictment (Jacobsen 1979). However, his alleged associations with Nazism ensured that after the war, geopolitics was largely ostracized by geographers in general, and by political geographers in particular. In a 1948 textbook for example, under the sub-heading of 'Perversion', G. Etzel Pearcy, Richard Fifield and Associates (1948: 23) revealed tortuous attempts to distinguish 'Political geography [which] is a sane, cautious and — above all else — honest science' from the geopoliticians, 'in no normal state of mind' who:

> had fallen victim to the psychosis of 'all or nothing'. To achieve their ends they were willing to convert political geography into *total geographical nonsense* if need be . . . it enabled the German geographers themselves to escape the moral censorship of their science and to side-step their own scientific consciences.
>
> (Etzel Pearcy, Fifield and Associates, 1948: 23)

Other histories of geography chose to virtually ignore geopolitics altogether. In his *Background to Political Geography* (1967), Crone mentioned German geopolitics in only one paragraph of 239 pages; and this only because he felt himself 'required' to do so (Crone 1967: 106). Likewise, in Johnston's *Geography and Geographers* (1987),

Haushofer and geopolitics merit only two lines (Johnston 1987). A reluctance to engage with this troubled past was thus evidenced in attempts to isolate the topic, and exclude it from geographical histories. At the same time, academics who tried to reassess 'geopolitics' in the post-war period often risked fierce criticism. When Ladis Kristof (1960) tried to rehabilitate the term at the height of the Cold War, he suffered immediate censure (Alexander 1961).

Of course there were exceptions to this rule. As Hepple points out (1986), the notion that the post-war taboo of geopolitics in the Anglophone world was absolute is wide of the mark. Several writers (mainly non-geographers) still used the term (Chubb 1954; Roucek 1955, 1956, 1962), others continued to investigate geopolitical themes (Cohen 1964), while some attempted to develop a 'geopacifics' of peace (Taylor 1957). Additionally, English-language experiences are not the whole story: in some parts of the world, geopolitics thrived. In Latin America, geopolitics became closely linked to the violent, militaristic and expansionist regimes of countries such as Brazil, Argentina and Chile (Dodds this volume; Hepple 1992). With dictators such as Augusto Pinochet having taught geopolitics in military academies, these ideas found application in foreign and domestic policy initiatives. Geopolitics under the generals appeared at best partial in so far as national security doctrines only identified 'threats' to the state; at worst, they were bloody and awful (Child 1985; Hepple 1992). Certainly, the experience did little to improve the intellectual reputation of geopolitics.

Regardless of this, within Anglophone geography, geopolitics seems to have been largely proscribed and demonized as a malign aberration for much of the post-war period. By 1969, when Brian Berry castigated political geography as a 'moribund backwater', geopolitics was surely the most stagnant reach of this lifeless sub-discipline (Berry 1969).

New perspectives

In the last fifteen years or so, understandings of the histories and development of geopolitics have improved markedly. For example, studies of German *Geopolitik* have allowed far more subtle understandings of the emergence, contexts and significance of *Geopolitik* in Weimar and Nazi Germany (Fahlbusch, Rössler and Siegrist 1989; Heske 1986, 1987; Heske and Wesche 1988; Jan van Pelt and Dwork 1996; Korinman 1990; Kost 1989; Murphy 1994; Sandner 1988; Sandner and Rössler 1994); the ways that *Geopolitik* was appropriated for intellectual legitimation by Nazism; and also the eventual marginalization of the movement by the suspicions of the Third Reich's hard-line racial theorists (Bassin 1987b; Heske 1987).

There also exist more nuanced understandings of the distribution of geopolitical ideas throughout the inter-war world, and their re-negotiation in a series of different countries. For example, geopolitics met varying degrees of success in

Finland (Paasi 1990), Sweden (Holdar 1992) and Japan (Fukushima 1997; Takeuchi 1980, 1994). There are studies of French contestations of German *Geopolitik*, and efforts to forge a *Géopolitique* (Desfarges 1996; Heffernan 1998; Parker 1985, 1987). Recent studies relate how both French and German geopolitics were read in Fascist Italy, where geographers developed an 'Italian geopolitics' in the light of, but distinct from, these extant geopolitics (Antonsich 1997, Atkinson 1995, 1996; Raffestin, Lopreno and Pasteur 1995). Similarly, geopolitics were re-worked in Franco's Spain (Bosque-Maurel *et al.* 1992; Sidaway, this volume), and chapters in this volume (by Chaturvedi, Dodds, and Sidaway) demonstrate the emergence of geopolitics in Indian, Argentine and Portuguese contexts. Reactions to geopolitics in the United States (Smith 1984, 1999; Ó Tuathail 1996) and the Soviet Union (Hauner 1992) have been addressed, while there are also studies of early geopolitical thinkers such as Ratzel (Bassin 1987a), Kjellén (Holdar 1992) and Mackinder (Ó Tuathail 1992, 1996). There is coverage of the geopolitical assumptions, theories, and practices of post-1945 American 'security intellectuals' (Dalby 1988, 1990a, 1990b, 1991; Ó Tuathail and Agnew 1992), and the popular elements of Cold war America's 'geopolitical cultures' (Sharp 1993, 1999). Finally, the 'anti-geopolitics' that resisted the connections between geopolitical expressions and statescraft are also addressed (Ó Tuathail, Dalby and Routledge 1998; Wittfogel 1985).

In the light of this body of work, the history of geopolitics is far less partial than the outline reproduced earlier. Certainly German *Geopolitik* played a significant role in the development and diffusion of geopolitics throughout the modern world. However, the German experience is not the entire story. Rather, it is now clearer that the geopolitical tradition entails a varied and complicated set of experiences, and a series of different 'geopolitics' negotiated and filtered through many different contexts.

Renewed responsibilities

In the light of this more sophisticated understanding of geopolitical histories, we argue that geographers have an ever-greater responsibility to reflect upon the consequences of geopolitical knowledges past and present. Recent scholarship on the connections between geography, the state and the military only emphasize this point. These connections provided a powerful incentive for the establishment of academic geography in the late nineteenth century. State patronage was welcomed by geographers who, in turn, often sought to reinforce their embryonic discipline by demonstrating the practical utility of their 'science' to the nation. The collection, collation and circulation of geographical knowledge in the form of maps, charts, surveys and reports had much practical significance for the execution of state power (Bell, Butlin and Heffernan 1995; Godlewska and Smith 1994; Harley

1988, 1989; Livingstone 1992; Jacob 1999). Thus, in addition to the intellectual armoury that geography provided for imperialism, geographers also offered practical assistance to political elites.

It was amidst this context that Rudolf Kjellén's original invocation of geopolitics explicitly recognized the territorial basis of the modern state and the geographical elements of governance. Geopolitical thought therefore has its roots planted firmly in this context. It may frequently have been a murky business grounded in the practicalities of governance, but any serious intellectual engagement with the history of geopolitics has to recognize that geopolitics is simply one of the more problematic consequences of geography's entangled historical connections to state power.

This collection therefore advocates an intellectual responsibility that acknowledges the situated, contingent and troubled past of geopolitics, while contributing an appreciation of how geopolitical knowledges have unfolded in Europe, North America, Latin America, Asia and the Middle East. The book adopts a critical perspective that situates and contextualizes these knowledges. It also admits a greater degree of pluralism, diversity and uncertainty into extant debates about geopolitics. While it is recognized that western scholars have yet to appreciate the rich geopolitical literatures that exist in China, Japan, Eastern Europe and parts of Africa, this book argues that even the increasingly well known histories of European and North American geopolitics can be re-interpreted with considerable intellectual profit. To such ends, this collection draws upon recent developments in human geography that enable the development of more nuanced, rigorous and critical understandings of geopolitical thought. These include developments in the writing of critical histories of geography, and the themes of critical geopolitics. Both approaches are outlined in the following pages. In the light of these, the idea of a 'geopolitical tradition' is developed. Finally, the structure of the collection is discussed.

Geopolitical traditions and critical histories of geography

Our title *Geopolitical Traditions* consciously refers to David Livingstone's seminal text *The Geographical Tradition* (1992). A key statement in the emergence of 'critical histories of geography', Livingstone's fluent narrative was the first significant history of geography to deliberately avoid whiggish historiography and disciplinary exceptionalism in favour of *situating* the production of geographical knowledges in the social, cultural, economic and political contexts whence they came (Haraway 1991; Livingstone 1992; Smith 1987; Smith and Godlewska 1994). In contrast to previous histories that often cast geography as a neutral, disembodied 'science' (Rose 1993), Livingstone recognized geographical knowledge as 'a social and cultural construction, and a political resource' (1992: 3).

6

The debate prompted by Livingstone's book highlighted some of the *different* aspects of geography elided in Livingstone's tradition (Driver 1994; Driver *et al.* 1995; Withers *et al.* 1996). Some critics perceived the exclusion of non-western geographies (Sidaway, 1997), of the embodied practices of geography (Matless, 1995), or of gender issues (Rose, 1995). Others observed that political and economic contexts (Withers *et al.* 1996), or technical and physical geographies (Werrity and Reid 1995) might have been emphasized more. Some simply questioned the relevance of histories (Barnett 1995). But in the light of these debates, and following Livingstone's (1992: 358) assertion that geography has 'connoted rather different things to different people at different times and in different places', it gradually became clearer that rather than privilege one, singular geographical tradition, it is more constructive to think of plural geographical *traditions* (Livingstone 1995a). Similarly, rather than search for an immanent, essential core of 'geography', it is more productive to recognize the different, negotiated geographical knowledges that are produced in each distinct context.

The importance of contexts demands that they are not merely considered as backdrops to the constitution of geopolitical knowledges. For instance, the debates within Nazi Germany about *Lebensraum* and expansionism into Eastern Europe (Clarke and Doel 1998; Jan van Pelt and Dwork 1996), inevitably informed *Zeitschrift für Geopolitik's* representations of the region as Germany's imagined 'living-space'. Simultaneously, *Geopolitik* reinforced these broader cultures by proposing the same ideas. Likewise, in Latin America, the articulation of geopolitical discourses in the 1960s and 1970s was shaped profoundly by the contested politics of the Cold War and anti-communist paranoia which gave rise to the national security doctrines of military regimes in Argentina, Brazil, Chile and Peru. These contexts actively contributed to the construction of geographical imaginations and knowledges about places and societies: it is therefore crucial that we take proper account of contexts.

At the same time as these modern horrors in Latin America were unfolding however, Quentin Skinner warned that an emphasis upon the context of knowledge had to be balanced by a concern for the specific texts which *also* informed that knowledge (Skinner 1969). In other words, text *and* context were symbiotically linked and mutually constitutive. Over twenty years later, geographers such as Felix Driver (1992, 1994) and David Livingstone (1992) also reminded readers that contextual histories of the discipline are not necessarily synonymous with a critical appreciation. Understanding how ideas migrate across time and space requires more than just half-hearted appeals to context. Texts, contexts and intellectual cultures must all be addressed in order to investigate how, and with what consequences, ideas are negotiated and situated.

Finally, the critical re-examination of geographical traditions also provoked interest in the varied texts and practices that have constituted geography. Scholars no longer limit themselves to orthodox, formal academic treatises, but increasingly investigate cartographies (Harley 1988, 1989), and a range of 'popular' sources such as film, exhibitions, photography and art, all of which have helped constitute geographical imaginations and knowledges (Cosgrove 1999; Matless 1998; Ryan 1994, 1997; Schwartz 1996). This tranche of research sustains and expands critical work upon the construction of geographical knowledges within European empires and the modern state system. Gearóid Ó Tuathail's deconstruction of Halford Mackinder, for example, explicitly recognizes how multiple contexts – ranging from personal background and overseas experiences to the machinations of late imperial politics – could shape an individual's geopolitical writings and their subsequent reception (Ó Tuathail 1992, 1996; GoGwilt 1998). Finally, at the end of this volume, Nigel Thrift calls for the extension of such work to encompass the mundane practices and structures that enable the realization of geopolitics in the 'real' world.

This collection draws upon these critical histories to inform the histories of geopolitics. The recognition that geopolitical knowledges are situated complements an acknowledgement of the plurality of geographical traditions and the lack of any essential, unchanging core to geopolitical thought (Ó Tuathail 1997). This abandons the long-standing desire of many geopolitical writers to unlock the nature of 'the geopolitical', or to secure the boundaries of a geopolitical tradition. In some respects, this was a pointless exercise: figures such as Mackinder shunned the term (Mackinder 1942; Gilbert 1972), while its doyen, Karl Haushofer, seldom settled on any definition of the geopolitical. It is clear that the meaning of geopolitics has always been historically ambiguous and contingent. Hence Ó Tuathail's simple warning:

> All concepts have histories and geographies, and the term 'Geopolitics' is no exception. The word 'geopolitics' has had a long and varied history in the twentieth century, moving well beyond its original meanings . . . Coming up with a specific definition of geopolitics is notoriously difficult, for the meaning of concepts like geopolitics tends to change as historical periods and structures of world order change. Geopolitics is best understood in its historical and discursive context of use.
>
> (Ó Tuathail 1998a: 1)

This inescapable fact was recognized by earlier commentators such as Hans Weigert who noted in 1942 that:

> There is no such thing as a general science of geopolitics which can be subscribed to by all state organizations. There are as many geopolitics as there are conflicting state systems struggling under geographic conditions.

8

...There is a *Geopolitik*, a *Géopolitique*, there are different geopolitics' for
the United States and England. Each nation has the geopolitics it deserves.
(Weigert 1942: 22–3)

Yet while these ideas found distinct and negotiated expressions in different
countries, they were not hermetic discourses. The history of inter-war geopolitics
in Europe, for instance, was characterized by a series of intellectual exchanges and
debates between 'geopoliticians' in Germany, France, Italy, Spain, and Scandinavia.
For example, Kjellén's writings on the state were translated into German in 1924
and attracted attention from French and German scholars including Karl
Haushofer (Heffernan 1998). In the 1930s, Haushofer's own ideas would reach
France, Italy, Spain, Japan and the United States, to be debated by still more
geographers. As a result, it is impossible to talk of discrete, exclusive geopolitical
traditions. Indeed, many of the contributors to this collection (including Atkinson,
Claval, Hepple, Sidaway and Takeuchi) demonstrate how geopolitical ideas and
practices were exchanged across the territorial boundaries of states.
Consequently, the circulation, exchanges, and historical geographies of these ideas
(Livingstone 1994, 1995b) are one of the dimensions of geopolitics addressed by
this collection. Similarly, the importance of social and political contexts, and the
popular knowledges and practices that have all constituted geopolitics at different
times, are also acknowledged by the book. And complementing the approaches of
critical histories are those of 'critical geopolitics'.

Common ground? Critical geopolitics and histories of geopolitics

The recent interest in critical histories of geography coincided with a renewed
interest in geopolitical ideas. A body of writing labelled 'critical geopolitics'
sought to combine the interrogation of contemporary political change with critical evaluations of geopolitical reasoning and representations. These concerns can
be traced to isolated critical studies of the geopolitics of the Cold War (O' Sullivan
1982, 1986; Ó Tuathail 1986), and accounts of corresponding geo-economic
disorder (Agnew and Corbridge 1989). However, throughout the 1990s, such
studies have multiplied and critical geopolitics has propelled geopolitical themes
to the heart of debates in contemporary human geography. Consequently, many of
the chapters in this collection draw upon the insights and approaches of critical
geopolitics while constructing their histories of geopolitical thought.

Underpinning the emergence of critical geopolitics was the recognition that
the geographies of global politics were neither inevitable nor immutable, but were
constructed culturally and sustained politically by the discourses and representational practices of statescraft (Dalby 1990a, 1990b; Ó Tuathail and Agnew 1992).

To this, end Ó Tuathail and Agnew (1992) re-conceptualized geopolitics as a discourse comprising two overlapping components. First, the 'practical geopolitics' of everyday statescraft, whereby the world is spatialized into regions with imagined attributes and characteristics – leading to a mosaic of places of 'danger', 'threat', or 'safety' that underpins foreign policy. Second, the 'formal geopolitics' created by 'security intellectuals' who produced theories and strategies to guide and justify the statescraft of practical geopolitics. The military-political establishment of the United States in the Cold War period proved particularly fertile ground for study. Simon Dalby (1988, 1990a, 1990b), and Gearóid Ó Tuathail and John Agnew (1992) critiqued the American geopolitical discourse that portrayed the Soviet Union as dangerous, threatening and expansionist. In turn, this legitimated US defence budgets, and policies of 'containment' and intervention in other parts of the world. The irony was that these simplistic representations, while supposedly based upon 'expert' knowledges, were profoundly a-geographical and one-dimensional (Ó Tuathail 1993; Ó Tuathail and Agnew 1992). Neither did they need to be terribly sophisticated in order to be politically effective, for in the absence of significant critiques, geopolitical discourses remained largely unchallenged.

In the introduction to *Rethinking Geopolitics*, a recent book that explores the scope of critical geopolitics still further, Ó Tuathail and Dalby (1998b) outline four more points that characterize critical geopolitics. After re-emphasizing the importance of interrogating the spatialization of the world by statescraft, their second point highlights how these representational practices impact upon the politics of identity. They argue that the distinctions drawn between the domestic Self and external Others are sustained by geopolitical imaginations and the moral and physical boundaries that divide the world into 'our' space and 'their' spaces (Shapiro 1998).

Third, Ó Tuathail and Dalby state that geopolitical knowledges are not the exclusive preserve of political elites, their advisors, and their policy documents. Rather, 'geopolitical cultures' also find expression in the everyday realms of television and films, novels and newspapers, the formal education system and the routine politics of banal nationalism (Billig 1995; Sharp 1999). And just as critical histories of geography now address popular cultures, so too 'popular' geopolitical knowledges are important precisely because they contribute to the production and circulation of 'common sense' geopolitical reasoning which impacts upon public opinion. In Argentina, for example, the formal education system and state-sanctioned geography textbooks have played a significant role in encouraging the geographical imaginations of its citizens, and attendant beliefs that territories such as the *Islas Malvinas* must be recovered in order for Argentina to secure development, progress and political justice (Dodds 1998).

Fourth, the intellectual practices and epistemological assumptions inherent to geopolitical thought and practices need to be investigated rather than reified uncritically. One element of this critique has been the interrogation of the politics of vision. Geopolitical texts have frequently appealed to a particular synoptic-vision and insight, while seldom problematizing their own right to declare places as 'heartlands', 'rimlands' or 'shatterbelts'. A more embodied approach to geopolitics demonstrates that these supposedly privileged gazes fail to acknowledge their own assumptions and conceits. Recent research on gender and geopolitics, for example, has illustrated how the gendered and racialized gazes of European and North American geopolitical observers ensured that knowledge of the political world was never (as some claimed) universal or free from particular prejudices (Dalby 1994). Critical geopolitics should thus interrogate the visualization of the world by geopolitical 'experts', to ensure geopolitical claims to truth and privileged insight are exposed as partial and subjective. Finally, Ó Tuathail and Dalby discuss the 'situated reasoning' of geopolitics within the social, spatial, political and technological parameters of the modern state-system. Given that geopolitical thought frequently addresses the pressing questions of governance that face states in their administration and regulation of territory, the recognition and critique of these contexts is also an essential task.

At the core of critical geopolitics, therefore, is the belief that these geopolitical representations of global politics deserve serious attention, for it is such 'scripting' of the world that helps constitute and legitimate foreign policies. But while these studies began to expose the spatial assumptions and representations of statescraft (Dodds and Sidaway 1994), critical geopolitics also became progressively more sophisticated in terms of its analytical frameworks and subject matter. Some critics complain that the emphasis upon discourse analysis, representation, meaning and identity is pursued at the expense of questions of political economy. Others worry that the term itself is insufficiently problematized, or the analysis of geopolitical discourse is defined too narrowly (Ó Tuathail 1996; Thrift, this volume). For some, Enloe's (1989) emphasis upon the roles of women that underpin practical geopolitics is still to be addressed thoroughly. It is also worth remembering that Yves Lacoste foreshadowed many concerns of critical geopolitics in the French journal *Hérodote* in the 1970s (see Claval and Hepple, this volume; Lacoste 1976; Ó Tuathail 1996).

However, there is little doubt that the insights of critical geopolitics have helped re-invigorate interest in geopolitical themes. For instance, several of the chapters that follow − including those by Atkinson, Sidaway and Dodds − consider the spatialization of world regions. Equally, the construction of boundaries between Self and Other informs the contributions of Kirby, Ó Tuathail, Chaturvedi, and Sharp amongst others. Likewise, the growing awareness of the popular dimensions of geopolitics, and the positioned contexts that produce geopolitical knowledge is addressed by many of the subsequent essays. Although not all the authors represented in this book might associate themselves with the phrase 'critical geopolitics', there is no doubt that this literature has helped to change the intellectual and political agendas surrounding the interrogation of geopolitical thought. Therefore, the combination of these insights with the concerns of critical histories offers the potential for fruitful engagement (Dodds and Sidaway 1994). It also promises opportunities to develop further our understandings of geopolitical traditions throughout the twentieth century.

Geopolitical traditions?

Notwithstanding the plurality of geopolitical knowledges and the situated and contingent nature of their production, this collection also suggests that a further, significant element of the phenomenon of geopolitics has been the continuity, resilience and endurance of geopolitical thinking over the years, despite the controversies that surround it. Given this, the book suggests that a loose 'geopolitical tradition' has connected geopolitical ideas across the last hundred years. For while other authors have traced the development of geopolitics through the century (Raffestin, Lopreno and Pasteur 1995; Desfarges 1996; Ó Tuathail 1996; Ó Tuathail, Dalby and Routledge 1998), and several invoke the phrase 'geopolitical tradition' (Muir 1997; Ó Tuathail and Agnew 1992; Ó Tuathail 1996; Parker 1985, 1998), few have problematized the concept that there might be traditions of geopolitical thought (Ó Tuathail 1996).

Livingstone's *Geographical Tradition* and its attendant debates posed a number of questions about intellectual traditions. Several critics pointed out that certain versions of the geographical tradition were often partial (Stoddart 1986). In particular, they highlighted how women and alternative forms of knowledge were frequently excluded from these versions of the 'tradition' (Domosh 1991a, 1991b; McEwan 1998; Matless 1995; Rose 1995). Gillian Rose (1995) argues that the construction of *any* tradition is inevitably an exercise in inclusion and exclusion; and thereafter, the *analytical* spaces territorialized by these traditions inevitably reproduce these exclusions. She proposes that one solution to this problem may be the recognition of the multiple spatialities of traditions (Rose 1995).

In this collection we have tried to provide analytical spaces within which different expressions of geopolitics can be considered. There is no attempt to demarcate the 'geopolitical', or to prescribe its boundaries beyond recognizing traces of a *self-selecting* tradition in which individuals and groups have *consciously* adopted, used and appropriated the label 'geopolitics' over time. This does not imply any special degree of certainty and incremental progression in the 'tradition' (cf. Parker 1998), but suggests that there are *some* connections between geopolitical expressions that voluntarily position *themselves*, however vaguely, in relation to other ideas, texts or movements, historical or contemporary, that also labelled

12

INTRODUCTION: GEOPOLITICAL TRADITIONS

themselves 'geopolitical'. However amorphous the inter-connections and recurrent themes of geopolitics, it might be helpful to think about geopolitical traditions in this loose sense.

The philosopher Alisdair MacIntyre (1985, 1990) talks about 'traditions of enquiry' that are historically extended and find periodic expression in specific sites. Livingstone (1992, 1995a, 1995b) adopts this idea, and emphasizes that contestation, debate and pluralism evidence a healthy, ongoing tradition. Thus, critics of geopolitical ideas, including the popular sensationalism of 1940s American journalism, the critiques of Wittfogel or Lacoste, the 'anti-geopolitics' of protest, or the developments of critical geopolitics, are all part of this loose geopolitical tradition (Ó Tuathail 1996; Wittfogel 1985). Likewise, those who adopt geopolitical reasoning, from Nazi Germany or Fascist Italy, to Argentina or India, equally engage with this tradition. The intention of this collection is to nuance our sense of these myriad geopolitical expressions, their overlaps and their trajectories through a series of different contexts. But it also suggests tentatively that, on occasions, geopolitical ideas are re-worked and re-negotiated within a broad and plural 'tradition', however ill-focused and ill-formed this may be. It is the contributions of our colleagues that have enabled the consideration of these themes. The final section of this introduction outlines their essays.

Discussion of the chapters

The book is divided into four sections. The chapters are arranged in loose chronological order across these. The first section concentrates upon the emergence and development of geopolitical ideas in various different countries around the world. The second addresses notions of spirituality, national identities and geopolitical thought. The third considers various attempts to re-conceptualize and re-cast geopolitical themes in the more recent past. The final section comprises two short, suggestive commentaries that speculate upon possible futures for geopolitical thought as the term commences its second century. These sections are not intended to signal significant divisions between the essays; indeed, there is a series of recurrent themes that are braided through the contributions.

Rethinking geopolitical histories

The opening chapter by Michael Heffernan affirms a tendency for geopolitics to surface especially in periods of instability. Heffernan returns to the origins of European geopolitics and situates these ideas amidst the uncertain academic and political climate of the late nineteenth century. He argues that re-conceptualizations of early geopolitics ought to focus upon the significant concerns for the ideological and geographical extent of 'Europe' that characterized British, French and

13

German scholarship in the era. For instance, when writers concentrate on Mackinder's 'Heartland' concept, most marginalize his wider contribution to debates surrounding Europe's cultural and political geography. Heffernan's discussion of these debates not only reflects the uncertainties of the late nineteenth century, but also highlights the possibility of re-reading early geopolitical sources for the purpose of recovering different angles upon the 'geopolitical tradition'.

In the same period, national anxieties about the newly-'closed world' also influenced the less-formal texts that Andrew Kirby examines. His concerns are the geopolitical motifs that suffused popular English novels in the first decades of the twentieth century. As a consequence of the broader geopolitical cultures of the day, literature such as Erskine Childers' *Riddle of the Sands*, John Buchan's *The Thirty Nine Steps*, and Captain W. E. Johns' 'Biggles' adventures can be interpreted in distinctly geopolitical terms. They are replete with fears of the 'encirclement' of England's cultural 'heartland' by German submarines, of the dangers posed to Britain by European eccentrics, or of the heroism of plucky Major James Bigglesworth, sustaining the empire against an international series of foes. With his close reading of these texts, Kirby demonstrates that popular realms of geopolitical reasoning should not be divorced from the formal pronunciations of statescraft, for alongside contemporary geopoliticians, many popular authors were predicting the rise of new geographical spaces in world politics, or articulating common but informed concerns about the international arena.

The interactions of ideas, individuals and institutions in Japanese geopolitical discourse is the guiding theme in Keiichi Takeuchi's essay. During the 1920s and 1930s, the writings of Kjellén and Haushofer were translated into Japanese and their ideas prompted a critical engagement from geographers such as Cikao Fujisawa and Taro Tsujimura. An interesting dimension of this exchange resides in the development of ideas about state and environment occasioned by Japanese debates on nationalism and imperial expansion. While urging Japanese geographers to take responsibility for their discipline's historical associations with imperialism, Takeuchi reveals how some geographers demonstrated considerable intellectual courage to resist certain currents of thought. Methodologically, this chapter illustrates the value of a rigorous examination of concepts such as environmental determinism and territorial politics, and their negotiation in Japanese universities, research institutions, and government departments. It also reveals the importance of key geopolitical movements such as German *Geopolitik*, and crucial figures like Kjellén and Haushofer in the global circulation of geopolitical ideas.

A sense of the migration of geopolitical ideas is also evident in David Atkinson's account of geopolitical journals in modern Italy. He discusses the emergence of a self-consciously Italian version of geopolitics amidst the instabilities of the 1930s world. The desire of the Fascist regime to increase the geographical imagination of the Italian people ensured state support for the production of *Geopolitica* from 1939 to 1942; Atkinson discusses the journal's perspectives upon the world through its claimed modern, synoptic powers of mastering survey. And although geopolitics suffered from association with *Geopolitik* in post-war Italy, the fall of the First Italian republic in 1992-1994 coincided with the emergence of a new Italian geopolitical journal entitled *Limes*. Atkinson speculates that it is periods of instability and crisis that encourage the development of putative encompassing, explanatory geopolitical gazes.

James Sidaway provides one of the few English language accounts of Spanish and Portuguese geopolitics. Again, geopolitical discourses emerged in Iberia thanks to the intellectual engagement with German, British, American and French geography in the 1930s and 1940s. However, in a striking parallel with the South American experience, the emergence of military governments stimulated exchanges between scholars such as Jaime Vicens Vives and military academies. As with some other geopolitical movements, Iberian geopolitics produced striking visual images to articulate this presumed insight. Likewise, its geopolitical discourse was intended to enrich and expand national geographical imaginations that seemed ill-equipped to deal with the instabilities of decolonisation and the Cold War period. Concluding his survey in the present, Sidaway reveals how the integration of Spain and Portugal into the European Union has provoked new geopolitical debates and controversies over national identity.

Finally in this section, Klaus Dodds situates Argentine geopolitical writings within the emergence of the post-colonial state in the early nineteenth century. Yet again it was geographical and cultural uncertainties within the state-building project that prompted patriotic forms of education, including sustained interests in Geography and History. With the active support of the military, geopolitical discourses filtered into twentieth century Argentine political life. Inspired by German, French and American geopolitical writings, Argentine geographers and military authors produced tomes concerned with national geography and development programmes. Most disturbingly, the military regimes of the 1960s and 1970s precipitated interests in national security doctrines and geopolitical security. Dodds argues convincingly that geopolitics enabled an intellectual context that legitimated expansionist foreign policies and repressive domestic politics. Finally, he discusses the revisionist reassessments of Argentine geopolitical debates that emerged after the fall of the military in the 1980s.

Geopolitics, nation and spirituality

While questions of national identity occupy many discussions of geopolitics, few accounts engage with spiritual or religious identities. The second section of this book seeks to remedy this situation with two chapters that problematize the 'spiritual' amidst questions of nationhood, geopolitics and international relations. In this respect, the essays touch upon the long-standing tensions between idealism and realism that have characterized practical and formal geopolitical debates. Many geopoliticians have reproduced the simple dualism that perceives these categories as mutually exclusive. These two chapters demonstrate how these perspectives can coincide to underpin geopolitical thought.

The interconnections between religion and geopolitics inform Gearóid Ó Tuathail's investigation of Father Edmund Walsh. Trained as both a Jesuit Priest and a political scientist, Walsh's extraordinary career included positions as Director General of the Papal Relief Mission in Moscow in the 1920s, and special inter-rogator to Karl Haushofer before the Nuremburg War trials. Later, as a senior member of Georgetown University in Washington DC, he influenced and cajoled a generation of American political leaders including Senator Joseph McCarthy. His formidable religious zeal combined with his extensive experience of Russia to produce several high-profile works which displayed considerable antipathy to the Soviet Union and communism. Ó Tuathail demonstrates how the intellectual and political trajectories of individuals can help us unravel the histories of geopolitics. Additionally, he emphasizes the place of religion in traditions that are frequently self-styled as modern, secular and materialist.

Religious affiliations were also involved in the construction of inclusive and exclusive geopolitical imaginations in post-colonial India. Sanjay Chaturvedi demonstrates that after 1947, India's governing elites constructed a national-polit-ical community based upon exclusive conceptions of citizenship. This was because Hindu nationalists reconstituted India's geography and history in ways that failed to recognize ethnic and religious diversity. Indian politicians subsequently employed practices and discourses associated with national security and territorial borders to justify military spending and aggressive posturing in disputed territo-ries in the Kashmir and Punjab. While neighbouring states such as Pakistan also played their part in sustaining communal violence, Indian geopolitical reasoning remains rooted in fears that the borders of the state must be protected against all perceived 'threats', regardless of any human costs. Recent tensions over nuclear testing in May 1998 highlight the persistence of these geopolitical imaginations in the Indian sub-continent, and point to the challenges of formulating more inclusive political communities.

Reclaiming and refocusing geopolitics

The penultimate group of essays considers the reformulations of geopolitical ideas in the changing contexts of the late twentieth century. They also demonstrate the continuing recourse to geopolitics in attempts to explain global transformations. Perhaps the most consistent source of geopolitical analysis in the post-war period – and one of the most important attempts to re-work the 'geopolitical' – has been the French left-wing journal *Hérodote*. Accordingly, two essays in this book provide complementary accounts of *Hérodote* and its impacts.

First, Paul Claval provides a fascinating and personal portrait of French geography in the post-war period. As Professor of Geography at the Sorbonne, he is a long-standing commentator on French geography and its relationship to Anglo-American academia. Claval demonstrates how, in the 1970s, Yves Lacoste's journal *Hérodote* gradually engaged with geopolitics, initially through a recognition of the politics of geographical knowledge. Thereafter, this re-constituted geopolitics of the left has seen *Hérodote* extending the boundaries of the geopolitical. Claval outlines some of the results in his paper. He concludes with an honest appraisal of the *Hérodote* group and identifies some lacunae in this impressive body of thought.

Leslie Hepple's interpretation of post-war French geopolitics complements Claval's analysis with an English-language perspective upon *Hérodote*. In contrast to earlier geopolitical conversations and debates, Hepple argues that there has been a *lack* of exchange between French and Anglophone political geographies in recent decades. Historically, French political geography enjoyed important connections with Italian and Hispanic geographers. Anglophone geographers, however, were largely absent from these conversations due to poor language skills and their commitments to Anglo-American contacts. Like Claval, Hepple describes some of the significant themes of *Hérodote*. However, he claims that the potential connections between *Hérodote* and critical geopolitics – including their shared concerns with the politics of geographical knowledges and the importance of popular geopolitics – are compromized by this lack of communication. Hepple thus provides a contemporary argument for the re-building of networks of critical thinking about geopolitics.

The varied geopolitical imaginations of Israel and the Jewish Diaspora concern David Newman's timely chapter. In the aftermath of the 1993 Oslo Peace Accords, many Jewish and Arab commentators are seeking political and cultural solutions that might enable peaceful co-existence between the communities of Israel and the West Bank. But to make progress in this direction, Newman argues that Israelis have to address the cultural and geographical identities of the Jewish people. To this end, he outlines competing conceptions of the Israeli state. To some, Israel has been been defined as the *Jewish* state. But alternative geopolitical imaginations emphasize either Israel's special relationship with the United States, or situate the country as 'European'. Both these approaches marginalize other groups, notably Arabs, from Israeli mainstream political and social life. Other Jewish groups introduce a more spiritual angle and talk of Israel and Jerusalem as the centre of the Arabic, Christian and Jewish worlds. Whichever interpretation is favoured, Newman demonstrates that these geopolitical imaginaries are integral to daily life in Israel. Similarly, they are further evidence that we should not lose sight of the diffusion of geopolitical cultures throughout societies.

Joanne Sharp's account of the *Reader's Digest* in the post-Cold War period provides an insight into the huge significance of popular media in the translation and reproduction of geopolitical ideas. Reading the *Reader's Digest* as a critical source, Sharp explores how a rich vein of popular geopolitical ideas has informed and shaped contemporary cultural debate in America. In Sharp's analysis, the magazine also casts light upon America's attempts to identify a new Other to replace the vanished threat of Communism. This search encompasses drug traffickers, terrorism, and Islamic fundamentalism, but also touches upon a new range of domestic fears including excessive federal power and anxieties about social breakdown within America. As new dangers are inscribed onto the globe, these popular geopolitical debates demonstrate that the territorial and moral boundaries of the United States are likely to be quite different from the Cold War order of things.

Finally, Timothy Luke's examination of US environmental discourse in the 1980s and 1990s similarly reflects recent anxieties about global well-being. It also demonstrates the rise of environmental issues in geopolitics (Ó Tuathail, Dalby and Routledge 1998; Dalby 1996), and possibilities for re-reading the Anglo-American geopolitical tradition. By returning to Halford Mackinder's commentary on the closed system and time-space compression at the last *fin-de-siècle*, Luke traces ecological resonances within aspects of geopolitical thought, from the start of the century to its close. To illustrate these continuities, he outlines the goals, operational philosophies and policy orientations of the American environmental lobby organization, the Worldwide Watch Institute. These ideas are subjected to careful analysis which considers the institution's claims that the natural world can be controlled within a social network of capitalist and managerial regimes. Ultimately, Luke concludes that contemporary green debates in the United States revisit several of the geopolitical and ecological assumptions of Mackinder and others at the close of the nineteenth century: thus providing another angle upon the re-casting of geopolitical ideas.

Futures and possibilities

As the twenty-first century looms, it is clear that geopolitical thought, in whatever form, is unlikely to disappear. Indeed, during the Cold War period when geographers tried to pretend that it might, their actions simply surrendered geopolitical discourse to other, less scrupulous 'geopoliticians'. These ideas and traditions may well prove to be as problematic in the twenty-first century as they have for the past hundred years. However, they remain a intrinsic part of our wider geographical traditions, and as such, geographers have a continuing responsibility to engage with these ideas critically and rigorously.

In certain respects though, geopolitics will *not* be operating within the same matrices of politics, economies and cultures as throughout this past century. John

Agnew (1998) has argued that the 'modern geopolitical imagination' is founded upon the nation-state as its basic unit of analysis; the geopolitical traditions traced in these pages all reflect this. However, the accelerating erosion of orthodox political boundaries by processes such as globalization and information and financial flows undermines this traditional framework. Ó Tuathail (1998b) emphasizes that this is not simply a transition from the modern to the post-modern, but argues that contemporary geopolitics develop amidst a more complicated and changing milieu as a consequence of these transformations. What is not disputed is that a fundamental transformation is underway. To conclude the collection while leaving the broader problematic of geopolitics open-ended, the final section of the book comprises two brief commentaries upon the prospects and future possibilities for geopolitics.

Peter Taylor's reflections on the position of geopolitics within the social sciences provide a sobering conclusion. As a minor intellectual movement or cluster of ideas, geopolitics has always been on the periphery of university disciplines. Moreover, the contribution of geopolitics to exploring dominant themes of the twentieth century such as the nuclear threat to humanity has been slight. However, there are grounds for optimism and Taylor outlines the contribution of geopolitics to the reformulation of research on social and political change. He suggests that geopolitics has to engage with the bewildering changes of the contemporary world, from the restructuring of states and inter-state relations to the rise of religious fundamentalism. Ironically, the eclectic traditions of geopolitics might then prove a source of strength in these changing times.

Finally, Nigel Thrift challenges the over-emphasis upon narrowly-defined discourse and representation in critical geopolitics. Instead, he draws upon Billig's *Banal Nationalism* (1995) to suggest it is the practical, mundane, everyday 'little things' that actually enable the functioning of a broader sense of geopolitical discourse. This ranges from the everyday language of 'we' and 'us' that condones 'our' (usually national) formulations of political space, to the embodied geopolitical practices of clerks (usually women), files and archives that permit the practical governance of space. As the book concludes, Thrift's challenge for the future is the development of broader conceptualizations of geopolitical discourse that encompass the ways that geopolitics actually 'work'. The collection is thus left open-ended to allow these debates to continue.

Bibliography

Agnew, J. (1998) *Geopolitics*, London: Routledge.

Agnew, J. and Corbridge, S. (1989) 'The new geopolitics: the dynamics of geopolitical disorder', 226–88 in R. J. Johnston and P. J. Taylor (eds) *A World in Crisis? Geographical Perspectives*, 2nd edition, Oxford: Blackwell.

Alexander, L. (1961) 'The new geopolitics: a critique', *Journal of Conflict Resolution* 5: 407–10.

Antonsich, M. (1997) 'La geopolitica Italiana nelle Rivista 'Geopolitica', 'Hérodote/Italia' ('Erodoto'), 'Limes', *Bollettino della Società Geografica Italiana* series XII, 2: 411–18.

Atkinson, D. (1995) 'Geopolitics, cartography and geographical knowledge: envisioning Africa from Fascist Italy', 265–97 in M. Bell, R. A. Butlin and M. Heffernan (eds) *Geography and Imperialism, 1820–1940*, Manchester: Manchester University Press.

—— (1996) *Geopolitics and the Geographical Imagination in Fascist Italy*, unpublished Ph.D. Thesis, University of Loughborough, Loughborough.

Barnett, C. (1995) 'Awakening the dead: who needs the history of geography?', *Transactions of the Institute of British Geographers* 20: 417–19.

Bassin, M. (1987a) 'Imperialism and the nation state in Friedrich Ratzel's political geography', *Progress in Human Geography* 11: 473–95.

—— (1987b) 'Race contra space: the conflict between German *Geopolitik* and National Socialism', *Political Geography Quarterly* 6: 115–34.

Bell, M., Butlin, R. A. and Heffernan, M. (eds) (1995) *Geography and Imperialism, 1820–1940*, Manchester: Manchester University Press.

Berry, B. (1969) Book review, *Geographical Review* 59: 450–1.

Billig, M. (1995) *Banal Nationalism*, London: Sage.

Bosque-Maurel, J., Bosque-Sendra, J. and Garcia-Ballesteros, A. (1992) 'Academic geography in Spain and Franco's regime, 1936–55', *Political Geography* 11: 550–62.

Child, J. (1985) *Geopolitics and Conflict in South America*, New York: Praeger.

Chubb, B. (1954) 'Geopolitics', *Irish Geography* 3: 15–25.

Clarke, D. B. and Doel, M. A. (1998) 'Figuring the Holocaust: singularity and the purification of space', 39–61 in G. Ó Tuathail and S. Dalby (eds) *Rethinking Geopolitics*, London: Routledge.

Cohen, S. B. (1964) *Geography and Politics in a World Divided*, Oxford: Oxford University Press.

Crone, G. R. (1967) *Background to Political Geography*, London: Museum Press.

Cosgrove, D. (ed.) (1999) *Mappings*, London: Reaktion.

Dalby, S. (1988) 'Geopolitical discourse: the Soviet Union as Other', *Alternatives* 13: 415–22.

—— (1990a) *Creating the Second Cold war*, London: Pinter.

—— (1990b) 'American security discourse: the persistence of geopolitics', *Political Geography Quarterly* 9: 171–88.

—— (1991) 'Critical geopolitics: discourse, difference and dissent', *Environment and Planning D: Society and Space* 9: 261–83.

—— (1994) 'Gender and critical geopolitics: reading security discourse in the new world disorder', *Environment and Planning D: Society and Space* 12: 595–612.

—— (1996) 'The environment and geopolitical threat: reading Robert Kaplan's "Coming Anarchy"', *Ecumene* 3: 472–96.

Desfarges, M. (1996) *Introduzione alla geopolitica*, Bologna: Il Mulino.

de Seversky, A. P. (1942) *Victory through Airpower*, New York.

Dodds, K. (1998) 'Towards rapprochement: Anglo-Argentine relations and the Falklands/Malvinas in the late 1990s', *International Affairs* 74: 617–30.

Dodds, K. and Sidaway, J. (1994) 'Locating critical geopolitics', *Environment and Planning D: Society and Space* 12: 515–24.

Domosh, M. (1991a) 'Towards a feminist historiography of geography', *Transactions of the Institute of British Geographers* 16: 95–104.

—— (1991b) 'Beyond the frontiers of geographical knowledge', *Transactions of the Institute of British Geographers* 16: 488–90.

Driver, F. (1992) 'Geography's empire: histories of geographical knowledge', *Environment and Planning D: Society and Space* 10: 23–40.

—— (1994) 'New perspectives on the history and philosophy of geography', *Progress in Human Geography* 18: 92–100.

Driver, F., Matless, D., Rose, G., Barnett, C. and Livingstone, D. N. (1995) 'Geographical traditions: rethinking the history of geography', *Transactions of the Institute of British Geographers* 16: 403–22.

Enloe, C. (1989) *Bananas, Beaches and Bases: Making Feminist Sense of International Relations*, Berkeley: University of California Press.

Etzel Pearcy, G. E. , Fifield, R. H. and Associates (1948) *World Political Geography*, New York: Crowell.

Fahlbusch, M., Rössler, M. and Siegrist, D. (1989) 'Conservatism, ideology and geography in Germany, 1920–50', *Political Geography Quarterly* 8: 353–67.

Fushukima, Y. (1997), 'Japanese geopolitics and its background: what is the real legacy of the past?', *Political Geography* 16: 407–21.

Gilbert, E. W. (1972) 'Mackinder and Haushofer', in *British Pioneers in Geography*, Newton Abbot: David and Charles.

Godlewska, A. and Smith, N. (eds) (1994) *Geography and Empire*, Oxford: Blackwell.

GoGwilt, C. (1998) 'The geopolitical image: imperialism, anarchism, and the hypothesis of culture in the formation of geopolitics', MODERNISM/*modernity* 5: 49–70.

Harley, J. B. (1988) 'Maps, knowledge and power', 277–312 in D. Cosgrove and S. Daniels (eds) *The Iconography of Landscape: Essays on the Symbolic Representation, Design and Use of Past Environments*, Cambridge: Cambridge University Press.

—— (1989) 'Deconstructing the map', *Cartographica* 26: 1–20.

Haraway, D. (1991) *Simians, cyborgs and women: the reinvention of nature*, London: Free Association Books.

Hauner, M. (1992) *What is Asia to us? Russia's Asian heartland Yesterday and Today*, London: Routledge.

Heffernan, M. J. (1998) *The Meaning of Europe. Geography and Geopolitics*, London: Arnold.

Hepple, L. W. (1986) 'The revival of geopolitics', *Political Geography Quarterly* vol. 5 supplement: 21–36.

—— (1992) 'Metaphor, geopolitical discourse and the military in South America', 136–54 in T. Barnes and J. Duncan (eds) *Writing Worlds. Discourse, text and metaphor in the representation of landscape*, London: Routledge.

Heske, H. (1986) 'German geographic research in the Nazi period: a content analysis of the major geography journals', *Political Geography Quarterly* 5: 267–81.

—— (1987) 'Karl Haushofer: his role in German geopolitics and Nazi politics', *Political Geography Quarterly* 6: 135–44.

Heske, H. and Wesche, R. (1988) 'Karl Haushofer, 1869-1946', *Geographers Biobibliographical Studies* 12: 95–108.

Holdar, S. (1992) 'The ideal state and the power of geography: the life and work of Rudolf Kjellén', *Political Geography* 11: 307–23.

Jacob, C. (1999) 'Mapping in the mind: the earth from ancient Alexandria', 24–49 in D. Cosgrove (ed.) *Mappings*, London: Reaktion.

Jacobsen, K-A.(1979) *Karl Haushofer: Leben und werk* vol. I and II, Boppard am Rhein: Boldt.

Jan van Pelt, R. and Dwork, D. (1996) *Auschwitz: 1270 to the present*, London: Yale University Press.

Johnston, R. J. (1987) *Geography and Geographers*, 3rd edition, London: Arnold.

Kearns, G. (1983) 'Closed space and political practice: Frederick Jackson Turner and Halford Mackinder', *Environment and Planning D: Society and Space* 2: 23–34.

—— (1993) '*Fin-de-siècle* geopolitics: Mackinder, Hobson and theories of global closure', 9–30 in P. J. Taylor (ed.) *Political Geography of the Twentieth Century*, London: Belhaven.

Kern, S. (1983) *The Culture of Time and Space, 1880–1918*, Cambridge: Harvard University Press.

Korinman, M. (1990) *Quand l'Allemagne pensait le monde: grandeur et décadence d'une géopolitique*, Paris: Masparo.

Kost, K. (1989) 'The conception of politics in political geography and geopolitics in Germany until 1945', *Political Geography Quarterly* 8: 369–85.

Kristof, L. (1960) 'The origins and evolution of geopolitics', *Journal of Conflict Resolution* 4: 15-51.

Lacoste, Y. (1976) *La Géographie, ça sert, d'abord, à faire la guerre*, Paris: Maspero.

Livingstone, D. N. (1992) *The Geographical Tradition*, Oxford: Blackwell.

—— (1994) 'Science and religion: foreword to the historical geography of an encounter', *Journal of Historical Geography* 20: 367–83.

—— (1995a) 'Geographical traditions', *Transactions of the Institute of British Geographers* 20: 420–2.

—— (1995b) 'The spaces of knowledge: contributions towards a historical geography of science', *Environment and Planning D: Society and Space* 13: 5–34.

McEwan, C. (1998) 'Cutting power lines within the palace? Countering paternity and eurocentrism in the "geographical tradition"', *Transactions of the Institute of British Geographers* 23: 371–84.

MacIntyre, A. (1985) *After Virtue: A Study in Moral Theory*, London: Duckworth.

—— (1990) *Three Rival Versions of Moral Inquiry: Encyclopaedia, Genealogy and Tradition*, London: Duckworth.

Mackinder, H. J. (1942) 'The round world and the winning of the peace', *Foreign Affairs* 21: 595–605.

Matless, D. (1995) 'Effects of history', *Transactions of the Institute of British Geographers* 20: 405–9.

—— (1998) *Landscape and Englishness*, London: Reaktion.

Meilinger, P. S. (1995) 'Proselytiser and prophet: Alexander P. de Seversky and American airpower', *The Journal of Strategic Studies* 18, 1: 7–35.

Muir, R. (1997) *Political Geography. A New Introduction*. London: Macmillan.

Murphy, D. T. (1994) 'Space, race and geopolitical necessity: geopolitical rhetoric in German colonial revanchism, 1919–33', 173–87 in A. Godlewska and N. Smith (eds) *Geography and Empire*, Oxford: Blackwell.

O'Sullivan, P. (1982) 'Antidomino', *Political Geography Quarterly* 1: 57–64.

—— (1986) *Geopolitics*, London: Croom Helm.

Ó Tuathail, G. (1986) 'The language and nature of the New Geopolitics – the case of US –El Salvador Relations', *Political Geography Quarterly* 5: 73–85.

—— (1992) 'Putting Mackinder in his place: material transformation and myth', *Political Geography* 11: 100–18.

—— (1993) 'The effacement of place? US foreign policy and the spatiality of the Gulf Crisis', *Antipode* 25: 4–31.

—— (1996) *Critical Geopolitics*, London: Routledge.

—— (1997) 'At the end of geopolitics? Reflections on a plural problematic and the century's end', *Alternatives* 22: 35–55.

—— (1998a) 'Introduction: thinking critically about geopolitics', in G. Ó Tuathail, S. Dalby and P. Routledge (1998) *The Geopolitics Reader*, London: Routledge.

—— (1998b) 'Postmodern geopolitics? The modern geopolitical imagination and beyond', 16–38 in G. Ó Tuathail and S. Dalby (eds) *Rethinking Geopolitics*, London: Routledge.

Ó Tuathail, G. and Agnew, J. (1992) 'Geopolitics and discourse: practical geopolitical reasoning in American foreign policy', *Political Geography* 11: 190–204.

Ó Tuathail, G. and Dalby S. (eds) (1998a) *Rethinking Geopolitics*, London: Routledge.

—— (1998b) 'Introduction: rethinking geopolitics. Towards a critical geopolitics', 1–15 in G. Ó Tuathail and S. Dalby (eds) *Rethinking Geopolitics*, London: Routledge.

Ó Tuathail, G., Dalby, S. and Routledge, P. (1998) *The Geopolitics Reader*, London: Routledge.

Paasi, A. (1990) 'Political geography around the world VIII: the rise and fall of Finnish geopolitics', *Political Geography Quarterly* 9: 53–65.

Parker, G. (1985) *Western Geopolitical Thought in the Twentieth Century*, London: Croom Helm.

—— (1987) 'French geopolitical thought in the interwar years and the emergence of the European idea', *Political Geography Quarterly* 6: 145–50.

—— (1998) *Geopolitics: Past, Present and Future*, London: Mansell.

Raffestin, C., Lopreno, D. and Pasteur, Y. (1995) *Géopolitique et Histoire*, Lausanne: Éditions Payot.

Rose, G. (1993) *Feminism and Geography*, Cambridge: Polity Press.

—— (1995) 'Tradition and paternity: same difference?', *Transactions of the Institute of British Geographers* 20: 414–16.

Roucek, J. S. (1955) 'The geopolitics of the United States', *American Journal of Economics and Sociology* 14: 185–92 and 287–303.

—— (1956) 'The geopolitics of Yugoslavia', *Social Studies* 47: 26–9.

—— (1962) 'The geopolitics of the Congo', *United Asia* 14: 81–5.

Ryan, J. (1994) 'Visualising imperial geography: Halford Mackinder and the Colonial Office Visual Instruction Committee, 1902-1911', *Ecumene* 1: 157–76.

—— (1997) *Picturing Empire. Photography and the Visualization of the British Empire*, London: Reaktion.

Sandner, G. (1988) 'Recent advances in the history of German geography, 1918–45: a progress report for the Federal Republic of Germany', *Geografische Zeitschrift* 76: 120–33.

Sandner, G. and Rössler, M. (1994) 'Geography and empire in Germany, 1871–1945', 115–29 in A. Godlewska and N. Smith (eds) *Geography and Empire*, Oxford: Blackwell.

Sauer, C. (1927) 'Recent developments in cultural geography',154–212 in *Recent Developments in the Social Sciences*, NewYork: Lippincott.

Schwartz, J. M. (1996) 'The geography lesson: photographs and the construction of imaginative geographies', *Journal of Historical Geography* 22: 16–45.

Shapiro, M. (1998) *Violent Cartographies*, Minneapolis: University of Minnesota Press.

Sharp, J. (1993) 'Publishing American identity: popular geopolitics, myth and *The Reader's Digest*', *Political Geography* 12: 491–503.

—— (2000) *Condensing Communism: The Readers Digest and American Identity 1922–94*, Minneapolis: University of Minnesota Press.

Sidaway, J. (1997) 'The (re)making of the western "geographical tradition": some missing links', *Area* 29: 72–80.

Skinner, Q. (1969) 'Meaning and understanding in the history of ideas', *History andTheory* 8: 3–53.

Smith, N. (1984) 'Isaiah Bowman: political geography and geopolitics', *Political Geography Quarterly* 3: 69–76.

—— (1987) '"Academic war over the field of geography": the elimination of geography at Harvard', *Annals of the Association of American Geographers* 77: 155–72.

—— (1999) 'The lost geography of the American century', *Scottish Geographical Journal* 115: 1–18.

Smith, N. and Godlewska, A. (1994) 'Introduction: critical histories of geography', 1–12 in A. Godlewska and N. Smith (eds) *Geography and Empire*, Oxford: Blackwell.

Sondern, F. (1941a) 'Hitler's scientists', *Current History* 53: 10–18.

(1941b) 'The thousand scientists behind Hitler', *The Reader's Digest* 38, 7: 44–8.

Stoddart, D. (1986) *On Geography and its History*, Oxford: Blackwell.

Takeuchi, K. (1980) 'Geopolitics and geography in Japan re-examined', *Hitosubashi Journal of Social Studies* 12: 14–24.

—— (1994) 'The Japanese imperial tradition, western imperialism and modern Japanese geography', 188–209 in A. Godlewska and N. Smith (eds) *Geography and Empire*, Oxford: Blackwell.

Taylor, G. (1957) 'Geopolitics and geopacifics', 587–608 in G.Taylor (ed.) *Geography in the Twentieth Century*, 3rd edition, London: Methuen.

Walsh, E. A. (1944) 'Geopolitics and international morals', 12–39 in H. Weigert, V. Stefansson and R. Harrison (eds) *Compass of theWorld*, London: Harrap.

Weigert, H.W. (1942) *Generals and Geographers:The Twilight of Geopolitics*, Oxford: Clarendon.

Werrity, A. and Reid, L. (1995) 'Debating the geographical tradition', *Scottish Geographical Magazine* 111: 196–8.

Whittlesey, D., Hartshorne, R. and Colby, C. (1942) *The German Strategy ofWorld Conquest*, London: F. E. Robinson.

Withers, C.W. J., Camerini, J., Heffernan, M. J. and Livingstone, D. (1996) 'Conversations in review', *Ecumene* 3: 351–60.

Wittfogel, K. (1985) [1929] 'Geopolitics, geographical materialism and Marxism', trans. G. L. Ulmen, *Antipode* 17: 21–72.

Part 1

RETHINKING GEOPOLITICAL HISTORIES

2

FIN DE SIÈCLE, FIN DU MONDE?

On the origins of European geopolitics, 1890–1920

Michael Heffernan

Introduction

The assemblage of ideas and theories that have been gathered under the curious label 'geo-politics' has a complex and often disturbing history. The different phases in this history are discussed in the other chapters in this volume and in several recent texts (Ó Tuathail 1996; Ó Tuathail, Dalby and Routledge 1998; Parker 1998; Raffestin, Lopreno and Pasteur 1995). My concern here is with points of departure, with intellectual roots rather than subsequent trajectories. What follows is a personal reading of why and how something called 'geo-politics' emerged as a distinctive, albeit amorphous, intellectual project about one hundred years ago. I am primarily interested in geopolitical writings from Britain and Germany and I hope to demonstrate how these formative texts can be read as commentaries on the future of Europe and 'European civilization'.[1]

The 'transformation of the year 1900': the birth of a 'speculative science'[2]

The term 'geo-politics' was coined in the last year of the nineteenth century by the Swedish political scientist Rudolf Kjellén. Like many newly invented words of the 1890s, 'geo-politics' was a portmanteau expression devised to convey a sense of novelty. This was to be a new kind of study for a new century, one which offered the prospect of understanding the world's nation-states, their borders and territorial capacities, and their relations with other states (Holdar 1992; Ó Tuathail 1996: 43–5; Raffestin, Lopreno and Pasteur 1995: 77–102).

Needless to say, geopolitics did not emerge as a fully-formed, coherent body of fact and theory. The discipline (in so far as it ever warranted such a term) was little more than a loosely defined perspective, an attempt to reveal textually and carto-graphically the complex relationships between geography and politics at a variety of spatial scales from the local to the global. The invention of the word was, in this sense, relatively unimportant and marked only a terminological modification of an existing intellectual agenda, previously labelled 'political geography'. This older subject had long been taught in universities and was widely discussed by academics, diplomats and professional politicians. The first problem we encounter is, therefore, a simple definitional one: what was geopolitics?

There is little to be gained by adopting a tightly restrictive definition for we are dealing with what might be termed a 'discourse', a constellation of writings and images produced by a varied constituency operating in several loci, including the universities, the media and government ministries. By no means everyone involved in this discourse would have recognized or used the term geopolitics (or even political geography). However, they all shared an interest in the kind of study which Kjellén sought to stake out under this label; they were all concerned to make sense of the often unexamined geographical dimension in world politics.

The idea that this project required a new name in 1899 reflected a widespread belief that the changes taking place in the global economic and political system were seismically important. The shift from an older industrial capitalism based on steam, coal and iron to a newer version based on gas, oil and electricity seemed to change the ground rules by which the world economy functioned. The USA was poised on the brink of a Fordist revolution of intensive industrial production geared to a new, rapidly expanding mass market and seemed perfectly placed to dominate the new age. America had already supplanted Britain as the global economic hegemony by the turn of the century, and on the eve of the First World War American factories were producing one-third of the world's industrial goods, a dominance Britain had claimed a mere forty years earlier.

These changes seemed to foreshadow comparable transformations in the global political order. The fact that the USA was a continental-scale land power, with unprecedented rail and road connections linking major cities on both the Atlantic and the Pacific coasts, pointed towards a new relationship between space and state politics that was entirely at variance to the traditional European world order. This older system involved relatively small European states with distant, often chaoti-cally scattered empires, 'spatch-cocked' together by the fragile sinews of maritime trade. Many believed that the future would be dominated by three or four spatially extensive land confederations (comparable to the USA) which would emerge in Asia, Africa and Europe itself.

This was the context within which 'geopolitics' was born: an era of apparently dramatic global economic change with equally dramatic political implications. Not surprisingly, this vision of the future provoked fear, unease and uncertainty, a kind of 'geopolitical panic', particularly in Europe where an entire generation of politicians, diplomats and intellectuals expended much of their energy devising strategies to cope with the apparently imminent collapse of their familiar world order.

There were at least three dimensions of this 'geopolitical panic' which together opened a space for the supposedly new, 'scientific' project of geopolitics. The first was an upsurge in economic nationalism and a general clamour for tariff reform and protectionism. This flew in the face of the cherished ideal of free trade that an earlier generation of European political economists had espoused. To some extent, protectionism was a despairing attempt by traditional nation-states to limit the disruption caused by an increasingly global and integrated world economy. But it also reflected the growing conviction that the future would be dominated by large, spatially cohesive and economically self-sufficient geopolitical units. The USA was the exemplar here for its enormous economic power (and even greater potential) was based on vast internal resources and a rapidly expanding domestic market. The quest for economic autarky (self-sufficiency) was initiated by Bismarck's protectionist policy for the newly-united German Empire from the late 1870s and was followed by successive French governments, culminating with the Méline tariff of 1892 that sealed off the French agricultural sector from the world market. In Britain, Joseph Chamberlain's ill-fated campaign for imperial reform was likewise designed to strengthen the economic ties between the imperial core and the colonial periphery and make the British imperial system more profitable.

This last example points towards a second European reaction to the real and imagined economic and political transformations of the late nineteenth century. Fearful that geographical size would determine national power and aware that they would be unable to expand on a European stage, the principal imperial powers embarked on an unprecedented 'scramble' for imperial space from the 1880s onwards. Over 16 million square kilometres (20 per cent of the earth's surface) and 150 million people (10 per cent of the world's population) were added to the European empires during the last thirty years of the nineteenth century, notably in Africa which was divided between the 'great' powers at the Berlin Conference of 1884–5. Precise objectives varied, of course, but this final surge of imperial expansion was arguably the product of despair rather than triumphant optimism or cynical calculation. Colonies were coveted not for immediate economic gain (still less for moral reasons) but as symbols of an otherwise vulnerable national pride. No self-respecting power could be without its 'place in the sun'.[3] Colonial expansion became an attempt to acquire comparative territorial advantage outside Europe in the hope that this would allow small European states to survive in the coming world order.

Imperial optimists insisted that these vast empires would eventually generate huge incomes but most colonial systems remained vexingly unprofitable. Only

10 per cent of French exports and an even smaller percentage of French overseas investments were directed towards the country's overseas empire. Less than half of French imperial exports ended up in *la mère patrie* (Andrew and Kanya-Forstner 1981: 14–17). Maintaining a German colonial presence in Africa and Asia cost the German taxpayers £6 million in 1913 (Fischer 1967: 102–4).

Competition between the older, smaller European states, initially deflected into the colonial arena, soon reverberated back onto Europe itself. This produced the third dimension of Europe's late nineteenth-century 'geopolitical panic'. From the 1890s, the European inter-state system underwent a funda-mental transformation. The complex web of bi-lateral treaties and pacts that had characterized the middle decades of the century was suddenly replaced by a simpler, and more dangerous, bi-polar arrangement. While the former was by no means peaceful, the latter ensured that an otherwise containable war would escalate into the kind of general European conflagration that engulfed the continent in August 1914.

The bi-polar system was the direct reaction to Germany's economic and polit-ical ascension. The new manufacturing industries of electrical goods, chemicals and metal processing played a more dominant role in Germany than elsewhere in Europe and, partly as a result, German industrial productivity was second only to that of the USA by the eve of the First World War. This new-found economic dominance was matched by a more aggressive imperial stance both within Europe – particularly towards the south, where Germany's ally, Austria-Hungary, acted as a kind of 'cat's-paw' for German territorial ambitions, and towards the east where expansion was sought at the expense of the sprawling but vulnerable Russian Empire.

Anxious to contain Germany, the other European imperial powers – Britain, France and Russia – suspended their traditional antagonisms and established an encircling alliance to counteract German expansion, an arrangement secured by the Franco-Russian treaty of 1894, the Anglo-French Entente Cordiale of 1904 and the Anglo-Russian accord of 1907. This was accompanied by an unprece-dented arms race which further exacerbated Europe's economic difficulties. The main European armies increased by an average of 73 per cent between 1880 and 1914 while warship tonnage grew by a factor of four, the latter due primarily to a head-to-head Anglo-German contest for control of the seas (Herrmann 1996; Kennedy 1988: 249–354; Massie 1991).

These changes in the global economic order, together with the fear of compa-rable political shifts, created the intellectual environment in which the new 'science' of geopolitics arose. While this partially explains the timing of the new project, it does not explain that project's characteristics. For this, we need to look more carefully at the cultural assumptions and preoccupations of those who pioneered this new perspective.

Endgames

There is no reason why the turning of a century should carry any particular historical significance. The idea that history falls neatly into hundred-year blocks and can be assessed in these units is obviously ludicrous. Yet in the supposedly Christian parts of the world where the larger sweep of time is measured in such terms, the passing of a century tends to give rise to wistful reflection on the past as well as intense speculation about the future.[4] The arrival of 'landmark' dates engenders a heightened sense of temporality, of time passing. Changes which are actually taking place at these junctures tend to acquire extra (sometimes mystical) layers of meaning.[5] This was certainly the case in the 1890s, a decade of 'semiotic arousal' when everything, it seemed, was a sign, a harbinger of some future radical disjuncture or cataclysmic upheaval.

The emergence of geopolitics (which soon lost its hyphen) was one aspect of this *fin-de-siècle* mentality, a point which Gerry Kearns (1993) has eloquently demonstrated. The original French expression, meaning simply 'end of century', became a catch-all phrase to describe everything from the architectural and artistic styles of the period through the developments in fashion, design and technology to the wider, often impassioned debates about the past, the present and the future on the eve of a new century (Laqueur 1996; Schorske 1979; Stokes 1992; Teich and Porter 1990; Weber 1986).

Geopolitics echoed these wider concerns about the future. While masquerading as a new, rational and scientific approach, most early geopolitical texts reflected the anxieties, fears and hopes of the last *fin-de-siècle* and were highly – sometimes wildly – speculative. In common with much *fin-de-siècle* writing, geopolitical texts tended to assume that the passing of the nineteenth century would represent a fundamental historical discontinuity, a clear break with the past. As we have seen, there was some evidence to support this dramatic claim but this conviction reflected a deeper cultural conviction that centennial endings must, by definition, be accompanied by profound upheavals.

The same texts also reflected another *fin-de-siècle* trait: the conviction that the new world of the twentieth century would need to be understood in its entirety, as an integrated global whole. Technology and global transportation by steamship, road, rail and telegraph would make the world of the future a 'smaller' place, one that could be interpreted as a single system. The rise of the USA suggested lessons about the future world order but some optimists predicted that US-style, continental-scale systems would be supplanted by a new planetary consciousness, a utopian 'one-worldism', that would arise on the back of the new space-defying technologies.[6] But not everyone was optimistic about the future. Indeed, fearful ambivalence was the dominant mood of the *fin-de-siècle*. While some welcomed the prospect of a new age and saw rapid change as energizing and liberating, many others

31

worried about the dislocation and disorientation these imagined upheavals would create and lamented the destruction of cherished traditions and values that seemed destined to ensue. The debate between optimists and pessimists was at its most intense in the major European capitals, citadels of traditional imperial nation-states whose future prospects seemed less than assured. The fate of Europe (or 'European' civilization) was a principal concern in this debate. Those who embraced change tended to see a new (and perhaps united) Europe emerging to occupy a central place in the global order; those who feared change foresaw a dangerous era in which European hegemony would be undermined and ultimately destroyed by the rise of very different empires to the west (where the USA seemed destined to dominate the entire American land mass, north and south, and perhaps also the vast expanse of the Pacific Ocean) and to the east (where some predicted the emergence of an Asiatic power arising from the ashes of the still quasi-feudal empires of Russia or Japan).

In the remainder of this chapter, I want to illustrate some aspects of the geopolitical debate about Europe at the last *fin-de-siècle* by reference to the foundational British geopolitical writings by Halford Mackinder (1861–1947), the leading English geographer of the period, and a variety of German texts, including those produced by Friedrich Ratzel (1844–1904), doyen of the German geographical community.

Europe and the new world order: from London . . .

One of the more pessimistic readings of Europe's potential was developed by the founder of British geopolitics, Halford Mackinder. Much has been written on Mackinder, a towering figure in the early history of British geography as a university and school subject. He was appointed Reader in Geography at the University of Oxford in 1887, the first modern acknowledgement of the subject's significance by the country's senior university. He went on to become Principal of the fledgling university at Reading and Director of the London School of Economics (LSE). This was followed by a political career as Unionist MP after 1910 (Blouet 1987; Kearns 1985; Parker 1982). Although he retained a belief in the value of travel and exploration (he made the first ascent of Mount Kenya in 1899) Mackinder is best known as the leading prophet of the 'new geography' in Britain (Kearns 1997). Drawing on the ideas of French and German geographers, especially Friedrich von Richthofen, he blazed a trail for a more theoretically-informed, synthesizing subject, a conceptual bridge between the natural and social sciences (Mackinder 1887). As an influential advocate of imperial reform, Mackinder's ideas were shaped by his desire to preserve and enhance Britain's position as a global power. His was a planetary political geography, an attempt to understand how the different nations and regions of the world inter-related as elements in a holistic geopolitical structure. Education lay at the heart of his concerns and geography lay at the core of his educational strategy. An imperial nation like Britain demanded the cultivation of an 'imperial race' trained in the art of 'thinking imperially'. This depended on a spatial and visual sensibility gleaned from map and landscape interpretation, the central preoccupations of the geographer (Mackinder 1911; Ó Tuathail 1996: 75–110; Ryan 1994).

Mackinder's desire to build a 'new geography' was informed by his belief that the world was changing in fundamental ways. His views were eloquently expressed in a famous lecture, delivered before the Royal Geographical Society in London in January 1904 (Mackinder 1904). Like Penck, Mackinder argued that geographical exploration was a finite task which would soon be completed. The age when the basic geographical facts about the world were fully appreciated was at hand. Geography, hitherto driven by the necessary but conceptually straightforward challenge to map and identify the 'unknown' places of the earth, needed to respond to the challenges of the new century. '[G]eographical exploration is nearly over', Mackinder insisted, 'and . . . geography must be diverted to the purpose of intensive survey and philosophic synthesis'.

Failure to develop explanatory geographical theories carried dire consequences for the 'old' imperial states. The closing of the nineteenth century marked the end of the 'Columbian age' of European expansion, the era when Europeans had escaped their Medieval limits and expanded 'against almost negligible resistances'. Europe's 'colonial frontiers' had now been pushed to their limits: 'there is scarcely a region left for the pegging out of a claim of ownership' (Mackinder 1904: 421). This was a worrying development. Like many imperialists, Mackinder believed that the colonization of the 'wide-open spaces' of Africa and Asia had acted as a necessary 'safety-valve' allowing Europe's restless ambition to be deflected away from territorial conflict in Europe and towards the useful development of empires. The 'closure' of the world to further European expansion would generate social and political tension within Europe while increasing the likelihood of conflict along imperial frontiers which were now shared with rival imperial powers and hence static rather than unfolding. 'Every explosion of social forces', he wrote, 'instead of being dissipated in a surrounding circuit of unknown space and barbaric chaos, will be sharply re-echoed from the far side of the globe, and weak elements in the political and economic organism of the world will be shattered in the consequence' (Mackinder 1904: 422).

Mackinder's Eurocentrism, his refusal to consider the rights of colonized peoples, was characteristic both of his class and of the era in which he wrote. But his analysis was undeniably correct in one respect. By the late 1890s, European expansion into Africa and Asia had became a source of conflict between the great powers rather than a means of reducing tension. The British desire to establish a north-south 'Cape-to-Cairo' link through east Africa, for example, clashed directly with French objectives to open a route across the continent from west to east. The two countries came perilously close to war in 1898 when Kitchener's troops confronted Marchard's at Fashoda on the banks of the White Nile. Russia and Britain were also embroiled in a long-running 'Great Game' of espionage and counter-espionage, the 'Cold War' of the imperial age, along the flanks of their respective Asian empires in the mountains and foothills of the Hindu Kush and the Khyber Pass. While these contests would be partially resolved by the emergence of a bi-polar European system after 1907, the potential for European crisis across this divide was ever-present. France and Germany twice came to the brink of war, in 1905 and 1911, over their rivalry in Morocco.

The end of the 'Columbian age' represented, Mackinder asserted, a crisis for the old imperial powers of Europe, particularly Britain. With a minuscule 'domestic' territory and widely-scattered imperial possessions, Britain was vulnerable both to rival imperial powers and to gathering resistance within its empire (as the Boer War demonstrated). 'Imperial overstretch' would be exacerbated, Mackinder claimed, by the eclipse of sea-power in the post-Columbian era.[7] Sea-power, insisted Mackinder, had been the basis of Europe's hegemony in the Columbian age but global dominance in the future would depend on control and exploitation of the vast resources of the world's land masses (Mackinder 1902). In the new age, a revolution in overland transportation by rail would mean that continents could be bound together by networks of iron, radiating from great transcontinental railways. The Trans-Pacific railroad across North America and the Trans-Siberian across Russia's Asian empire were the first, probing tentacles in a new transport system destined to encircle the globe (Hauner 1992: 142–5).[8]

Asia, the largest land mass on the planet, was the key to Mackinder's world view. This was his 'geographical pivot of history'. The idea that a new railway system could conquer the immense expanse of Asia was an awesome, and deeply disturbing, prospect. 'The century will not be old', he predicted, 'before all Asia is covered with railways' (Mackinder 1904: 434). The failure of this system to materialize does not detract from the provocative impact of Mackinder's prediction in 1904. European civilization, 'the outcome of the secular struggle against Asiatic invasion' in Mackinder's estimation, was directly threatened by a new Eurasian land empire occupying the geostrategic centre of the globe. Europe's global hegemony, established by small, seafaring nations which had embraced the rule of law and democracy (Britain, France and the Netherlands), had prevailed in the age of sail and steam. The world they created was unlikely to survive into the age of continental railways and long-distance travel. Asia was poised on the brink of global dominance. The question was which of the existing states around the edge of the Asian land mass was destined to control this crucial arena?

Imperial Russia, the only existing 'Eurasian' power, was best placed to dominate the Asian land mass. However, this awesome task might prove too much for an unreformed Tsarist Russia which seemed incapable of embracing the modern age (the Trans-Siberian railway notwithstanding). Other powers, located in the 'crescent' surrounding the Asian land mass, offered alternative threats. The emergence of Japan, whose forces were poised to defeat the Russian army as Mackinder was preparing his lecture in 1904, raised the spectre of an Asian empire developing from the east rather than the west. But Germany, a state apparently obsessed with territorial expansion, seemed the obvious non-Russian source of a Eurasian empire. The dread scenario, Mackinder believed, was an alliance between Russia and Germany. This would establish an irresistible force: the modern, machine-efficiency of Germany coupled with the massive demographic bulk of peasant Russia. The Asian land mass, 'the world island', would be swiftly overwhelmed by such an alliance. This would herald an entirely new era in the world's geopolitical system. Whereas previous episodes of European colonial expansion had created peripheral, maritime empires which had sustained the dominance of the European core, the conquest of the 'world island' would represent not the expansion of Europe but its virtual abandonment in favour of the huge expanse of Asia. European civilization, the maritime empires, and the traditional balance of power would collapse. The 'Columbian age' of European dominance would end and a new, Eurasian era would begin. If Britain and Europe were to survive into the twentieth century, the fearful prospect of a Germano-Russian alliance must be prevented at all costs (Figure 2.1).

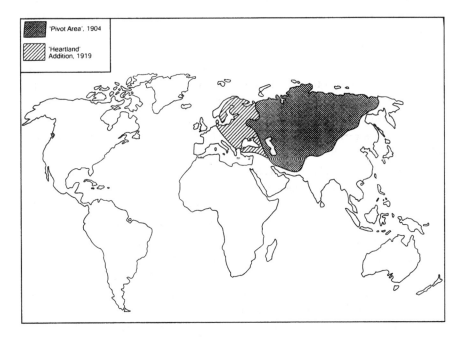

'Pivot Area', 1904

'Heartland' Addition, 1919

Figure 2.1 Mackinder's Heartlands, 1904–1919

Source: Mackinder 1904: 435; Mackinder 1919: 135 (redrawn for this volume to improve clarity)

The historical context within which Mackinder wrote was all-important. His lecture was delivered during the preliminary discussions for the Anglo-French Entente Cordiale, signed ten weeks later. As we have seen, this accord was a key element in the creation of the so-called 'Triple Entente' which encircled Germany and its central European allies and which created thereby the bi-polar European order. Mackinder's analysis of the Eurasian threat to the existing European order was an elaboration of Britain's new role in European power politics, an exercise in 'shock' tactics designed to underscore the need for an 'encircling' European alliance which would contain Germany on all sides, bind Russia to France and Britain, and remove the prospect of an alliance between the two potential Eurasian powers. The British lion, Mackinder implied, must awaken from its complacent imperial slumber. Any residual belief in Britain's 'splendid isolation' from European political intrigues was no longer tenable. The world had changed, claimed Mackinder, and Britain had to respond to protect its position and to ensure the survival of European civilization. Mackinder's 1904 lecture was a defining moment in the history of geopolitics; a foundation statement about the dangers to Britain and Europe in the twentieth century. It was both a reflection of, and a contribution to, a global territoriality which has grown steadily stronger during the twentieth century. For Mackinder, and those who followed his reasoning, geographical size was all-important. Land, resources and Mackinder's own 'portmanteau word' 'man-power' were assumed to determine national significance and prestige (Mackinder 1905).

In his *Democratic Ideals and Reality: A Study in the Politics of Reconstruction*, published on the eve of the post-First World War peace conferences in early 1919, Mackinder offered an extended analysis of his 1904 paper and a rather gloomy interpretation of Europe's future (Mackinder 1919). The war had 'established, and not shaken' his earlier views. The enhanced role of overland transportation had been demonstrated by Germany's ability to fight a war on two fronts and the eclipse of sea power had been revealed by Britain's failure to exploit its naval superiority. But the Bolshevik takeover in Russia and the likely changes to the political landscape of east-central Europe forced Mackinder to re-think the details of his theory. Although the war had raged around the world and reached its peak of savage intensity in northern France and Belgium, it had really been a struggle for control of eastern Europe, Mackinder claimed, for this was the gateway to the Asian land mass, a region he now dubbed the 'Heartland'. Control of the 'Heartland' was still the key to world power. However, the western limits of this sprawling area now 'intruded' into the core of 'old' Europe.

The idea of an expansive Europe stretching from the Atlantic to the Russian steppelands, what Mackinder called 'the real Europe', no longer existed. This had previously been limited by a line running east from St. Petersburg to Kazan and then south along the valleys of the Volga and Don to the Black Sea. This was 'the Europe of the European peoples, the Europe which, with its overseas Colonies, is Christendom'. But the 'real Europe', for so long 'a perfectly definite social conception', had been gradually dividing into two divergent and antagonistic regions, east and west.

The zone-of-transition between these two areas ran from the shores of the eastern Baltic to the Gulf of Venice, a remarkable prediction of the Cold War division after 1945 (Figure 2.2). The regions on either side were environmentally, culturally and geopolitically opposed; two different worlds. To the west, in the 'Coastland', a balance now existed between Britain and France, the great powers which had preserved the heritage of 'real' European civilization. To the east, on the western fringes of the Heartland, lay Russia and much of Germany, the most likely Eurasian powers. Russia's destiny as an Asian power was sealed but Germany, lying athwart this great divide, was the more problematic and unpredictable power; it looked in two directions at once. The division of Europe meant that Germany could not claim full European credentials and it seemed likely to turn its back on the Atlantic world of the 'Coastland'. This would spell the end of 'real Europe', the collapse of European civilization.

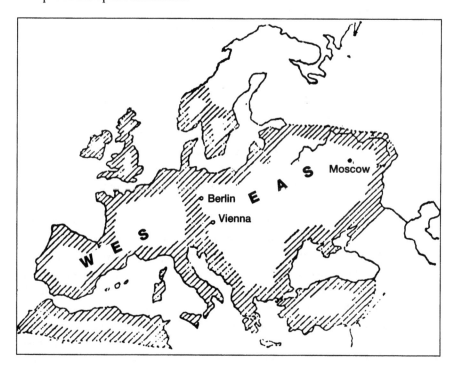

Figure 2.2 Mackinder's Divide

Source: Mackinder 1919: 154 (redrawn for this volume to improve clarity)

The threat of a future alliance between Germany and Russia was still the ultimate danger, particularly if Bolshevism were to take root in Germany. Equally worrying was the prospect of a German conquest of the Slavic world. This was unlikely in the immediate future but could not be ruled out in the long term. The encircling strategy of the pre-1914 era had failed to prevent Germany waging a war of expansion and had been in any case 'an incongruous . . . alliance between Democracy and Despotism' (Mackinder 1919: 161). An alliance between Democracy and Bolshevism would be even more incongruous, Mackinder implied. Germany's future expansion into the Heartland was thus a distinct possibility. The pulse of the Heartland might eventually beat to rhythms laid down in Berlin rather than Moscow or Petrograd.

The lessons were clear and unaltered by the war: 'West Europe, both insular and peninsular, must necessarily be opposed to whatever Power attempts to organize the resources of East Europe and the Heartland' (Mackinder 1919: 178). A few pages later, Mackinder coined his famous dictum to sum up his geopolitical vision:

> Who rules East Europe commands the heartland:
> Who rules the Heartland commands the World-Island;
> Who rules the World-Island commands the World.
>
> (Mackinder 1919, 194)

Preventing an alliance between Germany and Russia in the east, within the 'Heartland', should be the basis of Allied policy (Mackinder 1919: 155–60). The only way to achieve this in 1919 was to establish a tier of east European 'buffer states', supported by the western powers, which could separate Germany from Russia (Figure 2.3) (Mackinder 1919: 209). Mackinder's vision of a Europe divided into two, mutually-exclusive realms reiterated, in the language of twentieth-century geopolitics, the traditional east-west conceptualization of the continent which can be dated back to the eighteenth century if not earlier (Wolff 1994).

Mackinder's 1919 analysis was intended as a lesson in geopolitical realism, a counterblast to the 'naïve idealism' he detected in the proclamations of the American President, Woodrow Wilson, and amongst the advocates of a new League of Nations which might facilitate the peaceful resolution of international conflicts. If such an organization was to be a force for good, it must avoid unworkable concepts such as 'internationalism' or 'one-worldism'. Nor should its agenda be set by lawyers concerned with abstract legal principles. Mackinder's conclusion was that naïve 'democratic ideals' ran counter to harsh geographical realities. Appeals to 'universalism' or 'the equality of nations' were futile in a world where access to space and resources was inherently unequal. 'There is no such thing as equality of opportunity for nations', claimed Mackinder,

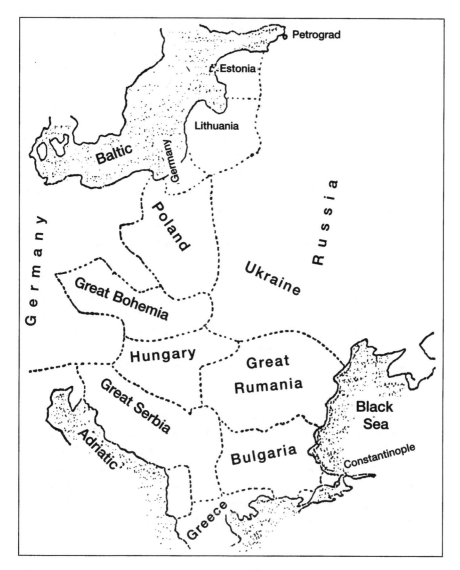

Figure 2.3 Mackinder's Buffers

Source: Mackinder 1919: 207 (redrawn for this volume to improve clarity)

Geography, the grouping of lands and seas, and of fertility and natural pathways, is such as to lend itself to the growth of empires, and in the end to a single World Empire. If we are to recognise our ideal of a League of Nations which shall prevent war in the future, we must recognise these geographical features and take steps to counter their influence.

(Mackinder 1919: 2–3)

... to Leipzig

Mackinder's predictions generated widespread interest though not everyone was impressed by his reasoning. George Chisholm, Professor of Geography at Edinburgh University, dismissed Mackinder's famous dictum about Eastern Europe, the Heartland and the World Island as 'a somewhat syllogistic chain of affirmation', while the American geographer, Charles Redway Dryer, criticized his failure directly to consider the future role of non-European powers, particularly the USA (Chisholm 1919: 250; Dryer 1920). Leo Amery (then a journalist on *The Times* and subsequently a prominent Conservative politician) argued in response to Mackinder's 1904 paper that a new era of land power and intercontinental railways was unlikely to materialize because another transport revolution was about to occur. The prospect of a world bound together by aeroplanes and global telegraph systems made the geopolitical distinction which Mackinder drew between land and sea power redundant (Mackinder 1904: 441).

Others were attracted by Mackinder's underlying logic, if not his conclusions. The American geographer Frederick Teggart transformed Mackinder's analysis into a quasi-racist denunciation of all things Russian and a spirited defence of Germany as a 'natural' ally of the 'West' in its age-old struggle against the 'East'. Germany, Teggart asserted, was Europe's bulwark against the westward expansion of Russian despotism, now manifested in the new guise of Bolshevism.

> [I]n the opposition to the power of the Heartland, the interests of the Western powers (England, France and Germany) were identical. In actively opposing the Russian menace to the marginal lands, Germany and Austria were carrying out, for good or ill, the historic mission which Western Europe had entrusted them. Germany is an integral part − the defensive frontier − of Western Europe. If her politico-military organization is a menace, it is the logical product of Western ideas.
>
> (Teggart 1919: 240)

A 'United States of Europe', including a fully-integrated Germany, was the only solution to 'the new menace of Russia', claimed Teggart (Teggart 1919, 241).

Such views were common currency in Germany, where Mackinder's ideas were widely discussed. Max Scheler also claimed that the war was a struggle between Europe and Asia, a spiritual crusade which would ultimately ennoble and reinvigorate the European life-force. Unfortunately, matters had become unnecessarily complicated, Scheler argued, by the unnatural alliance which Britain and France, Germany's natural European allies, had forged with alien, Asiatic Russia. Germany had been obliged to carry the sole burden of European culture in this historic struggle, while suffering murderous attacks by those peoples who shared a common European interest (Scheler 1915).

Mackinder's belief that cohesive land empires were likely to replace unsustainable maritime empires scattered haphazardly across the globe also struck a chord in Germany where an emerging school of 'pan-regionalism' likewise predicted that existing European states would be unviable in a future world of rapid, long-distance transport. Europe's political geography was destined to change as its states coalesced to form larger, land-based economic and political units, self-sufficient regional confederations. But the suggestion that a new global configuration would split Europe in two was firmly rejected by most German critics. This was scarcely surprising, of course, for this scenario directly challenged both the territorial integrity of a united Germany and the European credentials of the German people.

A different perspective, based on the idea of *Mitteleuropa* (Central Europe), was advanced as an alternative vision of a future Germany in a new Europe. Unlike Mackinder's 'Heartland', an emerging *Mitteleuropa* would preserve Europe's (and Germany's) historic integrity, or so its proponents claimed. The idea that Europe could be divided into three 'natural' regions ranged from west to east, with a central belt, dates back to the mid nineteenth century (Brechtefeld 1996; Schultz 1989; Stirk 1994; Szücs 1988). The debate about *Mitteleuropa* and its variants (*Zentraleuropa* or *Zwischeneuropa*) was immensely complex. There was never a clear consensus about its geographical limits nor its precise meaning and function. Initially, the concept was understood in economic terms. *Mitteleuropa* was invoked as a new economic space with low 'internal' and high 'external' tariffs which would protect the agricultural and industrial sectors of Central Europe and over-come the region's severe economic dislocations and imbalances. This was one manifestation of the late nineteenth-century belief in autarky and protectionism. The economist Friedrich List espoused a general Central European customs union, building on the German *Zollverein*, during the 1830s and 1840s. A similar idea was advanced by the Austrian Minister of Commerce, Karl von Bruck, through the late 1840s and 1850s. Von Bruck proposed a series of interlocking confederations – German, Austrian, Swiss and Italian – which would act as a coun-terbalance to the imperial economies to the west (Britain and France) and the east (Russia). The largely economic nature of the *Mitteleuropa* debate during the 1850s is underlined by Robin Okey's analysis of seventy-nine German atlases on sale in 1856. Only four showed the political geography of an area identified as *Mitteleuropa* whereas twenty-four included thematic economic maps organized on this basis (Okey 1992: 114).

Proposals for a customs union embracing Germany, Austria-Hungary and other south central European states re-surfaced on several occasions through to the early twentieth century, inspired in part by the establishment of the Pan-American League which suggested (misleadingly) that an economic 'pan-region' was about

41

to emerge, uniting the North and South Americas. An important figure in the campaign for a German economic 'pan-region' was the economist Julius Wolff, whose *Mitteleuropäischer Wirtschaftsverein* was an important focus for this debate (Stirk 1994: 12).

Lurking behind these economic arguments were ideological motives. These emerged more clearly towards the end of the nineteenth century. The idea of a political *Mitteleuropa* won some support on the left. A democratic federal system in Central Europe seemed an imaginative solution to the region's brooding ethnic tensions, a radical alternative to the moribund and inflexible autocracy of an Austro-Hungarian Empire in which under half the population were Austrian or Hungarian and where the dominant group of the Emperor's subjects, some 47 per cent, were Slavs. Here, ethnic communities, languages and religions intermingled in an un-mappable mosaic. The creation of ethnically homogenous micro-states seemed hopelessly impractical. Strategically vulnerable and economically unviable, such states would only exacerbate ethnic conflict. A new democratic confederation, on the other hand, in which all communities would have appropriate representation, offered an attractive and progressive alternative to the sclerosis of the existing geopolitical structure and the anarchy of an ethnic free-for-all. A federal *Mitteleuropa*, defined by its ethnic diversity rather than its geographical unity, offered a model from which a wider, federal Europe might emerge.

The German Marxist, Karl Renner, was one of the most articulate advocates of a federal *Mitteleuropa*. His scheme was based on a radical, and highly imaginative, re-thinking of how states, sovereignty and political territoriality might operate, and has some affinities with the modern notion of 'subsidiarity'. In place of discrete, bounded states, each developing their own economic and social policies, Renner advocated the functional separation of responsibilities within a federal system. Wider economic and political policies could be decided at federal level by a Parliament of elected representatives from the existing states but social questions and matters relating to ethnic or religious groups, including education, could be handled by a parallel organization comprising representatives of each geographically scattered community (Kann 1964, 157–67; Stirk 1994, 9).

The notion of a socialist Central Europe, with Germany at its core, was extended still further by a few radical thinkers who dreamed of an even larger socialist German empire which could challenge the capitalist empires of the British and French (Fletcher 1984). The concept of *Mitteleuropa* as a cosmopolitan core within a new, socialist Europe was by no means an orthodoxy in *fin-de-siècle* Germany though the idea that Central Europe could be identified by its very diversity was a common motif in political and academic debate. The geographer Joseph Partsch, for example, produced an important survey of the region's physical and human geography based on this theme (Figure 2.4) (Partsch 1904).

Figure 2.4 Partsch's *Mitteleuropa*
Source: Partsch 1904: 310 (redrawn for this volume to improve clarity)

The idea of a political *Mitteleuropa* was also appealing to conservative thinkers. Theodor Schieman, friend of the ultra-nationalist German Kaiser, Wilhelm II, listed the dominant objectives which right-thinking German patriots should insist upon at the *fin-de-siècle*: 'a central European customs and economic union, a settlement of the colonial question on generous lines, the humiliation of England, the preservation of peace with our allies Austria-Hungary and Italy and the containment of the powerful Russian influence' (Stirk 1996, 10–11). For Schieman, *Mitteleuropa*, based initially on a customs union, implied a Central European confederation dominated by a greater Germany in which all German-speaking peoples, 10 million of whom lived outside Germany in Austria-Hungary and numerous *irridenta* elsewhere, would finally be united. This was not a *Mitteleuropa* in which ethnic diversity would be celebrated or even tolerated. The 'smaller peoples' of the region would simply have to accept their gradual demise. Their destiny was to be assimilated into the dominant German *Kultur*. The call for a German *Mitteleuropa* featured as one of the published war aims issued by the German Chancellor Bethmann-Hollweg in September 1914. 'We must create', it was argued,

> a central European economic association through common customs treaties, to include France, Belgium, Holland, Denmark, Austria-Hungary, Poland, and perhaps Italy, Sweden and Norway. This association will not have any common constitutional supreme authority and all its

members will be formally equal, but in practice it will be under German leadership and must stabilise Germany's economic dominance over *Mitteleuropa*.

(Stirk 1996: 20).

Mitteleuropa was a complex and ambiguous concept which was invoked by different constituencies for mutually exclusive reasons. These contradictions came to the fore during the First World War when the idea of *Mitteleuropa* was widely debated as one alternative to the existing European geopolitical structure. Albrecht Penck wrote of a new Central Europe (*Zwischeneuropa*) under German patronage as a means of securing peace and a pathway towards a united Europe (Penck 1915). A similar argument was put forward the following year by Friedrich Naumann, a former Lutheran Pastor and member of the *Reichstag* (Naumann 1916). Naumann saw a self-sufficient Central Europe, dominated by a benign Germany, as the only solution to Europe's manifest geopolitical problems and the chronic incompatibility between German *Kultur*, Anglo-French *civilisation* and the 'Oriental despotism' of Russia. Initially this region would need to be physically isolated from its enemies to east and west by two great 'Chinese walls'. In the long term, however, *Mitteleuropa* would spawn a wider federal Europe. A united Europe at peace with itself would eventually emerge from its own core region. The essential characteristics of Central Europe were the essential characteristics of Europe as a whole.

Naumann's analysis opened a lively debate across the war-time division. His underlying belief in the incompatibility between German *Kultur* and Anglo-French *civilisation* was accepted by several critics in Britain and France. Henri Bergson interpreted the war in precisely these terms, as a clash between two fundamentally different world-views, the products of mutually exclusive national cultures. Peace required either the total eradication of one side or the complete separation of both (Bergson 1915). Some British critics offered a cautious welcome to Naumann's scheme. A German *Mitteleuropa* sealed off from the rest of the world would spell the end of the 'old' Europe for the foreseeable future but if that was the price of peace, it was perhaps worth paying (Chisholm 1917a; Chisholm 1917b; Lyde 1916–17). The more common Allied response to Naumann's *Mitteleuropa*, however, was angry rejection (Verosta 1977). Most French commentators dismissed the idea as a flagrant piece of geopolitical propaganda, a thinly disguised justification for German imperial domination. Central Europe was a myth, it was claimed, an invented region of diverse, mutually-antagonistic peoples whose 'unity' could only be maintained by force.

Paradoxically, the more aggressive proponents of German imperialism were also unconvinced by Naumann's scheme on the grounds that it was insufficiently ambitious and too conciliatory towards the smaller nations of Central Europe.

According to these critics, who included key military figures such as Generals Ludendorff and Hindenberg, *Mitteleuropa* might serve as a means to an end, a stepping stone towards a greater German Empire, but it could never be an end in itself. Naumann's *Mitteleuropa* would place 'unnatural' limits on Germany's potential growth, trapping the nation forever in a Central European zone.

These criticisms reflect the increasing scale of German geopolitical ambition after 1890 which developed its own terminology. The German concept of *Lebensraum*, or 'living space', was of central importance here. This idea was borrowed from mid-nineteenth century biology, having been coined by the German biologist Oscar Penschel in an 1860 review of Charles Darwin's *The Origin of Species* (1859) (Raffestin, Lopreno and Pasteur 1995, 31). Penschel used the term as an imprecise German alternative to the English 'habitat' and the French 'milieu'. Like many concepts developed in the natural sciences, however, the word soon began to feature in German geopolitical writings, part of a wider process by which a simplified Darwinism was used to explain and predict human behaviour (Dickinson 1943).

The writings of the German geographer Friedrich Ratzel, Professor at the University of Leipzig, were especially important in this process (Smith 1980; Hauner 1983; Wanklyn 1961). Ratzel had trained in the natural sciences and was influenced by Darwinian metaphors and ideas. The main weakness in Darwin's thesis, Ratzel argued, was its failure to consider the importance of space. According to Ratzel, the 'Darwinian' idea of the struggle for existence (*Kampf ums Dasein*) could be equated with a struggle for space. All life forms on the planet were involved in this ceaseless quest for *Lebensraum*. The application of this idea to European geopolitics implied a biological theory of state formation and development, a concept pioneered by Ratzel but further developed by the aforementioned Kjellén.

Both Ratzel and Kjellén saw the nation-state as a natural organism, greater than the sum of the individuals, communities and classes which it comprised. The state was a living geopolitical force rooted in, and moulded by, its soil. It was an organic entity, the physical embodiment of the popular will and the product of a centuries-old interaction between a people and their natural environment: *ein Stück Boden und ein Stück Menschheit* (Ratzel 1897: 2; Bassin 1987: 480). Kjellén, in particular, sought to challenge the narrow legal and institutional definition of the state in his path-breaking analysis, *Der Staat als Lebensform*, originally published in Swedish in 1916 and probably the conduit through which the term *Geopolitik* entered Germany. The legal concept of the state, claimed Kjellén, was derived from bourgeois republicanism and was therefore rooted in the ultra-rationalism of the French Enlightenment and Revolution. But the state should be far more than this; it should be vehicle for all the hopes, dreams and aspirations of a people, a measure of their demographic vitality (*Demopolitik*), their economic potential (*Wirtschaftspolitik*), their social vigour (*Soziopolitik*) and their political power (*Herrschaftspolitik*) (Herb 1997: 51). 'All that a man is', wrote Kjellén in 1916, 'he owes to the State. . . . This is where his spirit resides. His valour, his spiritual reality would have no meaning without the State. . . . The nation, and not the individual, is the true hero of history' (Kjellén 1924: 86, 117; Breuilly 1992; Raffestin, Lopreno and Pasteur 1995: 85–6).

For Ratzel, possession of space could be distinguished from desire for space. The existence of large, vulnerable states (such as the Ottoman and Russian Empires) demonstrated that power did not necessarily derive from the mere possession of space. If this were the case, large states would always grow at the expense of smaller states by virtue of their initial territorial advantage. Strong states, claimed Ratzel, were those which manifested a demographic, economic and cultural capacity which was greater than their existing territorial limits. Such states would inevitably develop expansive tendencies. A state's power was therefore determined by its territorial ambitions rather than by its spatial extent. The acquisition of 'living space', although an obvious manifestation of initial strength, often diminished the lust for territorial expansion and thus undermined a state's dynamism and energy, making it vulnerable to the ambitions of rival states. Ratzel's geopolitics therefore implied a ceaseless process of growth and decline, a permanent waxing and waning in the size of rival states. This was an inevitable consequence of human life, claimed Ratzel, the engine of progress. Ratzel's theory did not preclude the possibility of achieving balance between equally dynamic and vigilant states. Such a balance could only be sustained, however, by the permanent and general desire for greater *Lebensraum*. The endless struggle for 'living space' was thus a fundamental and inalienable geopolitical law (Raffestin, Lopreno and Pasteur 1995, 29–75).

Ratzel's ideas contain several conceptual confusions. He combined a characteristic German romanticism (derived from Hegel, the *Naturphilosophie* of Haeckel and the geographical writings of Carl Ritter which posited an essential life-force determining the nature and actions of all living things) with a Social Darwinism (which relied on fundamentally anti-Hegelian, materialist forms of explanation). Ratzel was thus able to imply an inherent human spirituality operating independently of external forces in some instances, and on other occasions to resort to crude environmental determinism. These intellectual contradictions were mirrored by Ratzel's ideological confusion. His romanticism manifested itself in profoundly anti-modern beliefs. Urban industrial life represented, he believed, a spiritual crisis. Modern cities were soulless places dominated by greed, cynicism and pernicious individualism. Industrial labour crushed the human spirit and destroyed social cohesion by breaking the necessary dialectic between the soil and the people. A rural existence, he implied, in which human beings lived in harmony with the rhythms of the natural world, was the only legitimate form of human existence. Yet his geopolitics depended on the very industrial prowess he claimed to abominate. He was, for example, a passionate advocate of an overseas German empire policed by a hugely expanded navy which could rival that of the British (Ratzel 1900).

Ratzel's *Lebensraum* was a theoretical rather than a geographically specific concept (Ratzel 1901). However, the idea could obviously be applied to legitimize German territorial ambitions, particularly in the east. The ancient Germanic myth of the *Drang nach Osten*, the drive of the Teutonic knights to the east, was thus re-born in the 'scientific' language of Ratzelian geopolitics. Germany's historic destiny was deemed to lie not merely in Central Europe, its immediate arena, but in Eastern Europe and beyond. This eastern realm, currently part of the Slavic world, was where a German *Lebensraum* would emerge. Here was the space for a new German Empire in the coming century; the zone where the burgeoning German population could begin their new era of colonization.

Conclusion

The *fin-de-siècle* geopolitical writings of Mackinder, Ratzel and the others are often represented as warrants for unfettered imperial expansion, as a pseudo-scientific rationale for state aggression. However, they can equally be read as nervous commentaries on Europe's uncertain fate in the changing conditions of the twentieth century. There were many differences of opinion here, from cautious optimism to dark pessimism, but the dominant impression was that Europe was bound to change dramatically in the future and that the traditional hegemony of 'European civiliza-tion' was under serious threat. The rise of continental-scale land-based powers (such as the USA) seemed to change the relationship between space and power and this carried unsettling implications for the traditional European states whose power had derived from, and produced, a different spatial structure. The world was changing, or so it appeared, and if these transformations were to be understood and predicted, a new language and a new mode of analysis was required.

Geopolitics was a child of its time, a product of the hopes, fears and anxieties of the last *fin-de-siècle*. Its early advocates wanted to establish a new, modern and rigorously scientific discipline with its own concepts, theories and language. But these early geopolitical writers were irrationally convinced that their era was fundamentally important. While there was some evidence to support such a view, the belief in a dramatic disjuncture reflected the heightened sense of history that accompanied the closure of the nineteenth century and the onset of the twentieth. Most early geopolitical writings also reflected this assumption and were, as a result, highly speculative and even prophetic in tone.

It is important to emphasize in conclusion that early geopolitical texts covered a wide range of themes and topics. It is misleading to interpret geopolitics as a singular, readily identifiable school of thought. It was, rather, a disparate, cross-disciplinary perspective and a modification of earlier forms of academic analysis. It flourished briefly in the very particular economic, political and intellectual circumstances which prevailed in the decades before the First World War. As we shall see in subsequent chapters in this volume, this project acquired new and more disturbing connotations after 1918.

Notes

1 For an extended analysis, see Heffernan (1998), especially pp. 49–110.
2 My sub-title is borrowed, with obvious modification, from the elegant model proposed by Guy Bois (1992) to explain the social and economic changes which accompanied the transition from antiquity to feudalism at the end of the first millennium. For a wider Marxist reading of the same transition, see Anderson (1974).
3 Even the USA, a nation with more than enough domestic space and with a political culture which drew on a powerful anti-imperial rhetoric born of its own struggle for independence, joined the clamour for imperial expansion, acquiring Puerto Rico, Cuba and the Philippines after the Spanish-American war of 1898. See Le Feber 1993, 129–82.
4 Even in our own de-Christianised, post-modern times, many of us still believe that rare calendar events are anything but arbitrary. The intensifying debate about the importance of the coming millennium – from discussion about how national governments should commemorate the new era in 2000 (or is it 2001?) to the wilder outpourings of internet 'millennial cults' – makes this abundantly clear. The fearful spectre of a 'millennial' computer virus, for example, lurking in the global communications system and threatening to do untold damage echoes comparable anxieties and phobias at the turn of the last millennium and those that re-emerged in different forms at the end of each intervening century. See Briggs and Snowman (1996), Cohn (1970) and Heffernan (1999).
5 Again, this can be detected in recent debates about the significance of the 1990s. The collapse of European communism and the associated splintering of the USSR has yet to produce much in the way of detailed explanation but it has spawned several highly speculative 'end-of-history' accounts, initiated by Francis Fukuyama (1992). This is recognizably part of a post-modern millennialism which denies the validity of historical explanation while being itself an obvious historical manifestation of our own *fin-de-millennium* moment. See Mestrovic (1994).
6 There were many examples of this 'one-worldism'. In geography, it partially inspired the 1891 proposal from the renowned German geographer, Albrecht Penck, for an international world map to be based on a 1:1 million scale and a standard set of conventions and symbols. Four centuries after Columbus, argued Penck, on the eve of the twentieth century and with only the polar regions still unexplored, the time had come for the great powers to unite in a common project to construct a single cartographic image of the world, an international tribute to the age of European expansion and exploration. Enthusiastically endorsed by leading geographers and initially by several governments, this intriguing scheme fell foul of the international tension it sought to overcome. Penck himself was later to repudiate internationalism in favour of a more conservative nationalism. See Heffernan (1996) and Robic (1996).
7 This challenged the orthodox view, re-stated in 1890 by the American military historian A.T. Mahan in his *The Influence of Seapower upon History,* that control of the seas was the key to strategic and economic power. See Ó Tuathail (1996, 38–43).
8 The *fin-de-siècle* enthusiasm for ambitious railway schemes, such as the German Berlin to Baghdad project, the British scheme for a Cape-to-Cairo railway and the French project for a north-south Trans-Saharan route, added force to Mackinder's view.

Bibliography

Anderson, P. (1974) *Passages from Antiquity to Feudalism*, London: NLB.

Andrew, C. and Kanya-Forstner, A. J. (1981) *France Overseas: The Great War and the Climax of French Imperialism*, London: Thames and Hudson.

Bassin, M. (1987) 'Imperialism and the nation state in Friedrich Ratzel's political geography', *Progress in Human Geography* 11: 473–95.

Bergson, H. (1915) *La signification de la guerre*, Paris.

Blouet, B. W. (1987) *Sir Halford Mackinder: A Biography*, College Station, TX: University of Texas A and M Press.

Bois, G. (1992) *The Transformation of the Year 1000: The Village of Lournand From Antiquity to Feudalism*, Manchester: Manchester University Press.

Brechtefeld, J. (1996) *Mitteleuropa and German Politics, 1848 to the Present*, London: Macmillan.

Breuilly, J. (1992) 'The national idea in modern German history', 1–28 in J. Breuilly (ed.) *The State in Germany: The National Idea in the Making, Unmaking and Remaking of a Modern Nation-State*, London: Longman.

Briggs, A. and Snowman, D. (eds) (1996) *Fins de Siècle: How Centuries End, 1400–2000*, New Haven: Yale University Press.

Chisholm, G. G. (1917a) 'Central Europe: a review', *Scottish Geographical Magazine* 33: 83–8.

—— (1917b) 'Central Europe as an economic unit', *Geographical Teacher* 9: 122–33.

—— (1919) 'The geographical pre-requisites of a League of Nations: a review', *Scottish Geographical Magazine* 35: 248–56.

Cohn, N. (1970 [1957]) *The Pursuit of the Millennium: Revolutionary Millenarians and Mystical Anarchists of the Middle Ages*, London: Paladin.

Dickinson, R. E. (1943) *The German Lebensraum*, Harmondsworth: Penguin.

Dryer, C. R. (1920) 'Mackinder's world island and its American satellite', *Geographical Review* 9: 205–7.

Fischer, F. (1967) *Germany's Aims in the First World War*, London: Chatto and Windus

Fletcher, R. (1984) *Revisionism and Empire: Socialist Imperialism in Germany*, 1897–1914, London: Allen and Unwin.

Fukuyama, F. (1992) *The End of History and the Last Man*, Harmondsworth: Penguin.

Hauner, M. (1992) [1990] *What is Asia to Us? Russia's Asian Heartland Yesterday and Today*, London: Rouutledge.

Heffernan, M. (1996) 'Geography, cartography and military intelligence: the Royal Geographical Society and the First World War', *Transactions of the Institute of British Geographers*, New Series 21: 504–33.

—— (1998) *The Meaning of Europe: Geography and Geopolitics*, London: Arnold.

—— (1999) 'Historical geographies of the future: three perspectives from France, 1750–1825', 139–64 in C. W. J. Withers and D. N. Livingstone (eds) *Geography and Enlightenment*, Chicago: University of Chicago Press.

Herb, G. H. (1997) *Under the Map of Germany: Nationalism and Propaganda 1918–1945*, London: Routledge.

Herrmann, D. G. (1996) *The Arming of Europe and the Making of the First World War*, Princeton: Princeton University Press.

Holdar, S. (1992) 'The ideal state and the power of geography: the life and work of Rudolf Kjellén', *Political Geography* 11: 307–23.

Kann, R. A. (1964) *The Multinational Empire: Nationalism and National Reform in the Habsburg Empire 1848–1918*, vol. 2: *Empire Reform*, New York: Columbia University Press.

Kearns, G. (1985) 'Halford John Mackinder 1861–1947', 71–86 in T. W. Freeman (ed.) *Geographers: Bio-Bibliographical Studies* vol. 9, London: Cassell.

—— (1993) '*Fin-de-siècle* geopolitics: Mackinder, Hobson and theories of global closure', 9–30 in P. J. Taylor (ed.) *Political Geography of the Twentieth Century*, London: Belhaven.

—— (1997) 'The imperial subject: geography and travel in the work of Mary Kingsley and Halford Mackinder', *Transactions of the Institute of British Geographers*, New Series 22: 450–72.

Kennedy, P. (1988) *The Rise and Fall of the Great Powers: Economic Change and Military Conflict from 1500 to 2000*, London: Fontana.

Kjellén, R. (1924) [1916]) *Der Staat als Lebensform*, Berlin: Leipzig.

Laqueur, W. (1996) '*Fin de siècle:* once more with feeling', *Journal of Contemporary History* 31: 5–49.

Le Feber, W. (1993) *The Cambridge History of American Foreign Relations. Volume II: The American Search for Opportunity, 1865–1913*, Cambridge: Cambridge University Press.

Lyde, L. W. (1916–7) 'Europe v. Middle Europe', *Sociological Review* 9: 88–93.

Mackinder, H. J. (1887) 'On the scope and methods of geography', *Proceedings of the Royal Geographical Society* 9: 141–60.

—— (1902) *Britain and the British Seas*, London: Heinemann.

—— (1904) 'The geographical pivot of history', *Geographical Journal* 23: 421–42.

—— (1905) 'Man-power as a measure of national and imperial strength', *National and English Review* 45: 136–43.

—— (1911) 'The teaching of geography from an imperial point of view, and the use which could and should be made of visual instruction', *Geographical Teacher* 6: 79–86.

—— (1919) *Democratic Ideals and Reality: A Study in the Politics of Reconstruction*, London: Constable.

Massie, R. K. (1991) *Dreadnought: Britain, Germany and the Coming of the Great War*, New York.

Mestrovic, S. (1994) *The Balkanisation of the West: The Confluence of Postmodernism and Postcommunism*, London: Routledge.

Naumann, F. (1916) [1915] *Central Europe*, London: P. S. King.

Okey, R. (1992) 'Central Europe/Eastern Europe: behind the definitions', *Past and Present* 137: 102–33.

Ó Tuathail, G. (1996) *Critical Geopolitics*, London: Routledge.

Ó Tuathail, G., Dalby, S. and Routledge, P. (eds) (1998) *The Geopolitics Reader*, London.

Parker, G. (1998) *Geopolitics: Past, Present and Future*, London: Pinter.

Parker, W. H. (1982) *Mackinder: Geography as an Aid to Statecraft*, Oxford: Clarendon.

Partsch, J. (1904) *Mitteleuropa: Die Länder und Völker von den Westalpen und dem Balkan bis an den Kanal und das Kurische Haff*, Gotha.

Penck, A. (1915) 'Politisch-geographische Lehren des Krieges', *Meereskunde* 9–10: 12–21.

Raffestin, C., Lopreno, D. and Pasteur, Y. (1995) *Géopolitique et Histoire*, Lausanne.

Ratzel, F. 1897, *Politische Geographie oder die Geographie der Staaten, des Verkehres und des*

Krieges, Munich and Leipzig.

—— (1900) *Das Meer als Quelle der Völkergrösse. Eine politische-geographische Studie*, Munich.

——. (1901) 'Der Lebensraum. Eine biogeographische Studie', 101–90 in K. Bücher and K. Fricker (eds) *Festgabe für Albert Schäffle zur siebzigsten Wiederkehr seines Geburtstags am 24. Februar 1901*, Tübingen.

Robic, M.-C. (1996) 'Les voeux des premiers Congrès: dresser la Carte du Monde', 149–78 in M.-C. Robic, A.-M. Briend and M. Rössler (eds) *Géographes face au monde*, Paris: Presses de l'Université de France.

Ryan, J. (1994) 'Visualizing imperial geography: Halford Mackinder and the Colonial Office Visual Instruction Committee, 1902–1911', *Ecumene* 1: 157–76.

Scheler, M. (1915) *Der Genius des Krieges und der deutsche Krieg*, Berlin.

Schorske, C. (1979 [1961]) *Fin-de-Siècle Vienna*, New York, Cambridge: Cambridge University Press.

Schultz, H. D. (1989) 'Fantasies of Mitte: Mittelage and *Mitteleuropa* in German geographical discussion in the 19th and 20th centuries', *Political Geography Quarterly* 8: 315–40.

Smith, W. D. (1980) 'Friedrich Ratzel and the origins of *Lebensraum*', *German Studies Review* 3: 51–68.

Stokes, J. (ed.) (1992) *Fin de Siècle, Fin du Globe: Fears and Fantasies of the Late Nineteenth Century*, London: Macmillan.

Stirk, P. M. R. (1994) 'The idea of *Mitteleuropa*', 1–35 in P. M. R. Stirk (ed.) *Mitteleuropa: History and Prospects*, Edinburgh: Edinburgh University Press.

—— (1996) *A History of European Integration since 1914*, London: Pinter.

Szücs, J. (1988) [1983]) 'Three historical regions of Europe', 291–332 in J. Keane (ed.) *Civil Society and the State: New European Perspectives*, London: Verso.

Teggart, F. J. (1919) 'Geography as an aid to statecraft: an appreciation of Mackinder's 'Democratic Ideals and Reality', *Geographical Review* 8: 227–42.

Teich, M. and Porter, R. (eds) (1990), *Fin-de-Siècle and its Legacy*, Cambridge: Cambridge University Press.

Verosta, S. (1977) 'The German concept of *Mitteleuropa* 1916–1918 and its contemporary critics', 208–14 in R. A. Kann, B. K. Kiraly and S. Fichtner (eds) *The Habsburg Empire in World War I*, Boulder, Co.: Greenwood Press.

Wanklyn, H. G. (1961) *Friedrich Ratzel*, Cambridge: Cambridge University Press.

Weber, E. (1986) *France, Fin-de-Siècle*, Cambridge, Mass.: Harvard University Press.

Wolff, L. (1994) *Inventing Eastern Europe: The Map of Civilization on the Mind of the Enlightenment*, Stanford: Stanford University Press.

3

THE CONSTRUCTION OF GEOPOLITICAL IMAGES

The world according to Biggles (and other
ficitional characters)

Andrew Kirby

The construction of geopolitical images

This chapter addresses a specific aspect of the evolution of geopolitics in the twentieth century, namely the ways in which images – of encirclement, say, or of national
spheres of influence – are constructed within the public sphere. It recognizes that
the varied assumptions and implications of geopolitical thought have now been well
dissected (Ó Tuathail 1996). It has also been demonstrated how such concepts have
played a part in the construction of foreign policy (Dalby 1990), and, from a different starting point, the poetics of the concept have also been explored (Jameson 1993).
However, there has been less consideration of the manner in which geopolitical
thought has been represented beyond the realms of statecraft and professional analysis, and even less discussion of the ways in which basic concepts became embedded
in popular discourse and subsequent political debate (but see Sharp 1996).

We must begin by confronting – and then avoiding – a definition of the term
'geopolitical'. Ó Tuathail's cautionary writings on the varied uses of the term, and his
preference for an alternative 'geo-politics', indicate that within the academy, it is a
complex and contradictory label (1996). Consequently, I start here with Ó Tuathail's
own tactic of hyphenating the term, in order to juxtapose geography and politics
(1996: 65). In the most general and unsurprising sense, then, I am taking geopolitics
to be a way of thinking about politics that takes geography into account (rather than
thinking about geography in a way that takes politics into account, a challenge which
has faced the discipline over the past fifty years). Such political thought can take place
at various scales (for example, internal colonialism), and we can use the term accordingly, but this discussion will restrict itself to relations between territorial states.

Drawing on Ó Tuathail further, I want to build on his claim that 'geopolitics cannot be abstracted from the textuality of its use' (Ó Tuathail 1996: 65). Indeed, I want to explore the construction of the concept at a pivotal moment within popular discourse, by looking at the intersection of international politics and geography. In this chapter I address these themes by focusing upon popular British fiction in the early part of the twentieth century, a period that is compelling for two reasons. First, this was a time of intense nationalism and state formation, that both reflected the material rivalries in the West and also fuelled further tensions. As Mann reminds us, this was the period when a formal state apparatus began to evolve, but it was as common for governments to spend large sums on war-making as it was to spend funds on state-making (Mann 1988). Relations between states, especially in Europe, were driven by a number of factors; economic rivalries were endemic in a time of industrial growth and imperial aspiration, but were also coloured by longer-standing tensions that had been evolving for decades or more (Overy 1996: 289). Competition and conflict between nations drew on archetypes of race and ethnicity and permanent historical markers such as the Prussian occupation of Paris in 1870 or the French Revolution and its aftermath. The frictions of the past lurked behind the conflicts of the present, and the strategic interpretations of trade routes, boundaries and other geographic realities were slow to change (Paasi 1996). This was a template of time and space that continued to define many conflicts, even until 1939, and it was, as we shall see, a template that was maintained via public discourse.

Second, because the state apparatus was still in many countries in a process of consolidation, it was also relatively rudimentary. For example, the second half of the nineteenth century was a period of massive migrations, from Europe (to both northern and southern America) and from China (to Southern Asia), for instance. Few restrictions were placed on movement between countries – American passports were not issued nationally until 1856, for example – in large measure because the technologies of surveillance were so poorly developed. The criteria for controlling immigration into the United States based on political ideology were not developed until 1918 (although skin colour and intelligence had been used for exclusion decades before). It is little exaggeration to suggest that just as the state's operatives did not exert complete control over the frontiers, so they could only exert partial control over the evolution of civil society within the country. The electronic media of communication, which would make that task so much more effective later in the century, were still in development, and their prestige was much less than that enjoyed by the written word; in consequence, both factual and fictional writing was influential.[1] The flood of words in this period was part of what Bhabha terms the 'daily plebiscite', the constant creation and recreation of a literal common sense that defines, through narration, what it is to be a nation. Part of that discourse has to do with what is to be remembered and what is to be forgotten, but a significant part is to do with the construction

53

of a geographical sense. Bhabha borrows from Foucault in seeing the territorial state facing an 'indefinite future of struggles', but these conflicts must be carefully chosen. This, then, is what many writers of fiction chose as a target, namely the definition of friends and foes, allies and enemies, zones of indifference and spheres of influence (Bhabha 1990: 291–322).

The chapter has three parts: in the next, I focus on the evolution of popular fiction in the early part of the century; the second provides some examples of such fiction and the third explores the geopolitical images that it contains; by way of conclusion, the chapter also analyses the congruence between the images and the evolution of popular thought and policy.

Mass readership

In his provocative study of public opinion, Claud Cockburn writes that:

> Historians and sociologists must examine innumerable sources when they are in search of the mood, the attitude, the state of mind of a nation or a class at this or that period of time. . . . They have to study what people said and wrote, as distinct from what they did. There are the public speeches, the leading articles . . . the personal letters and diaries. All these are indispensable. But of all indices to moods, attitudes, and above all, aspirations, the best-seller list is one of the most reliable. There is no way of fudging it.
>
> (Cockburn 1975: 9)

Cockburn's analysis of a number of best sellers has been influential in the development of this argument, not least in his focus upon popular literature (defined in terms of sales) versus academic or prize-winning literature *per se*. In this chapter, I have paid little attention to writers who come readily to mind in terms of defining the start of the century, such as Shaw, or Compton Mackenzie. While their names remain familiar to us, this is no guarantee of influence upon their contemporaries, who were more often in thrall of writers that we would be hard pressed to identify today. As Cockburn argues, attitudes to the Great War were for a long time shaped by writers like Ian Hay, whose novels (such as *The Right Stuff* (1908), *A Man's Man* (1909), *The First Hundred Thousand* (1915) and *The Last Million* (1918)) have long been out of print. Our enduring images come to us from Robert Graves and Siegfried Sassoon, but these were men whose critical texts achieved stature long after the war had been concluded. Nor should we underestimate just how long this critical reassessment can take: frank discussion of the Holocaust, for example, is in some ways only beginning fifty years after its cessation (Finkelstein and Birn 1998; Goldhagen 1996).

So what were the best-selling books of the new century? Cockburn's list includes many that are no longer even remembered: Hay's *The First One Hundred Thousand*, as noted; *When it was Dark*, by Guy Thorne, and *The Garden of Allah* by Robert Hichens. There are some that we recognize in an inchoate manner, such as *The Blue Lagoon* (de Vere Stacpoole 1908), and *Beau Geste* (Wren 1924): several films have been made of both of these chestnuts. And there is one that I will address later, namely *The Riddle of the Sands* (Childers 1903).

The themes chosen by these various novels are instructive, as are the unspoken tropes beneath the surface. In several, religion is dominant in a manner that could barely be sustained today: Cockburn provides an interesting analysis of *When it was Dark*, in which the basic theme is the impact of the discovery, by an archaeologist, of a document that reveals the Resurrection to have been a hoax. What we might today term 'orientalism' is also represented in both *The Blue Lagoon*, in which European children are cast away on a desert island and grow up without the constraints of community or society, and *The Sheik* (Hull 1921), in which a European woman is kidnapped by a Bedouin and held against her will until she has submitted to his advances.

Beneath these chosen themes are, inevitably enough, whole rafts of assumptions. There is the inevitable question of sex, of gender, of class, and of race. There is on display rape, nymphetism, the violent subordination of women by men, the demeaning of 'others', including both Jews and Arabs, criticism of workers who will not work and pacifists who will not fight. As a consequence, a veritable industry has developed in the last two decades to find examples of contemporary sin in the books of former generations. As many commentators have argued, it is easy to find stereotypical thinking in such best sellers; from this insight is often to be found the suggestion that particular books be banned so that future readers should not suffer their message (Livingstone 1998).

It should be stated categorically that there clearly are such dimensions to these books, just as there is in unrefined pulp fiction today. There is anti-Semitism; people of colour are often presented as villains; women are either absent or appear in positions of weakness, silliness or maliciousness. It is easy to find a form of situational ethic that excuses these failings on the ground that the author didn't mean the insult, didn't use it very often, or, most frequently, was only reflecting the mores of the day (e.g. Ellis and Schofield 1993; Winks 1988). In this chapter, however, I am less interested in the motives of the authors. When John Buchan wrote about 'Huns', 'Kaffirs' and 'Portuguese Jews', he may or may not have had virulent feelings about such characters (Webb 1994). More to the point is the fact that he wrote these lines with the expectation that such characterizations would be understood in specific ways by an audience (just as we can expect that they produce a different reaction today). This then forms the basis of the analysis that follows.[2]

The players

I have chosen a number of titles for examination and have included publications that were read by children and adolescents, on the grounds that these were books in which images were most simply and visibly presented to the reader. It can also be argued that these were books that tended to stay in print for a long period, and which were thus capable of influencing an even larger number of readers. It has been suggested that 'juvenile literature' is valuable in revealing both the norms of the day and the fantasies of its creators:

> Most societies clearly reveal both their moral norms and their political ideologies through their efforts to acculturate the young. . . . It is indeed one of the notable characteristics of the late nineteenth and early twentieth centuries that many European countries, their imperial territories and rapidly Europeanising imitators like Japan, established a powerful zone of intellectual, ideological and moral convergence in the projection of state power and collective objectives to children.
>
> (Mackenzie in Castle 1996: vii)

There are clearly very many sources that could be examined in this kind of study. The public school stories that featured Bunter and his pals, for example, are veritable open books on the subjects of class, race and sexuality (Castle 1996). There was a raft of comics that were popular through until the 1950s, that indeterminately mixed up contemporary issues (such as nuclear war) and long-standing imperial themes (Kline 1993).[3] There was also a segment of the market designed solely for girls, but this chapter follows Phillips in restricting its attention to work that can be thought of as 'adventures for men and boys', although it tends to deal with a later period than does his selection (Phillips 1997).

My motivation for restricting the discussion is a deliberate attempt to begin to interrogate the relation between gender and international relations in a manner which complements the feminist analysis done by Enloe and others but pursues somewhat different issues, notably the relations between men rather than those between men and women (Enloe 1990). From that standpoint, relatively little has been done to connect the tropes of imperialism and warfare on the one hand, and the presentation of the male within popular writing on the other. While the construction of masculinity and its relation to conflict may seem obvious, even a cursory examination of the examples below indicates that there exist complex shades of intimacy and allegiance, which in turn open up interesting avenues for further exploration.

Here I have chosen a number of writers and their characters solely on the basis of their longevity. For consistency, I have focused on their books published in

Britain, and I therefore do claim this as a representative sample, but the reverse holds; if there are no clear images on view in these works, then they were not visible in some of the most frequently read books of the period. The earliest source was published in 1903, the latest in 1946.

The propagandist: **The Riddle of the Sands**

Erskine Childers' novel was published early in the twentieth century and confronts very explicitly the likelihood of a naval war between Germany and Britain. Despite a title that is more redolent of the desert than the sea, the story is set amongst the Frisian islands close to Schleswig-Holstein and revolves about two upper middle-class Englishmen who are sailing a tiny boat for recreation. As Cockburn observes, the story is brilliant propaganda, from a man who became the information specialist for Sinn Fein during the Irish uprising. He was eventually captured, and executed on the express orders of Winston Churchill, who wrote 'such as he is may all who hate us be' (Cockburn 1975: 87).

By contemporary standards, the story is extraordinarily subtle. It takes a great deal of time to get under way, and there is never any likelihood of high speed chases, explosions or similar types of action. It is, in fact, a laborious exposition on the imminent war between the two countries, the possibilities of a 'coalition' surrounding and blockading the United Kingdom, and the role of small craft in defensive and offensive action:

> A word more as to our motive. It was Davies's conviction, as I have said, that the whole region would in war be an ideal hunting-ground for small free-lance marauders, and I began to know he was right; for look at the three sea-roads through the sands to Hamburg, Bremen, Wilhelmshaven, and the heart of commercial Germany. They are like highways piercing a mountainous district by defiles, where a handful of desperate men can arrest an army.
>
> (Childers 1903: x-xx)

The novel begins with a now-familiar device, that of claiming the account to be based on a 'true story'. As it unfolds, it develops a very explicit perspective on tactics and strategy that provides a prescient overview of guerrilla warfare:

> Follow the parallel of a war on land. People your mountains with a daring and resourceful race, who possess an intimate knowledge of every track and bridle-path, who operate in small bands, travel light and move rapidly. See what an immense advantage such guerrillas possess over an enemy which clings to beaten tracks, moves in large bodies, slowly, and does not 'know the country'. See how they can not only inflict disasters

on a foe who vastly overmatches them in strength, but can prolong a semi-passive resistance long after all the decisive battles have been fought. See, too, how the strong invader can only conquer his elusive antagonists by learning their methods, studying their country, and matching them in mobility and cunning. The parallel must not be pressed too far; but that this sort of warfare will have its counterpart on the sea is a truth which cannot be questioned.

(Childers 1903: 131–2)

It is worth restating that this was published more than a full decade before the outbreak of war, yet it is written as if it were common knowledge that Britain and Germany were destined to fight each other. Indeed, as I shall show below, Childers' book was the first in what became a torrent of such predictions. The suggestion that the latter would blockade Britain (albeit not with submarines) is also prescient.

The Riddle of the Sands has remained in print throughout much of this century, although predictably its content has been presented in different ways. In 1931, for example, a new edition written by one of Childers' relatives held a preface that explicitly removed all (geo)political aspirations from the novel: 'his profound study . . . convinced him that preparedness induced war . . . [it led] to international armament rivalries and bred in the minds of the nations concerned fears, antagonisms and ambitions that were destructive to peace' (Childers, M. in Childers, E. 1931, quoted by Cockburn 1975: 84).

Of course, Childers had always believed such things; his book was an explicit attempt to provoke such a process of rearmament in Britain in order that it would survive what he saw as an inevitable conflict. By 1931 very different, isolationist views were being promoted, and it was instead plausible to re-position the book as a mere sailing story: 'the book remains the cherished companion of those who love the sea and put forth in sailing ships in search of adventure' (Childers, M. , quoted by Cockburn 1975: 84). It was thus offered up as a precursor to *Swallows and Amazons* (Ransome 1930), despite the fact that this is a role in which it singularly fails to occupy the reader's attention.

The conformist: the adventures of Richard Hannay

Although John Buchan is best known for his story *The Thirty-Nine Steps* (in turn, now remembered as Hitchcock's first film), he wrote a full series of stories that involved his 'action hero' Richard Hannay. The books contain a number of recurring characters, including Hannay, a colonial who eventually becomes a British Brigadier during the Great War; Blenkiron, a mysterious American businessman and espionage agent; and Sandy Arbuthnot, an English peer who spends most of his time buried deep in the backstreets of Istanbul, or 'Llasa or Yarkand or Seistan'

(Buchan 1916: 94) It is hard not to pick up a Buchan story today without being struck by its profound 'incorrectness', but also by its revelations, most of which are unintended, dealing with sexuality. Without exception, its English heroes display predictable virtues: collegiality, pluck and so forth. However, without exception, they are also described in profoundly ambiguous terms. Arbuthnot, for instance, is described as 'tallish, with a lean, high-boned face and a pair of brown eyes like a pretty girl's' (Buchan 1916: 89).

As one moves further from Piccadilly, so Buchan's descriptions become more critical. One French character is described warmly but in large measure because he is clearly an Anglophile and dressed in tweeds. Some Germans are described in almost friendly terms, although Hannay recognizes the 'queer other side' of one arch foe, 'that evil side which gossip had spoken of as not unknown in the German army' (Buchan 1916: 129).[4] As one goes even further east, so the expectations drop further. Hannay's nemesis is one Rasta Bey, a foppish Turk, described as 'a wrathful Oriental with his face as fixed as a Buddha' (Buchan 1916: 156). The most interesting characterization is that of Hilda von Einem, a German spy whose depiction constitutes one of only two women to appear in these stories. Her character is the inverse of the gazelle-like males; she is voluptuous, dangerous, the female as a source of evil – 'mad and bad but principally bad' (Buchan 1916: 183). At one point Hannay finds himself in the back seat of a car with her, a scene which is portrayed as perhaps being his first sensual experience with a woman, but certainly the first time that he has sat alone with a woman in the intimacy of an automobile. He summarizes it as follows: 'this was something . . . as a cyclone or an earthquake is outside the decent routine of nature. Mad and bad she might be, but she was also great' (Buchan 1916: 190).

The Buchan stories are compelling because they attempt to portray the complexities of the world but they do so from an exterior position. Hannay is hardly in a position of subalternity (he has 'made his pile' in Rhodesia and South Africa), but he is a colonial and a sceptical outsider, a reality that finds him involved in a murder hunt at the outset of the first story without friends or a place to turn. Thus his conversations are always based on ignorance and wonder, while his informants tell him of the world that surrounds Britain; the Bolsheviks (evil), the Germans (dangerous), the Americans (naïve but friendly) and the many mysterious tribes and nations that stretch off to the east, places that inevitably draw him as the adventures unfold.

Biggles flies . . . everywhere

Major James Bigglesworth first appeared in 1932 as a fictional pilot in the Great War. Over a series of approximately ninety stories, he went to fight the Nazis, the Japanese and the Soviets before retiring during the depths of the cold war, while working for Interpol. The Biggles series has been described as one of the highest selling collections of adolescent volumes ever created (Ellis and Schofield 1993) and has returned to print in the 1990s with sales in the hundred thousands. Written by a former Flying Corps officer, the stories focus on a band of fliers, led by the eponymous Biggles, and include both a peer and an orphan. There is one brief appearance (early in the First World War) by a love interest who, once more, proves to be a source of danger to the honourable and innocent male. There are consequently few references to gender, although, on one occasion, the opinion is given that a princess is not quite the thing in a time of war: 'Biggles shook his head sadly. "You need a man in times like this."' (Johns 1938: 11). This apart, sex is never mentioned, although to the modern reader, some of the dialogue seems to drip with double entendres.[5]

Like other fictional characters (such as O'Brian's Aubrey and Maturin), Biggles has taken on the trappings of reality, including a biography and a newsletter. His adventures extend through some of the most explosive periods in European history and provide an overview of the shifting alliances and campaign strategies that occurred between the two world wars and after. Like other successful fictional teams — such as Dumas' *The Three Musketeers* — the world-view on display is energized through a series of character studies that clearly contrast Biggles and his group with a series of 'others': Germans, of course, but also French, Belgians, Russians, Norwegians, Arabs, Indians, Malays, Chinese and Japanese. What is fascinating is that Johns, Biggles creator, visited virtually none of these countries, a fact that, as we shall see, has some importance in the way in which he portrayed them in his writing.

The airmen's behaviour is not impeccable (drinking and smoking are endemic) but it does foreground a hierarchy of values. Most prominent of these is what an Australian would recognize as 'mateship' — that is, allegiance to a comrade. Beyond that is the service, that is to say, the Royal Air Force and the hierarchy of command. In turn, there is the camaraderie of the air, which leads occasionally to chivalrous acts between enemies. And finally, there is nationality, the ultimate category of allegiance, for which Biggles will turn to spying and other distasteful actions.

Inevitably, Biggles' exploits cover a great deal of territory, both literal and metaphorical. Some of the stories are little more than pot-boilers, but others are almost didactic. One of the novels published in 1938, entitled *Biggles Goes to War*, turns out to be an updated version of Childers' work in so far as it constructs a primer on how an unspecified enemy will attack in the new era of aviation and what steps should be taken for defence: this was, of course, still the period of denial and disarmament in Britain. The novel begins there, but the fliers are quickly recruited by an envoy and flown to Maltovia, in a remote part of Europe. The latter principality is threatened by its neighbour, Lovitznia, which is supported in turn by a 'great power' that is unspecified but is clearly Germany. Drawing on his observations of the Spanish Civil War and the Japanese campaigns in China,

Johns constructed a scenario for invasion, which involved both a traitorous 'fifth column' (as occurred in Spain) and an active role for aircraft in terms of espionage, bombing (as happened in Nangking) and urban defence. The key theme of the book is the importance of standing up to threats of military annexation, which recede once the Maltavians display air superiority; the quarrel between the two neighbours is then adjudicated by the League of Nations.

Biggles Goes to War was only one of the author's attempts to stimulate rearmament but it may have been one of the more vigorous. He was fired from the editorship of the magazines *Flying* and *Popular Flying* in 1939 due to his political stance, an outcome that indicates that even agitation within the juvenile market was taken seriously by publishers and other opinion makers.

The outlaw, William Brown

Of the narrow selection examined here, Richmal Crompton's hero William Brown is perhaps the least exotic and is certainly the youngest, being clearly a boy in short pants rather than an adult. Although his adventures are virtually non-stop, and are again almost always restricted to a band of males, they are of a very different type from those experienced by his older counterparts, in so far as they take place in the suburbs of 1930s and 1940s England. William thus represents the middle-class version of the boarding school stories that were so common during this period. As reportage on domestic arrangements, the books (of which there were several dozen) are essential reading on the now mysterious topics of housemaids, coal furnaces, potting sheds and the Anglican church.

Although the classic tropes of boyhood from this period are expressed (pirates, cannibals and so forth), there is also an explicit dose of realism in books such as *William and Air Raid Precautions* and *William and the Evacuees*. Indeed, it is surprising (to a reader at the end of the millennium, accustomed to fiction located in a virtual realm) just how seamlessly the contexts of international relations, war and national survival become incorporated into the stories. A bathing beauty describes the League of Nations as being 'too cute for words' (Crompton 1996 'A Spot of Heroism': 7) while a spinster declaims that 'the country had been full of their spies for years before the war began . . . tourists or students or even professors . . . and they'd each take a bit of the coastline and study it until they knew every inch of it' (Crompton 1940 *William and the Spy*: 123).

Geopolitical images

In his study of adventure stories, Phillips argues that the latter are typically set within a dual geographical structure of 'home and away'. In such stories, then, there exists a constant referencing between the exotic and the familiar, such that the distant location becomes a liminal space: 'a marginal, ambiguous region in which the elements of normal life are inverted' (1997: 13). While this is not entirely absent in the examples above, they do not simply operate as morality tales, nor does place exist only in a metaphorical way. Rather, these tales display evidence of what Pick terms 'national representations and cultural anxieties' (Pick 1994: 77). Without trying to create a rigid taxonomy, it is possible to identify four themes: encirclement; the protection of the Heartland; absence and erasure; and the erosion of place from narrative.

Encirclement

While it is important to characterize twentieth-century geopolitics in the context of emerging global thought (see for instance Kearns 1993), it was also the case that the European nation-states maintained a consciousness of the enemies that surrounded them. This reached its apotheosis before the Second World War, when maps alleging Germany's encirclement constituted common propaganda displays (see Herb 1997). It was however also true of the period before the Great War, when complex plans for mobilization and attack were widely known. An expressive example of this rests in the novel *The Riddle of the Sands*. This is a very early precursor of the Great War, that recognizes explicitly the conflict of interest that will ultimately bring Britain and Germany to strife. It also goes on to predict fairly accurately the course of the war, with Britain encircled and besieged by Germany. In parts, the novel becomes extremely didactic, developing a series of very explicit policy recommendations:

> Davies in his enthusiasm set no limits to its importance. The small boat in shallow waters played a mighty role in his vision of a naval war, a part that would grow in importance as the war developed and would reach its height in the final stages. 'The heavy battle-fleets are all very well,' he used to say, 'but if the sides are well matched there might be nothing left of them after a few months of war. They might destroy one another mutually, leaving as nominal conqueror an admiral with scarcely a battle-ship to bless himself with. It's then that the true struggle will set in; and it's then that anything that will float will be pressed into the service, and anybody who can steer a boat, knows his waters, and doesn't care the toss of a coin for his life, will have magnificent opportunities. It cuts both ways. What small boats can do in these waters is plain enough; but take our own case. Say we're beaten on the high seas by a coalition. There's then a risk of starvation or invasion. It's all rot what they talk about instant surrender. We can live on half-rations, recuperate, and build; but we must have time. Meanwhile our coast and ports are in danger, for the millions we sink in forts and mines won't carry us far. They're fixed – pure passive defence. What you want is boats – mos-

quitoes with sting – swarms of them – patrol-boats, scout-boats, torpedo-boats; intelligent irregulars manned by local men, with a pretty free hand to play their own game. And what a splendid game to play! There are places very like this over there – nothing half so good but similar – the Mersey estuary, the Dee, the Severn, the Wash, and, best of all, the Thames, with all the Kent, Essex, and Suffolk banks round it. But as for defending our coasts in the way I mean – we've nothing ready – nothing whatsoever!'

(Childers 1903: 132)

A somewhat similar trope is explored in *The Thirty-Nine Steps*, which revolves around the attempts to assassinate a prominent European statesman and German efforts to steal British battle plans. It was published in 1915, when the continental war was established, but is set seven weeks before the outbreak of the conflict. It has some similarities to *Mr. Standfast*, the third of the Hannay quartet which appeared in 1919 and constitutes an allegorical version of *The Pilgrim's Progress*. While there is less geostrategic writing, there is a great deal spoken by the characters on the subject of the individual's responsibility to the state during a time of war.

In both novels, the hero moves back and forth between London and the high-lands of Scotland. London is portrayed as the seat of imperial and national power but the wilds of Scotland are revealed as an area of undomesticated menace, in which opposing forces (notably various German spy rings) operate freely. Untamed nature is matched by uncontrolled enemies. In the first story, the latter are poorly organized, but in the next tale, there are enemy submarines to be found hiding in deep-water inlets. Thus England is portrayed as a civilized but lonely power, encircled by the wild, Gaelic countries, Scotland to the north and Ireland to the West, both of which are implied to have dangerous connections with the enemy (a partial truth that accounts for Churchill's enmity for Childers, of course).

The struggle for the heartland

The most fully realized geopolitical discussion is presented in *Greenmantle*, which Buchan published in 1916; this was perhaps the lowest point in the War for Britain and France, who did not yet have the United States as an ally. At a time when most attention was placed on the Western Front, Buchan chose to send his characters further and further east, until they reach the scene of fighting between the Turkish and the Russian armies. This was not accidental, in so far as one of his characters offers long and complex speeches on the geopolitical designs of Germany.

The last dream Germany will part with is the control of the Near East. That is what your statesmen don't figure enough on. She'll give up Belgium and Alsace-Lorraine and Poland, but by God! She'll never give up the road to Mesopotamia till you have her by the throat and make her drop it. The worst happens, the Kaiser will fling overboard a lot of ballast in Europe, and it will look like a big victory for the Allies but he won't be beaten if he has the road to the East safe. Germany's like a scorpion: her sting's in her tale and that tale stretches way down into Asia.

(Buchan 1916: 171)

Interestingly, these perspectives do not disappear with the Peace of Versailles. In 1938, Biggles is made to question the location of his future employer. "'In Europe?'" he asks. "'Of course, but only just'", comes the response. "'Still we can claim to be Europeans, and there is much difference between Europe and Asia, which is not far from our eastern frontier'. "'As you say, there is much difference'", agreed Biggles' (Johns 1938: 10).

Erasure

When Biggles finds himself posted to another country, or follows his foes to a foreign setting, he frequently acts as though he remains on British territory. To all intents and purposes, this is true, as he lives and works in what Enloe terms an 'artificial society', an airbase that is an extension of British sovereignty (Enloe 1990). The virtual absence of all traces of indigenous peoples from the desert in *Biggles Flies East*, for example, may be helpful for the development of a claustro-phobic narrative that places the Germans versus the British, but it also serves to erase the indigenous inhabitants. This has been a common phenomenon in texts dealing with exotic locations. Pratt, for instance, writes at length about Mr John Barrow's sojourn in Africa (published in 1801), which excluded all inhabitation of the landscape with the exception of occasional traces: 'abstracted away from the landscape that is under contention, indigenous peoples are abstracted away from the history that is being made, a history into which Europeans intend to reassert them as an exploited labour pool' (Pratt 1992: 64–5).

In the Biggles story, there are also little but traces to be found. On the one occasion when natives are encountered – typical Bedouins of the desert – they are incomprehensible: 'a medley of guttural voices had broken out but [Biggles] could not get the hang of the conversation'. Again, echoes of *Beau Geste* are raised. While it was not the case that the Bedouin were to become a pool of labour, it was to be their fate to lose their homeland. The views of the airmen are redolent of a *mentalité* that would later see Palestine as a *tabula rasa* across which new social and political relations could be written without problems.

'Gosh this place gives me the creeps' he muttered. 'Give me France every time.' He was far too much of a realist to be impressed by the historical associations of the ground over which he walked, land which had once been trodden by Xenophon at the head of his gallant ten thousand, Alexander the Great, Roman Generals, and Crusaders . . . but he was conscious of the vague depression that is so often the result of contact with remote antiquity. . . . It was all part of the scheme, the moving of the relentless finger of fate that had lain over Palestine like a blight for nearly two thousand years and left a trail of death in its wake.'

(Johns, *Biggles Flies East*; 90, 155, 208)

By the time of the Second World War, Biggles appears to have assumed some different attitudes. In *Biggles in the Orient*, set in India, he warns his cohort 'we don't use the term natives . . . it's discourteous' (Johns 1945: 34). None the less, Biggles talks later about the Oriental mind, while when addressing a mess steward, his salutation is reduced to '"hi, you; that'll do", he said curtly' (28). And when, in another novel set in the Pacific, Malays turn on the Japanese, Biggles is appalled. '"You'll never stop them now, they're berserk," said Marling casually. "The trouble with these chaps is they tend to get out of hand." "Tend to get out of hand" cried Biggles with bitter sarcasm. "They're like a lot of wild animals"' (Johns 1946 *Biggles Delivers the Goods*: 170). This apart, there is little commentary on the Chinese or the Indians, who provide only a backdrop for a struggle between the British and the Japanese. Although he claimed to have been posted to the Levant, Johns in fact never left Europe and all his stories are projections of European attitudes.[6]

Inversion

The Spanish Civil War and the Sino-Japanese conflict ushered in a new era in which there was an inversion of near and far. In an era of total warfare made possible by aircraft, adventure (if it could be termed that) was now to be found in the backyard. In the later *William* stories, there is a pervasive sense of an enemy that has not only moved closer but has even penetrated the domestic sphere. The hero becomes focused on the evacuation of children away from the threat of air raids, the air raids themselves and even the search for spies. This sense of imminent violation was not new: as we shall see below, plans to build a channel tunnel initiated an explicit sense of threat, but the subsequent development of flying machines confronted the ways in which many authors thought of 'this Fortress, this England' (Beer 1990: 265).

By the 1940s and 1950s there were few open spaces left to explore or appropriate in the ways that had been possible in Africa and South America during earlier periods. In consequence, the oceans and the planets came under greater scrutiny, and under such circumstances a different geopolitical conception began to emerge. As the cold war evolved, comics too created new spaces. In part, some of these reflected an interest in hidden dimensions, populated by mutant creatures who had been irradiated (something of an extension and juxtaposition of stories such as the *Phantom of the Opera* or *Twenty Thousand Leagues under the Sea*).[7] Others represented a liberation from the earth entirely, and science fiction moved into outer space, where new empires and new colonies awaited.

The role that science fiction and extraterrestrial locales play in this story are only tangentially relevant, but it is interesting that there is a clear projection outwards that occurs once the European nations settle into economic integration, stability and military eclipse. Outer space took on the role of the frontier and aliens became the natives across whose territories fought the new heroes (Dan Dare or Flash Gordon) and villains (The Mighty Mekon or Ming the Merciless).

When fiction becomes fact

Having been offered these simple tropes, the reader might reasonably ask for some conclusive evidence that these stories were, in effect, something more than mere stories. It is usually difficult to trace explicit connections between fiction and fact, although it is rarely hard to show how specific tropes enter literature. But did the perceived reconfiguration of time and space implicit in the construction of a channel tunnel or the manufacture of Zeppelins then have an impact on the evolution of British foreign policy? Is it possible that H. G. Wells' 1908 novel *The War in the Air*, in which mass civilian bombing is perpetrated by the German air force, actually brought such ideas into existence? The book was written at a time when such weapons simply did not exist: so was their development inevitable, or did Wells really provide the blueprints that Goering later developed? By the most rigid criteria of causality, the interpretations made here must remain speculative. This notwithstanding, it has been my intention to show that there was a broad consistency between the fiction of this period and the evolution of geostrategy, and this can be plausibly done. None the less, there is also some harder evidence that such writing could be influential, and this penultimate section explores this material in greater depth.

This chapter began with *The Riddle of the Sands* and used that as an example of fiction that was created to serve a geo-political purpose. Childers' book was, however, only the most successful of a veritable freshet of volumes that appeared in the late Victorian era. In a period when traditional rivalries within Europe remained sharply drawn, imaginative and speculative fiction served as an arena within which international relations could be debated and hostilities expressed in a relatively painless manner. Until the end of the century, these stories were essentially a reprise of the usual antipathy between France and Britain, and titles such as *La Guerre Anglo-Franco-Russe* (unknown 1900); *L'Agonie d'Albion* (Demolder n.d.) and *La Guerre Fatale: Franco-Angleterre* (Danrit 1901) were matched by *The Invaders* (Tracy 1901); *The Battle of Dorking* (Chesney 1871); *The Great War in England* and *The New*

Battle of Dorking (Maude 1900). As Pick has shown, the genre was so large that there was even an identifiable sub-field of pot-boilers that focused explicitly on the threats implicit in the construction of a channel tunnel (Pick 1994).

It was in this context that the enormity of Childers' work has to be understood. While Prussian militarism and the construction of the Kiel canal in 1895 were hardly secrets, the traditional foe remained the most threatening. Even in 1901, Pemberton's novel *Pro Patria* had French troops pouring from a tunnel and invading England. *The Riddle of the Sands*, coming a mere two years later, was thus in every sense 'novel'. Members of the government were driven to deny its predictions, and as Clarke observes 'several hundred thousand copies of the cheap edition were sold and in Germany the book was ordered to be confiscated' (Clarke 1965: 110).

Once Childers had established the genre, it was followed by *The Invasion of 1910* (le Queux 1894), *An Englishman's Home* (du Maurier 1909) and *When William Came* (Saki 1929), while German authors responded with *Der deutsch-englisch Krieg* and *Mit deutschen Waffen über Paris nach London*. Nor did these publications occur in a vacuum. A German novel such as *Der kommende Krieg* was followed by an editorial in the *Daily Mail* in 1906; while a serial, William le Queux's *Invasion of 1910*, was critically discussed by an article in *Marine Rundschau*. A high-ranking soldier wrote the drama *An Englishman's Home* (du Maurier 1909) that also dealt with invasion, and the lines between politics, policy-making and prognostication became fully blurred once an army recruitment office was opened in Wyndham's Theatre (Clarke 1965).

From these examples, we can see that adventure fiction played a central role in the creation and development of popular discourse in an era before the electronic media of communication. The numerous titles that followed Childers' work all served to develop the themes of encirclement and invasion such that the declarations of war in 1914 were a literal relief. But they did not exist in isolation. As Clarke shows in fascinating detail, each statement by a British author dealing with invasion was seen by the Germans as a provocation, while each German statement of national destiny was viewed as a threat to British sanctity. And these were not the only audiences for such fiction; le Queux's *Invasion of 1910* ultimately sold over a million copies and was translated into 27 languages, including a special boys' edition (*Der Einfall der Deutschen in England*) which removed the ending in which the Germans were defeated.

Conclusions

Ó Tuathail and Agnew have written that 'it is only through discourse that the building up of a navy or the decision to invade a foreign country is made meaningful and justified. It is through discourse that leaders act, through the mobilization of certain geographical understandings that foreign-policy actions are explained and through ready-made geographical reasoning that wars are rendered meaningful' (1992: 191). In this chapter I have demonstrated some of the ways in which such reasoning was assembled. First, we can see that there was a clear notion of something that we would today identify as a 'sphere of influence'. Childers, for instance, was making unambiguous statements about the threats that were facing Britain even as Germany was moving into the first phases of industrial expansion and mounting a challenge to British hegemony. Second, we trace a clear connection between popular sentiment and the more formal pronunciation of geopolitics on a world stage. Mackinder's address to the Royal Geographical Society, in which he reasserted the importance of the European and Asian land area as pivotal, took place in 1904; it was three years after Hannay is found identifying Turkey and Islam as key factors in the stability of the West that Halford Mackinder projected the concept of the Heartland into formal geographical publication. In other words, we can see that popular literature was perfectly timed to reflect these academic debates (Mackinder 1904; Mackinder 1919).

Third, and most critically, we can also infer from the silences that are expressed in these pages. The removal of most recognizable women is now well understood, as is a credible portrayal of class.[8] Much has been made of the manner in which racial thinking permeates this period's writing, but in some ways the more subtle erasure of indigenous people is more troubling. Once they cease to occupy their own territories, these political spaces cease to exist as political entities: they simply become spaces once again. There is a case to be made that the explorations of the nineteenth century generated profound changes in the colonial societies which recoiled from, and yet were profoundly influenced by, the natural and human diversity of the lands beyond Europe. In some measure, the views of the twentieth century are more perplexing, in so far as the diminution of the other is a reflection of a new economic orthodoxy and of the creation of the Third World as an underachieving social object to be experimented upon. The fourth dimension of this story is the way in which both conventional geographical and geopolitical imagery begin to erode. Adventure still exists: indeed, the persistence of such fiction constitutes an interesting insight into the contemporary construction of masculinity (Phillips 1997). None the less, it has changed dramatically. Desert islands have become the settings, not for adventure but for comedy (notably *Gilligan's Island*, still in syndication on America television after three decades), or soft pornography (such as *The Blue Lagoon*). It is only with the relocation of imperialism and expansionism to other worlds that the liminal spaces of adventure can be recreated, thus generating a 'whole new world' of fiction to match.

Acknowledgements

A preliminary version of this chapter was presented to the members of the Department of Geography at the University of Reading in November 1997, and their comments were very helpful. Other useful insights were offered by Jennifer Hyndman, Julie Murphy

Erfani, Kristin Koptiuch, Julia Patterson, an anonymous referee and, in large measure, the editors themselves. This work is dedicated to my grandfather, who bought me comics throughout my childhood, even when he couldn't afford them.

Notes

1 This is not to argue, of course, that only fictional writing is important in this context, a point well demonstrated with regard to magazines by Sharp (1996).
2 The authors did use the phrases they did and audiences clearly consumed them, often uncritically. Consequently, the books were not banned in their day, a reality that some see as a failing that should now be redressed. However, such belated censorship would be merely self-exculpating on the one hand, and another small extension of state control on the other (for further discussion on this, see Phillips 1997, chapter 8).
3 I have a distant recollection of a series that was still going strong in the early 1960s that featured a character named the Wolf of Kabul. He was engaged in constant violent activity in the company of a monoglot sidekick "Chung" who was armed solely with a "clicky-ba" or cricket bat. Given that the British had been attempting to subdue Afghanistan since the First Afghan War of 1829 and had of course left India in 1947, this was by any standards a very strange message to be sending to adolescent readers.
4 The great Eulenburg homosexuality scandal of the German officer class that erupted at the end of the century would have been well known to Buchan and his contemporaries: see Steakley 1993: 233.
5 In a story entitled *The Great Arena*, Biggles escapes death when a German pilot sportingly ceases firing on him when his guns become jammed. On returning to base, Biggles is asked where he has been: "'you didn't get those testing!" returned the Major, frowning, pointing to a row of neat holes in the fin of Biggles' Camel. "No, sir, I had a little affair – nothing to speak of – with a lad in a yellow-tailed Fokker", replied Biggles.' (Johns 1934 *Biggles of the Camel Squadron*: 123).
6 This leads, unsurprisingly, to some inconsistencies. Biggles spends several hours tramping across the scalding desert, but then munches some chocolate, miraculously unmelted, from his pocket (*Biggles Flies East*).
7 These hidden dimensions are only simple precursors to the much more complex hidden landscapes that are found in many computer games such as Myst and Riven.
8 While there were always working-class characters in British fiction, they were either orphans (i.e. redeemable) or strangely eccentric (such as the 'tough of the track' Alf Tupper, who smoked cigarettes and ate fish and chips before sprint races).

Bibliography

Primary sources

Buchan, J. (1915) *The Thirty-Nine Steps*, London: Nelson.
—— (1916) *Greenmantle*, London: Nelson.
—— (1919) *Mr. Standfast*, London: Nelson.
Chesney, G. (1871) 'The Battle of Dorking', *Blackwood's Magazine*, May.
Childers, E. (1903) *The Riddle of the Sands*, Oxford: Oxford University Press.

Crompton, R. (1939) *William and Air Raid Precautions*, London: Macmillan.
—— (1940) *William and the Evacuees*, London: Macmillan.
—— (1940) *William and the Spy*, London: Macmillan.
—— (1996) *Just William on Holiday*, London, Macmillan.
Danrit, Capitaine (1901) *La Guerre Fatale: Franco-Angleterre*, Paris.
de Vere Stacpoole, H. (1908) *The Blue Lagoon*, London: Unwin.
Demolder, E. (n. d.) *L'Agonie d'Albion*, Brussels.
du Maurier, G. L. B. (1909) *An Englishman's Home*, London: Harper.
Hay, I. (1908) *The Right Stuff*, London: Blackwood.
—— (1909) *A Man's Man*, London: Blackwood.
—— (1915) *The First Hundred Thousand*, London: Blackwood.
—— (1918) *The Last Million*, London: Blackwood.
Hichens, R. (1904) *The Garden of Allah*, London: Methuen.
Hull, E. M. (1921) *The Sheik*, London: Eveleigh Nash.
Johns, W. E. (1934) *Biggles of the Camel Squadron*, Oxford: Oxford University Press.
—— (1935) *Biggles Flies East*, Oxford: Oxford University Press.
—— (1938) *Biggles Goes to War*, Oxford: Oxford University Press.
—— (1945) *Biggles in the Orient*, London: Hodder and Stoughton.
—— (1946) *Biggles Delivers the Goods*, London: Hodder and Stoughton.
le Queux, W. (1894) *The Invasion of 1910*, London: Nash.
Maude, F. N. (1900) *The New Battle of Dorking*, London: G. Richards.
Pemberton, M. (1901) *Pro Patria*, London, Ward: Lock.
Ransome, A. (1930) *Swallows and Amazons*, London: Jonathan Cape.
Saki (Munroe, H. H.) (1929) *When William Came*, New York: Viking.
Thorne, G. (1903) *When it was Dark*, London: Greening.
Tracy, L. (1901) *The Invaders*, London: Pearson.
Unknown (1900) 'La Guerre Anglo-Franco-Russe', *Le Monde Illustré*, March.
Wells, H. G. (1908) *The War in the Air*, London: Bell.
Wren, P. C. (1924) *Beau Geste*, London: Murray.

Secondary sources

Beer, G. (1990) 'The island and the aeroplane', 265–90 in H. Bhabha (ed.) *Nation and Narration*, London: Routledge.
Bhabha, H. (1990) 'DissemiNation', 291–322 in H. Bhabha (ed.) *Nation and Narration*, London: Routledge.
Castle, K. (1996) *Britannia's Children: Reading Colonialism Through Children's Books and Magazines*, Manchester: Manchester University Press.
Clarke, I. F (1965) 'The shape of wars to come', *History Today*, February: 108–16.
Childers, M. A. (1931) 'Foreword', in Childers, E. *The Riddle of the Sands*, London: Edward Arnold.
Cockburn, C. (1975) *Bestsellers: The Books that Everyone Read, 1900–39*, Harmondsworth: Penguin.
Dalby, S. (1990) *Creating the Second Cold War*, London: Pinter.

Ellis, P. B. and Schofield, J. (1993) *Biggles! The Life Story of Captain W. E. Johns*, Godmanstone: Veloce.

Enloe, C. (1990) *Bananas, Beaches and Bases*, Berkeley: University of California Press.

Finkelstein N. G. And Birn, R. (1998) *A Nation on Trial*, New York: Henry Holt.

Goldhagen, D. J. (1996) *Hitler's Willing Executioners: Ordinary Germans and the Holocaust*, New York: Knopf.

Herb, G. (1997) *Under the Map of Germany*, London: Routledge.

Jameson, F. (1993) *The Geopolitical Aesthetic*, Minneapolis: University of Minnesota Press.

Kearns, G. (1993) 'Prologue', 9–30 in P. J. Taylor (ed.), *Political Geography of the Twentieth Century*, London: Belhaven.

Kline, S. (1993) *Out of the Garden*, London: Verso Press.

Livingstone, D. (1998) 'Reproduction, representation and authenticity', *Transactions of the Institute of British Geographers*, 23 (1):13–20.

MacKenzie, J. M. (1996) 'General editor's introduction', in K. Castle *Britannia's Children: Reading Colonialism Through Children's Books and Magazines*, Manchester: Manchester University Press.

Mackinder, H. (1904) 'The geographical pivot of history', *Geographical Journal* 23: 421–37.

—— (1919) *Democratic Ideals and Reality*, London: Constable.

Mann, M. (1988) *States, War and Capitalism*, Cambridge: Blackwell.

Ó Tuathail, G (1996) *Critical Geopolitics*, Minneapolis: University of Minnesota Press.

Ó Tuathail, G. and Agnew, J. (1992) 'Geopolitics and discourse: practical geopolitical reasoning in American foreign policy', *Political Geography Quarterly* 11: 190–204.

Overy, R. (1996) *Why the Allies Won*, New York: Norton.

Phillips, R. (1997) *Mapping Men and Empire: A Geography of Adventure*, London: Routledge.

Pick, D. (1994) 'Pro Patria: blocking the tunnel', *Ecumene* 1 (1): 77–93.

Paasi, A. (1996) *Territories, Boundaries and Consciousness*, Chichester: Wiley.

Pratt, M. L. (1992) *Imperial Eyes*, London: Routledge.

Sharp, J. (1996) 'Hegemony, popular culture and geopolitics: *Reader's Digest* and the construction of danger', *Political Geography* 15 (6/7): 557–70.

Steakley, J. (1993) 'Iconography of a scandal', 233–63 in M. Duberman, M. Vicinus and G. Chauncy (eds) *Hidden from History: Reclaiming the Gay and Lesbian Past*, New York: New American Library.

Webb, P. (1994) *A Buchan Companion*, Dover N.H.: Sutton.

Winks, R. W. (1988) 'John Buchan, stalking the wilder game', v-xxii in J. Buchan *The Four Adventures of Richard Hannay*, Boston: Grodine.

4

JAPANESE GEOPOLITICS IN THE 1930s AND 1940s

Keiichi Takeuchi

The introduction of geopolitics in Japan

The concept of geopolitics according to Swede Rudolf Kjellén was first introduced to Japan in 1925, when his work of 1916 appeared in a book review in a
Japanese journal of international law and diplomacy (Kjellén 1916; Fujisawa
1925). The reviewer, Chikao Fujisawa, rightly pointed out that 'this new
approach opens up new horizons in the real study of the phenomena of the state,
casting off the erstwhile prevailing old, abstract, theoretical and conventional
approach'. Perhaps Fujisawa was not aware of the geopolitical movement in
Germany, already established in Munich due to the initiative of Karl Haushofer.
Some months later, Taro Tsujimura, then head of the recently-created
Department of Geography at the Imperial University of Tokyo, discussed the
term 'geopolitics' in his review of Otto Maull's book (Maull 1925). Tsujimura
and other geographers who had occasion to refer to geopolitics in the 1920s
considered it to be merely an application of geography to real politics; they were
critical of the standpoint that considered it as either a branch or a development
of political geography (Tsujimura 1925). Rather, they believed that political
geography was firmly based on the recognition of the interdependence or interaction between the state and the physical and cultural landscape. Takuji Ogawa
(Ogawa 1930) and Goro Ishibashi (Ishibashi 1927), leading geographers of the
Imperial University of Kyoto, criticized geopolitics for its lack of a precisely
defined object of study. Their criticism, however, never touched upon the essential character of geopolitics, that is, an organicist (*sic*) view of the state.
Somewhat of an exception was Hikosaburo Sasaki, who criticized German
geopoliticians for their adherence to environmental determinism and direct
causality between politics and land, whilst failing to take into account the intermediate economic mechanism (Sasaki 1927). During that period, therefore,
only a few geographers actually applied this new approach to political science in

72

Japanese international politics. Nobuyuki Iimoto, then a young specialist in political geography at Tokyo, was one of the few who recognized the value of geopolitics in policy-making (Iimoto 1935: 1–13).

In 1931, Japan commenced her military invasion of Northeast China, marking the start of the fifteen-year war which ended in the defeat of Japan at the end of the Second World War in 1945. The year 1931 was also the turning point between the more liberal 1920s and the reactionary or imperialist decade of the 1930s in Japan. Early in 1932, Japan formed the puppet state of Manchukuo. In 1933, as a result of the adoption of the Lytton Report on Manchuria by the League of Nations, which strongly criticized the aggressive Japanese military actions in Manchuria, Japan withdrew from the League of Nations. The isolation of Japan in the new international political sphere led to an increasing interest in geopolitics. This was due in great part to the strengthening of the relationship between Japan and Germany and the ensuing formation of the Axis. In Germany, the rise to power of the Nazi party had strengthened links with the Munich-based geopolitical school. While in the 1920s the German term *Geopolitik* was translated into Japanese as *chiseijigaku*, literally geographical politics, it dropped out of use in the 1930s; instead, the term *chiseigaku* was adopted. The shift was due to a combination of the three ideograms involved in its written form which better reflected the term geopolitics rather than geographical politics.

The geopolitical tradition in Japan before the introduction of German geopolitics

The flourishing of geopolitics in Japan was not, however, prompted simply by the new, internationally isolated position of Japan in the 1930s. In order to obtain a clearer picture of the circumstances surrounding the establishment of German-style geopolitics, it is necessary to examine the background or tradition of geopolitical or geostrategic thought in modern Japan.

From the beginning of the 1870s, Japan expanded northwards into the Korean peninsula. In 1875, Japan dispatched a gunboat to the shores of the island of Kanghwado to force Korea into opening diplomatic relations, just as the United States had done to Japan two decades earlier. After the military attack at Kanghwado, the Japanese government sent an envoy, Kiyotaka Kuroda, to Korea to conclude a treaty of friendship. This marked the beginning of the Japanese intrusions into the Asian continent, culminating in the conflict with China in 1894–5, with Russia in 1904–5 and further, the so-called 'fifteen-year war' with China. Given these increasing encroachments in other territories, both military and economic, a considerable number of Japanese began to travel on missions of inquiry to China and Russia. They included people from all walks of life, ranging from prominent political and diplomatic figures to military inspectors and secret agents (Nagasawa 1973). Despite their diverse backgrounds, they were united in their belief in the appropriateness of Japan's northwards advance, referred to as *hokushin*, for geostrategic or geopolitical reasons. Obviously, due to the covert nature of some of the missions, not all the accounts of these journeys saw publication. It should be noted, however, that these accounts, confidential or otherwise, duly influenced the decision-making of the Japanese government and military groups; moreover, numerous protagonists of these investigative travels frequently gave public lectures catering to both academic and popular audiences, thus contributing to the gradual formation of Japanese geopolitical and geostrategic thought among the population at large. It is the published travel accounts or transcriptions of delivered lectures, such as those of Buyo Enomoto, sent to Russia on a diplomatic mission, Kiyotaka Kuroda, the prominent diplomat and politician, and Yasumasa Fukushima, who served as military attaché with the Japanese envoy at Berlin, that allow us a sense of their considerations and their specifically geopolitical nature (Takeuchi 1998, forthcoming).

Expansion was by no means limited to the North, for already by the 1870s many private Japanese concerns were to be found operating in Southeast Asia and the Pacific area. Meanwhile, in the 1880s, with the increasing German influence in Micronesia and Melanesia, publications in support of southward expansion (*nanshinron*) began to appear. In the 1880s and 1890s, when chairs of geography in higher education had not yet been created (Takeuchi 1974), a certain number of books dealing with political geographical considerations aimed at the general public were published. After pursuing studies in fishery science for three years in the United States, Kanzo Uchimura, a graduate of the Sapporo College of Agronomy, published a treatise of geography titled, in literal translation, *Considerations on Geography* (Uchimura 1894). In the preface, he considered the geographical position of Japan: facing the United States across the Pacific Ocean. This led him to emphasize the need for friendly relations with the United States.

Shigetake Shiga, another graduate of the Sapporo College of Agronomy, obtained passage on board a navy training ship and visited the South Pacific islands and Oceania in 1886. Subsequently, he began to insist that a Japanese advance to the south was required, both from the economic and military points of view. He argued that this move was appropriate due to Japan's geographical situation, allowing easy access to Southeast Asia and the South Pacific (Shiga 1887). Uchimura was a pacifist and, on the basis of his Protestant convictions, argued against the absolutist tennoism of Meiji Japan. He firmly opposed the outbreak of the Sino-Japanese war of 1894, whereas Shiga laid stress on the nation-building efforts of Meiji Japan, though admittedly, he was never at any time overtly prejudiced (Takeuchi 1988, 1994b; Minamoto 1984; Yamamoto and Ueda 1997). In spite of the ideological differences between these two graduates of the Sapporo College of Agronomy, through their studies they acquired a common understanding of the importance of geographical conditions. They were not academic geographers, yet they made pertinent observations about the international scene on the basis of their firsthand experience of the geographical position of Japan. It should also be noted that *nanshinron*, or southward expansionism, represented the interests of private sectors, business and immigration affairs, and that only in the second half of the 1930s did it become a governmental concern supported by naval interests.

In addition to these concerns, as in other nations around the inter-war world, the impact of German *Geopolitik* became evident in Japan. In order to analyse Haushofer's subsequent influence on Japanese geopolitical thought, it is necessary to take into account his personal connection with Japan. During his stay from the end of 1908 to the summer of 1910, he made the acquaintance of prominent Japanese political and military figures and availed himself of opportunities to delve into the traditions and culture of the country. After he became a professional geographer or geopolitician, he published not only the famous book on the geopolitics of the Pacific Ocean (Haushofer 1925) but repeatedly referred to Japanese affairs (Jacobsen 1979: 86–112). For Karl Haushofer and perhaps also for the Haushofer family, Japan retained a special meaning. In his Munich house hung (and still hang) numerous pictures and photographs of Japan. His eldest son, Albrecht Haushofer, geomorphologist and army officer, a child at the time the family lived in Japan, was executed by the Nazis in 1945, in the last days of Nazi Berlin, after being imprisoned for his part in the failed plot to assassinate Hitler. Afterwards, in his pocket, a handful of poems were discovered, two of which had Japanese titles, *Kami* and *Itsukushima*.

However, Japanese geopolitics were not solely derived from German *Geopolitik* and Haushofer's ideas. There were several trends of geopolitical thought and movements in Japan, and this chapter seeks to outline a few of these different versions of geopolitical thought. First, there was a geopolitical school of the Imperial University of Kyoto, directed by Saneshige Komaki; second, a group comprised of faithful followers of Haushofer or the German-type geopolitical school. Third, there was the Japan Association for Geopolitics (Nihon Chiseigaku Kyokai) (Takeuchi 1980, 1994a); and finally, the members of Hidemaro Konoe's 'brains trust' (Fukushima 1997). Although not all schools or groups blindly adopted Haushofer's doctrine, the latter none the less exerted an enormous influence which contributed to their development and success. This was partly because Haushofer was thoroughly acquainted with the circumstances of Japan, but also because the geographical position of Japan was highly conducive to the acceptance of Haushofer's doctrine of pan-regions (Abdel-Malek 1977). To some Japanese academics, who felt themselves otherwise incapable of employing their scientific achievements in the difficult situation facing Japan during that period, the German political movement appeared an archetype to which they could look for guidance in formulating their thinking (Yamaguchi 1943: 230–7). For other politicians and academics who felt antipathy towards the prevailing irrational, chauvinist and ultranationalist demagogy, Haushofer's mechanistic and apparently rational and realistic analyses came as a relief. It is not difficult to understand why, under such ambivalent circumstances, three separate Japanese versions of *Geopolitik des pazifischen Ozeans* (Geopolitics of the Pacific Ocean) were published in the 1930s and 1940s as well as two separate Japanese versions, respectively, of *Bausteine zur Geopolitik* (Building Blocks for a Geopolitics) and *Weltmeer und Weltmacht* (World Oceans and World Power). And it was through such publications that geopolitical thought spread throughout Japan.

Geopolitical practices of geographers before the mid-1930s

While the attitude of geographers towards geopolitics differed according to the 'school' of geopolitics and the individuals involved, the development of geopolitical ideas in Japan can be divided broadly into two phases: the first phase was from the 1920s through to the early 1930s when geographers such as Tsujimura and Sasaki criticized Kjellén, and the second period was from the mid-1930s onwards, when geopolitics came to be widely discussed in journalistic, political and military circles. Moreover, some contemporary geographers, who even during the second phase maintained a critical stance towards geopolitics, began to recognize the distinctions between the two phases (Watanabe 1942).

In the first phase, apart from the strictly scientific journals of geography such as *Chirigaku Hyoron*, *Chikyu* and *Chigaku Zasshi*, geographical journals aimed at a broader spectrum of readers, notably primary school teachers who aspired to the Teachers' Licence for the Teaching of Geography in Secondary Schools. *Chirigaku*, published by Kokon Shoin, and *Chirikyoiku*, issued by Chukokan, were just two of the main journals of this type. It should be noted that during the first phase critical introductions to geopolitics were originally confined to scientific journals. Up to the beginning of the 1930s, some degree of freedom of speech was still permitted in Japan, hence geopolitics came in for a certain amount of criticism due to its vindication of the fascist regime in Germany. For instance, Masakane Kawanishi, citing the Marxist critic of *Geopolitik* Karl Wittfogel (Wittfogel 1929), regarded geopolitics as 'an explanation neglecting the intermediate mechanism of the connection between existing natural conditions and political patterns'. Furthermore, Keishi Ohara published a series of papers in the first half of the 1930s, which were eventually included in his book of 1936. He clearly stated that:

> The fundamental method of a geopolitical approach still continues to be one involving an explanation of the nature of the state and the process of its political development, not in terms of the development of social

productive forces or other socio-economic factors such as the pursuit of profit or of capitalist economies, but directly and one-sidedly through natural conditions. This masks the socio-economic factors existing behind the activities of the state and justifies the claims and the acts of exploitation on the part of the state in regard to existing natural conditions. . . . Geopolitics and present-day political geography are thus based on an organic view of the state and on the geographical materialism of past times. Only the social and economic situations of present-day Germany have restored these conventional theories. . . . Political geography expressed in present-day Germany is an ideological reflection of the recovery of German capitalism and its nationalistic development, and serves as a scientific instrument for its development.

(Ohara 1936:335–6)

Not all Japanese geographers adopted this sort of fundamental theoretical criticism, but many considered geopolitics a mere application of political geography to state strategy. To some extent, this was reflected in contemporary geographical journals aimed at a wider readership: no papers appeared on geopolitics except for short mentions of the German geopolitical movement before the mid-1930s. However, the journals did contain a large number of papers on political geography, as these comprised mandatory reading for primary school teachers taking the examinations for secondary school teaching certificates. The negligent attitude of Japanese geographers towards geopolitics was not entirely without reason. German geopolitics under the Weimar Republic manifested a certain chauvinistic and patriotic character with reference to their own state and/or German-related lands as well as potential territories (*Lebensraum*) and potential enemy countries. However, in the case of a country beyond the frontiers of Germany's envisioned expansionism, the treatment accorded to such places was indistinguishable from a political geographical analysis. Japan's political geographical situation, as a victorious nation of the First World War and in possession of colonial lands such as Korea, Formosa and a number of Pacific islands, differed considerably from that of Weimar Germany, suffering under the restrictions of the Versailles Treaty. Consequently, the specific political geography produced in Weimar Germany was considered inapplicable to Japan.

During the second phase of geopolitical developments from the mid-1930s, geographers were rather slow to analyse the geopolitical situation relating to Japan and its roles in the 'fifteen-year war'. At the end of the 1930s, the Nazis achieved their target of *Kampf gegen Versailles*, or the breaking down of the terms of the Versailles Treaty; they commenced to flaunt a new slogan, the 'New European Order'. Directly or indirectly, this stimulated Japanese leaders to invent the 'New Order in East Asia' (*Towa Shin Chitsu-jo*), which later developed into the 'Great East

Asia Co-Prosperity Sphere' (*Daitowa Kyoeiken*). Only Joji Ezawa, a graduate in commerce of the Tokyo Commercial College where he later taught German, energetically took up the subject of German geopolitics of the 1930s. In spite of his efforts to reconstitute conventional human and economic geography, which was, according to him, based on the methodology of the natural sciences, he subsequently proposed an alternative anthropocentric and rational science of spatial organization (*Raumordnung*) (Ezawa 1938). Ezawa reified a curious mixture of romanticism with aspirations to a cool and mechanistic order of spatial organization. He was not recognized as a geographer in Japanese geographical circles of that period, but was nevertheless a vanguard geographer who advocated and popularized geopolitics in the journalistic world. He remained an epigone of Haushofer, as evidenced in his last geopolitical book (Ezawa 1943). In this, he paralleled the case of Ichigoro Abe, a former economist, who in 1933, had already published the first systematic treatise on German geopolitics (Abe 1933). Both men were not considered to be geographers by recognized geographical specialists.

Analyses of the geopolitical practices of geographers during the late-1930s second phase presented a number of difficulties for the following reasons. First, and until quite recently, most of the practitioners of geopolitics remained silent regarding this period of their lives because of the measures implemented by the Allied Forces authorities after the Second World War condemning geopolitics. Negative public opinion was shaped by events which took place during this second phase. Second, a large number of documents were destroyed by burning on the orders of the government and army authorities in August 1945, during the two-week hiatus between Japan's surrender and the arrival of the Allied Occupation Forces on mainland Japan. However, almost all of the surviving documents confiscated by the Allied Forces authorities have now been returned to Japan by the United States in the form of microfilms or microfiche but the analysis of these records, with their wealth of confidential material pertaining to geopolitical affairs, has only just begun. Therefore, the analysis of this author is out of necessity based chiefly on currently available printed material. Third, as a reaction to governmental 'thought control' and the severe censorship of printed matter, many authors felt compelled to adopt devious methods to express their thoughts, or even uttered the standard positions of the day, in order to escape governmental suppression.

Papers and reviews of geopolitics by geographers were rarely to be found in scientific journals in this second phase, except for *Chiri to Keizai*, organ of The Nippon Economic Geographical Society, which published several papers on topics that were problematic given the international circumstances of that period. Instead, the greater part of this material appeared in popular journals, albeit of a geographic nature, and in cultural journals of general interest. During 1941, the two main journals read by trainees for the Teaching of Geography at Secondary

Schools, *Chirigaku* and *Chirikyoiku*, began to use the term geopolitics in the titles and texts of many of the papers. In the 1942 volume of *Chirigaku*, four or five papers appeared every month under the column 'Geography of Great East Asia', and number four of that year was devoted to a special issue, 'Geopolitics of Great East Asia', containing papers by all the protagonists of the Kyoto school of geopolitics, starting with Saneshige Komaki. In contrast to *Chirigaku*, *Chirikyoiku* – having changed its title to *Chirigaku Kenkyu* in 1942 – adopted a more scientific approach to geopolitics. Most of the authors who published their papers on geopolitics in *Chirigaku Kenkyu* (Iwata 1942; Watanabe 1941,1942; Watanuki 1942) were inclined to be critical of geopolitics, yet all of them were compelled to acknowledge the usefulness of geopolitics in the construction of geostrategy and geotactics during a crucial phase of the Japanese empire. Watanabe, whilst recognizing the advantages of geopolitical discourse, wrote the frankest criticism of geopolitics in 1942 as follows:

> Geopolitics by definition appears to be a systematic discipline, but its content consists merely of policy discussions. . . . Its content and aims can be summarised substantially as follows: First, to provide politicians with guidelines for state policies on the basis of 'intuitive reasoning' and geographical considerations; Second, to justify the policies decided upon; and third, to convince the people of the validity of the foreign policies of the state and persuade them to collaborate with those policies on the basis of a moral conviction. Geopoliticians generally complemented their lack of logic with the shock quality of their discourses. People were deluded into believing in the logic or system paraded forth in these difficult discourses, which was the ulterior motive of the geopoliticians, or in other words, the purpose aimed at was a sensational effect.
>
> (Watanabe 1942: 8–9)

It could be construed that Watanabe, as a professor at a military academy and an individual in a privileged position, could afford to indulge in such straightforward criticism. In any case, all these popular journals had to cease publication in 1944 due to government-implemented measures to combat the paper shortage brought on by the destruction of paper mills in Allied air raids. The last remaining geographical journal, *Chirigaku Hyoron*, organ of the Association of Japanese Geographers, also had to interrupt publication for over a year from the beginning of 1945.

Some geographers who criticized geopolitics in the first phase began to recognise the practical and political validity of geopolitics in the later 1930s. In his thick treatise on political geography of 1941, Toshiyuki Iimoto, who in 1928 had refused to acknowledge geopolitics as an independent discipline, now devoted three whole chapters to a favourable consideration of geopolitics (Iimoto 1928, 1935).

Apropos the changing attitude of Iimoto, it should be noted that he became Secretary General for the Japan Association for Geopolitics founded at the end of 1941, a few weeks before the Pearl Harbor attack.

Meanwhile, Keishi Ohara who developed a fundamental and theoretical criticism of geopolitics in the first phase, published a paper in a leading intellectual review in 1940, thereby acknowledging the importance of geopolitical considerations, under the conditions of the prevailing totalitarian economic system or the so-called controlled economy (Ohara 1940). In 1942, he was compelled to resign from a teaching post at a higher commercial school in Yokohama. One of the reasons cited for his dismissal was the publication of the somewhat Marxist-oriented book of 1936 called *Shakaichiri-gaku no Kisomondai*. Around 1940, he was subjected to the stringent surveillance of the public security police and hence his argument of 1940 would seem a typical example of the distorted logic and convoluted phraseology that evolved out of a situation of limited freedom of speech. In contrast, Koji Iizuka wrote numerous papers for popular journals in this period. These appeared to reveal the conspiracies that the Western powers used to justify the Pacific war, and geographical analyses of the United States as enemy. At the same time he produced exceedingly penetrating critiques on *Geopolitik* in publications for the Faculty of Economics at the Imperial University of Tokyo, during 1942 and 1943 (Iizuka 1942/43). He amended and re-published almost all his collected works after the Second World War, but the writings published in popular journals in this period were not included.

Saneshige Komaki, third head of the Department of Geography of the Imperial University of Kyoto, was a brilliant specialist in historical and prehistorical geography, especially with regard to the reconstruction of historical landscapes and interpretation of past landscapes, a tradition which continues to this day. The circumstances under which Komaki's book, *Manifesto of Japanese Geopolitics* (Komaki 1940b), was published are not clear. In this work, he wrote that a new Japanese geopolitics had to develop on the basis of a geographical study of Japan which emphasized the traditional ethics and mentality of the Japanese. He also argued that:

> In this way Japanese geopolitics is different from the many world geopolitical currents imitating German geopolitics, from the colonialist in the British style and also from the old-fashioned type of Chinese geopolitics; it is a distinctly Japanese type which has existed since the beginnings of the imperial family and will develop in line with the prosperity of the imperial family as a truly creative science of Japan.
>
> (Komaki 1940a: 5)

Along these lines, he placed particular emphasis on the need for recognition of a proper national Japanese polity based on tennoism (Komaki 1942). This would appear to be a kind of divine inspiration on his part, influenced perhaps by his family belonging to the Shinto priesthood in Shiga Prefecture. It should also be noted that at the Department of Geography at Kyoto, the *Anthropogeographie* and *Politische Geographie* of Friedrich Ratzel were widely read and that some of Komaki's students had been extremely interested in geopolitical thinking since the mid-1930s. One of them was Jiro Yonekura, assistant of the department in the second half of the 1930s, who by the 1937 Japanese invasion of China had already carried out geostrategical examinations on behalf of the Japanese army, based on the documentary analysis of past wars on Chinese soil. These studies were later published (Yonekura 1942). It is impossible to judge the motivation behind Komaki's conversion to geopolitics. Whether it was due to divine inspiration or the impulse of the academic tradition of the Kyoto school remains to be ascertained. Regardless of his motivations, he managed to mobilize most of the graduates of the Kyoto geography department to research geopolitical themes. *Chiri Ronso*, the organ of the Kyoto geography department, in its eleventh volume commemorating the mythological 2,600th year of the foundation of Japan by the first tenno, published twenty-five papers relating to geopolitics including that of Komaki.

Under the directive of Komaki, a group of younger graduates at the geography department in Kyoto came to specialize in area studies of specific countries of the Greater East Asia Co-Prosperity Sphere; they later collaborated on an uncompleted multi-volume publication project involving a series of geopolitical works such as *Sekai Chiri Seijigaku Taikei*. Their division of labour was as follows: Atsuhiko Bekki in Southeast Asia, Jiro Yonekura in China, Nobuo Muroga in North America, Saburo Noma in Europe, Tokuichi Asa in South Asia, Masatoshi Mikami in Siberia, Shunji Wada in Australia, Yojiro Tomonaga in Africa, Kiyoshi Kawakami in the Polar regions and Tsugio Murakami in the Pacific islands. In 1942, Bekki and Asai were respectively dispatched as military administrators to Indonesia and Burma, then under Japanese occupation.

In numerous published writings, the geopoliticians of Kyoto remarked on the economic problems of Japan caused by the dominance of the Western powers in East Asia, and on the racial discrimination against the Japanese, which considerably affronted the Japanese public. Yet at the same time, these authors sensed that the mere exposure and condemnation of Western imperialism was insufficient to legitimize similar Japanese imperialist policies. As an alternative ideology, they were obliged to construct 'Asianism', a communal unity binding Asian people together. This was an extension of the idea of the communal state centred on the tenno family and applied to the Asian community as a whole. In order to exalt this communalism, they mobilized an indigenous ideology which underlined familial and pseudo-familial ties as the basis of social organization. In doing so, they cemented the logic of the apparently divinely-inspired discourses derived from Japanese mythology. Moreover, this logic was applied to the vagaries of competition among nation-states at the height of the imperialist era.

81

Thus Komaki emerges as fanatical, nationalistic and ethnocentric in the sense of being Asiacentric. According to Murakami's memoirs (Murakami 1993), which were privately published and deemed reliable as the author was by then already 82 years old and hence without qualms as to the effect of his revelations, Komaki and the Kyoto group exercised a marked influence on military decision-making and availed themselves of financial resources placed at their disposal by military and ministerial authorities. According to his memoir and fragmented statements made by Komaki to this author in the 1980s, and more recently by Yonekura, they rented an independent building complete with attendant janitors near the University of Kyoto, and held weekly study meetings there. The study group also received contracts from the General Staff Office, which was anxious to plan and elaborate military tactics in the light of the attack on Singapore via the Malaysian Peninsula in 1941, of operations in New Guinea in 1942, operations in Southern China in 1943, and finally in 1945, foreseeing the landing of the American forces, and the defence strategy pertaining to Kyushu. It is not clear to what extent the proposals of the Kyoto geopolitical group were adopted by the military authorities, but high-ranking military officers from the General Staff Office invariably attended the meetings in the building near the university, and there is no question that the relationship with the military group was a close one. Some of the young members of the group later obtained a salary from the University of Kyoto, and they proceeded to build up a collection of books and other materials in the building, all of which was sold immediately after the surrender of Japan in 1945. To this day, archival evidence confirming the connection of the Kyoto geopolitical school with the wartime military authorities has not been found. The discovery of such sources to substantiate these affairs and the eventual analysis of this material constitutes a further task for investigation.

The social relevance of geopolitical discourse

The Kyoto school of Japanese geopolitics certainly had a social relevance not only because Komaki's 'Manifesto of Japanese geopolitics' was widely read, as testified by Komaki himself (Komaki 1944), but also for its connection with the military as mentioned earlier. There is also no doubt that a certain rivalry existed between the Kyoto geopolitical school and the geopoliticians of Tokyo. In November 1941, the Japan Association of Geopolitics was founded in Tokyo, and from January 1942 to November 1944 published the monthly *Chiseigaku* (only five numbers in 1944 because of the shortage of printing paper). On the board of directors and counsellors appeared a number of names of geographers, economists, lawyers, politicians, journalists and some military authorities. The names included translators of Haushofer such as Joji Ezawa, as well as those who took a rather critical stand with regard to German geopolitics such as Hiroshi Sato, who introduced Wittfogel to Japan. Of the Kyoto geopolitical school, we find only the name of Goro Ishibashi as a nominal advisor. He was a former professor of Komaki's but was outside Komaki's geopolitical group because of his ill health and also because of his ideological differences with Komaki. There were no contributions to this journal from the Kyoto school of geopolitics, and in the numerous papers stressing the necessity of establishing Japanese geopolitics, no references were made to Komaki's writings nor to the writings of members of his school. The only paper published which refers to Komaki's 'Manifesto' was one discussing the economic revival of the remote Oki islands, published in 1943.

The founding declaration printed on the first page of every number of the Tokyo-based *Chiseigaku* proclaimed that: 'now the defence of our motherland and the war for survival of the peoples of Great East Asia has a world-wide significance. It is required that geopolitics, which makes up the fundamental base, blood and land of the people, be studied more profoundly and with greater dispatch' (Japan Association of Geopolitics 1941). As it was, however, the papers of this journal were mostly political geographical or historical geographical descriptions of areas that were crucial in the Second World War, and no contributions were imbued with the Shintoistic mysticism or intimations of divine inspiration so often found in the writings of the Kyoto geopolitical school. On the other hand, there is no evidence that the contributors to this journal actually exercised influence on the decision-making of the General Staff Office and governmental authorities. Although the president of the society was Admiral Yoshitake Ueda, the journal contained very few contributions from military men. In spite of their bold declarations, it would appear that, essentially, the members of this association were little more than 'hangers-on' in society (see Watanuki 1941).

School textbooks are generally some of the most efficient instruments of either nation-building or the indoctrination of state ideology. This is especially so in the case of Japan, where from 1903 school textbooks were prepared under the authorship of the Ministry of Education (Takeuchi 1998 and see Figure 4.2). In the textbooks of civic society and citizenship, emphasis on the peculiar character of Japanese polity based on tennoism was invariably forthcoming. Moreover, the geography books compiled during 1930–1 contained new descriptions of the ethnic minorities of Japan. It was written that:

> There are more than ninety million people holding Japanese nationality, of which about twenty million are Korean, four million three hundred thousand Chinese, more than one hundred thousand Formosan aborigines and a small number of Ainu in Hokkaido and Sakharin. All these differences notwithstanding, they are all loyal subjects of the Japanese emperor.
>
> (Fifth grade textbook, page 23)

The compilation of geography textbooks for primary schools for 1938–9 depicted the new situation arising from Japan's military invasion of Manchuria in 1932, and China in 1937. Yet all these new descriptions, rather than expanding the horizon of study for the students, reflected a *de facto* recognition of the imperialistic expansion of Japan. It was only in the 1943–4 editions of fifth- and sixth-grade school geography texts, though, that this imperial form of geopolitical manipulation became obvious. The geography of foreign countries was limited to that of East and Southeast Asia and Oceania, that is, the constituents of the Greater East Asia Co-Prosperity Sphere. At the front of the sixth-grade textbooks, instead of the usual Mercator projection map, a specific projection map deliberately centred on Japan was reproduced. Japan's geographical position was explained as being 'an apt one for extending her influence northward and southward'. Apart from the emphasis on Japan's geopolitical advantages due to her geographical location, there were neither environmentalist nor racial interpretations, but there was great admiration for Japanese achievements in colonial and occupied lands. It is interesting that the map of the Japan-centred Greater East Asia Co-Prosperity Sphere was a copy of the map printed on the last page of every issue of *Chiseigaku* of the Japan Association of Geopolitics as shown in Figure 4.1.

Eminent Japanese politicians were prone to surround themselves with sympathetic scholars, with the result that geopolitical discourses had a considerable impact on Japanese politicians in the second half of the 1930s. Prior to the onset of the activities of the Kyoto geopolitical school and the founding of the Japan Association of Geopolitics in Tokyo, in 1938, Premier Fumimaro Konoe proclaimed the advent of the 'New Order in East Asia' (*Towa Shinchitsujo*), which aimed to establish a new political system in Japan, Manchukuo and China and appealed to the 'Asianism' in the Japanese people (Hatano 1980). According to Miwa's analysis (Miwa 1981), at the time when Fumimaro Konoe presented his idea of a 'New Order in East Asia', he was extremely susceptible to the influence of his intellectual cabal, Showa Kenkyukai (Study Group of the Showa Period). Masamichi Royama, a leading figure of this group and professor of political science first at the Imperial University of Tokyo and later of the Imperial University of Kyoto, published two papers (Royama 1938, 1939) in which he explained defensively that 'Japanese expansion is not imperialism but regionalism for the purpose of defence or development'; moreover, he cites the concept of 'national living sphere' (*minzokuteki seikatsuken*) which was clearly a paraphrase for the *Lebensraum* of German geopolitics, and which according to him, was

> a geopolitical concept, not a phrase belonging to the terminology of international law, or political sciences . . . in real international politics, however, England and France cannot ignore the claims of Germany made along the lines of this concept.
>
> (Royama 1938)

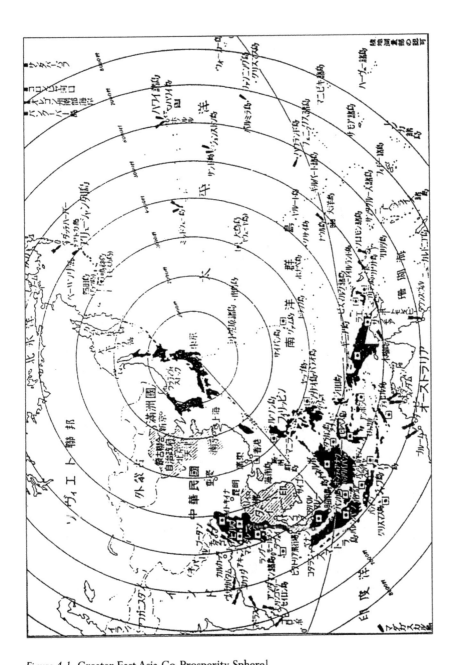

Figure 4.1 Greater East Asia Co-Prosperity Sphere[1]

Black areas show parts occupied after December 1941; progressive changes
were shown in maps for each issue. This map shows the maximum expansion
of Japanese-occupied areas.

Source: Chiseigaku, vol. 2, no. 9

初等科地理　下

Figure 4.2 Greater East Asia Co-Prosperity Sphere with Japan in the centre[1]
Source: Sixth-grade geography textbook, Ministry of Education, 1944

In his paper on the cultural aspects of the New Order in East Asia (Royama 1940a), he emphasized the relativism or pluralism of world culture, though admitting the superiority of Western material culture. Consequently, he was adverse to the Japanese cultural chauvinism which held Japanese culture to be the most superior in Asia and insisted on the need for a co-operative solidarity among East Asian nations. In the entry on geopolitics for an encyclopaedia of social sciences (Royama 1940b), he expressed himself more positively on the subject, acknowledging the validity of the geopolitical approach in the analysis of the reality of international relations, even though he denied the existence of geopolitics as an independent discipline. In August 1940, Foreign Minister Yosuke Matsuoka proclaimed the 'Greater East Asia Co-Prosperity Sphere' policy, as a development of the concept of the 'New Order in East Asia'. According to Fukushima (Fukushima 1997), with this new concept geopolitics was brought to the fore, in order to establish recognition of the inevitable linkage between East Asia and Southeast Asia. Miwa furthermore recognized that Royama's regionalist understanding of the 'economic community of East Asia' was always different from the emotional Asianist understanding of the Greater East Asia Co-Prosperity Sphere, being based as it was on a purely economic rationalism (Miwa 1981; Takahashi 1980).

In order to avoid criticism for being imperialist in the same way as the Western powers, the Kyoto geopolitical school turned away from rational reasoning and escaped into what can only be described as romantic discourses involving discussions of indigenous, divinity-related and spiritual traditions 'true' to Japan, and thereby receiving the support of the ultra-nationalists in the army. On the other hand, as a politician Konoe felt inclined to check the aberrant natures of the military group, whilst the 'brains trust' group surrounding him showed sympathy towards, and even adopted, geopolitical reasoning, apropos the pursuit of economic and political rationalism in the sphere of international relations as seen in the case of Royama. They subsequently encountered profound difficulty in distinguishing themselves from the Western imperialists and their politics. During the period leading up to the final catastrophe of 1945, when ultra-nationalist irrationalism came to a head, these intellectuals rapidly lost ground and many members of Showa Kenkyukai were arrested. One of them, Hidemi Ozaki, was executed for his part in the Sorge affair which involved the channelling of information to Moscow.

It is this author's belief that the main reason why the members of, and contributors of papers to, the Japan Association of Geopolitics failed to achieve social relevance as practitioners of geopolitics was the ambivalence in their attitude towards the fundamental contradiction inherent in Japanese geopolitics. On the one hand, they were compelled to rely on romanticist or fundamentalist ideals with sometimes irrational emphasis on ultra-nationalism or an emotional Asianism, whilst on the other hand they had to find a way out of the economic and political difficulties Japan faced in the pursuit of realism. Towards the end of the Pacific war, the pursuit of rationalism was already encountering setbacks and the romanticists, meantime, were being driven to mould their destiny with militarist Japan. The tragicomedy of Japanese geopolitics manifested itself in the mental acrobatics its adherents were obliged to perform in the course of distinguishing between these two positions, while they never reached a final decision.

Geopolitical affairs thereafter

After the defeat of Japan in the Second World War, the principal advocates of the Kyoto school of geopolitics and the representative figures of the Japan Association of Geopolitics were compelled to resign their offices, or were purged from public posts on the orders of the Allied Forces. This situation prevailed until the end of the occupation in 1952. Many problems pertaining to the purge measures require clarification. The Allied authorities were obviously more influenced by the social and political circumstances, positions and reputations of individuals in administrative institutions and academic circles. They seldom concerned themselves with any scientific evaluation of geopolitical practices. In some instances, protagonists of Japanese militarism who were particularly influential in certain spheres, for instance, managed to retain their posts. A type of quota was applied to every institute regarding the number of persons to be purged. Those who lacked social and political influence were made the scapegoats and lost not only their positions but also their livelihoods. Yet the real tragedy for Japanese geopolitics was that circumstances immediately after the Second World War led to the subject of geopolitics becoming the focus of a taboo. Consequently, Japan witnessed neither serious criticism of geopolitical theories *per se*, nor any analytical reflections on geopolitical practices in the country in the 1930s and 1940s (Uno 1981). The overwhelming impression in the intellectual world was that geopolitics was dead, and when 'geopolitical affairs' were referred to, they were considered events confined to the past. More generally, studies of political geography were also shunned to the point where whatever significance they had was negated, adding to the general impression of the demise of geopolitics.

With the end of the Allied occupation in 1952, most of the 'geopolitical figures' found posts in newly-created universities and other institutions, and continued to exert their influence both socially and academically. A veil of silence was drawn over their past involvement with geopolitics. They hardly ever wrote or spoke about the geopolitical practices of the 1930s and 1940s. Instead, surreptitious attempts were made to hinder the composition or publication of either critical or summary appraisals of Japanese geopolitics, of the kind carried out in Germany (e.g. Troll 1947) with regard to *Geopolitik*. None the less, Japanese geopolitics left its mark on the private and public activities of former geopolitical practitioners. Komaki obtained a professorship and then became president of the national university of Shiga, but he produced few academic contributions, apart from some writings on the local history and geography of Shiga Prefecture, before his death in 1990. Many of his former students who had been allotted specific area studies renewed their geographical work in foreign countries and became authorities in their respective areas of study, but without the former geopolitical overtones: Bekki in Southeast Asia, Mikami in the Soviet Unio and Yonekura, having changed his study area from China to India due to ongoing difficulties conducting field studies in China. Ezawa, translator of Haushofer and author of many papers in *Chiseigaku*, continued to personify the reification of the contradictory aspects of Japanese geopolitics: the shamanist-based mysticism on the one hand and the rational Escher-like spatial conception on the other. After his rehabilitation, he obtained a professorship in economic geography at a private university and became an authority on regional science in Japan. He was later elected president of the Japan Association of Economic Geographers and hence completed his intellectual transformation from geopolitics to mainstream economic geography. Other representative figures of Japanese geopolitics such as Hiroshi Sato and Nobuyuki Iimoto escaped the purges, for reasons unknown, but never published either geopolitical or political geographical papers after the end of the Second World War.

In 1957, the regional conference of the International Geographical Union (IGU) was held in Japan. It was the first time that Japanese geographers had organized this type of international event after two decades of cultural isolation. The IGU was a great contribution to the further development of geographical studies in Japan. The list of speakers and organizers of the conference and its excursions included several figures well-known for their role in 'geopolitical affairs', as well as geographical practitioners who worked during the fifteen-year war period: Fumio Tada, then vice-president of IGU, who joined several expeditions on the Asian continent, especially to Inner Mongolia; Ryuziro Isida and Taiji Yazawa, the two main promoters of the conference who were, respectively, a member of an investigative mission in Singapore attached to the Japanese military government, and a naval officer in charge of metallurgical work in Japan and Southeast Asia; and Soki Yamamoto, a specialist in hydrology in the research section of the Southern Manchurian Railway Company. Many rehabilitated geographers such as Jiro Yonekura read papers and led post-conference excursions to various parts of Japan. For those Japanese geographers concerned with geopolitical practices and past experiences in colonial and occupied lands, this constituted an occasion for the development of their international sensibilities.

By the mid-1950s, the Japanese economy had reached pre-war levels, and the following decade of the 1960s was considered the period of rapid economic growth, with the per capita GNP having more than doubled in real terms. This economic growth was realized by strengthening the export drive to the Asian continent. In the quarter-century after the Second World War, Japan actually realized in terms of economic expansion what imperialist Japan had not succeeded in accomplishing in military terms. In 1962, the Institute of Developing Economy was established as an extra-departmental organization of the Ministry of International Trade and Industry (MITI). The staff of this institute included several geographers, and it has played a leading role in the social and economic research of foreign countries, especially in the developing countries of the world. This research was not directly connected with Japanese economic expansion, nor did it directly support the country's economic expansionist system, yet it is necessary to recognize that regional studies by Japanese geographers in foreign countries have been influenced by the Japanese economic relationship with developing countries, and was due also to the differential allotment system of public and private grants.

As discussed earlier, during the 1930s and the 1940s a considerable number of Japanese geographers were critical of, or at least reluctant to collaborate with, the geopolitical practices that served the interests of the imperialists and militarists of Japan. None the less, they often found themselves forced to submit to these interests, and in order to maintain some semblance of intellectual critique they resorted to the mental acrobatics referred to above. During the past half century, the predominant state ideology, centred on the idea of the supremacy of economic interests in accordance with corporate interests, has admittedly been neither as fanatic nor irrational as the ultra-nationalist and militarist dogma of imperialist Japan. These days, freedom of speech is guaranteed, and the democratic political system works. Notwithstanding the opportunities afforded by these more conducive circumstances, no criticism has issued from Japanese geographers pertaining to the formation of the capitalist landscape of Japan and the new economic order brought about by the economic giant that is modern Japan. Yet the current state of affairs demands analysis, and constructive criticism of the hidden mechanisms of this new order by those either trained to disseminate it and/or whose task or duty it is to provide it.

There remains a great reluctance to confront the attitudes and discourses of Japanese geopolitics during the 1930s and 1940s. Contemporary geographers are inextricably enmeshed or submerged in a new brand of contemporary geopolitics tied, under some different name perhaps, into the new economic order. It would require a genuine desire and a certain degree of courage to objectively, and in all honesty, embark on a critical analysis. Ultimately the old adage of 'he who casts the first stone' is too well applicable.

Note

1   The only available versions of figures 4.1 and 4.2 are photocopies of poor quality. Despite this the figures have been retained as they give an indication of tendencies in Japanese education and geopolitical thought at that time.

Bibliography

Abdel-Malek, A. (1977) 'Geopolitics and national movements: an essay on the dialectics of imperialism', *Antipode* 9 (1) : 28–36.

Abe, I. (1933) *Chiseigaku nyumon*, Tokyo: Kokon-Shoin.

Ezawa, J. (1938) *Keizai chirigaku no kisoriron; shizen, gijutsu, keizai*, Tokyo: Nankosha.

—— (1943) *Kokudo no seishin*, Tokyo: Shinchosha.

Fujisawa, C. (1925) 'Rudolf Kjellén no kokka ni kansuru gakusetsu', *Kokusaiho Gaiko Zasshi* 24:155-175.

Fukushima, Y. (1997) 'Japanese geopolitics and its background: what is the real legacy of the past?', *Political Geography* 16: 407–21.

Hatano, S. (1980) 'Toa shinchitsujo to chiseigaku', in K. Miwa (ed.) *Nihon no 1930nendai*, Tokyo: Saikokusha.

Haushofer, K. (1925) *Geopolitik des pazifischen Ozeans. Studien über die Wechselbeziehungen zwischen Geographie und Geschichte*, Berlin: Kurt Vowinckel Verlag.

Iimoto, N. (1928) 'Iwayuru chiseigaku no gainen', *Chirigaku Hyoron* 4: 76–99.

—— (1935) *Seijichirigaku kenkyu*, Tokyo: Kokon-Shoin.

—— (1941) *Seijichirigaku kenkyu*, Tokyo: Chukokan.

Iizuka, K. (1942/43) 'Geopolitik no kihonteki seikaku(1),(2), (3)', *Keizai Ronso* 12: 816–44, 13: 288–314, 486–96.

Ishibashi, G. (1927) 'Seiji chirigaku to chiseigaku', *Chigaku Zasshi* 500: 611–14.

Iwata, K. (1942) 'Chiseigaku to senso', *Chirigaku Kenkyu* 1 (1): 1–15.

Jacobsen, H.-A. (1979) *Karl Haushofer: Leben und Werk, Band I Lebensweg 1869–1946 und ausgewählte Texte zur Geopolitik*, Boppard am Rhein: Boldt.

Japan Association of Geopoliticals (1941) 'Sengen', *Chiseigaku* 1 (1): 1.

Kjellén, R. (1916) *Staten som Lifsform*, Stockholm: H.Geber.

Komaki, S. (1940a) 'Nihon chiseigaku no shucho', *Chiri Ronso* 11: 3–6.

—— (1940b) *Nihon chiseigaku sengen*, Kyoto: Kobundo.

—— (1942) 'Daitowa no chiseigakuteki gaikan', *Chirigaku* 10 (4): 1–8.

—— (1944) *Nihon chiseigaku oboegaki*, Tokyo: Akitaya.

Maull, O. (1925) *Politische Geographie*, Berlin: Gebrüder Borntraeger.

Minamoto, S. (1984) 'Shiga Shigetaka, 1863–1927', *Geographers: Bibliographical Studies* 8: 95–105.

Miwa, K. (1981) '"Towa shinchitsujo" sengen to "Daitowa kyoeiken" Koso danso', in K. Miwa (ed.) *Saiko Taiheiyo senso zenya. Nihon no 1930nendai-ron to shite*, Tokyo: Soseiki.

Murakami, T. (1993) *Kaiso wa tsuzuku*, Kobe: private publication.

Nagasawa, K. (1973) *Nihonjin no boken to tanken*, Tokyo: Hakusuisha.

Ogawa, T. (1930) 'Jimmon chirigaku no ikka to shite no seiji chirigaku', *Chikyu* 9 : 239–47.

Ohara, K. (1936) *Shakaichirigaku no kisomondai*, Tokyo: Kokon-Shoin.

—— (1940) 'Geopolitik no hatten to sono gendaiteki kadai', *Shiso* 221: 391–404.

Royama, M. (1938) 'Towa kyodotai no riron', *Kaizo* 11: 6–27.

—— (1939) 'Sekai shinchitsujo no tenbo', *Kaizo* 11: 4–24.

—— (1940a) 'Toa shinchitsujo to shinbunka no sozo', 120–42 in *Kigen 2600nen kinen shin Toa kensetsu. Tokyo kondankai tokubetsu ronbunshu*, Tokyo: Tokyo Shiyakusho.

—— (1940b) 'Chiseigaku', 52–60 in I. Nakayama, K. Miki and K. Nagata (eds) *Shakaikagaku Shinjiten II*, Tokyo: Kawade Shobo.

Sasaki, H. (1927) 'Geopolitik to economic geography', *Chirigaku Hyoron* 3: 361–3.

Shiga, S. (1887) *Nanyo jiji*, in Shiga Shigetaka Kankokai (ed.) *Shiga Shigetaka zenshu*, Tokyo: Maruzen.

Takahashi, H. (1980) '"Toa kyodo tai-ron": Royama Masamichi, Ozaki Hidemi, Kada Tetsuji no baai', 50–79 in K. Miwa (ed.) *Nihon no1930nendai*, Tokyo: Saikosha.

Takeuchi, K. (1974) 'The origins of human geography in Japan', *Hitotsubashi Journal of Arts and Sciences* 15:1–13.

Takeuchi, K. (1980) 'Geopolitics and geography in Japan: re-examined', *Hitotsubashi Journal of Social Studies*, 12: 14–24.

Takeuchi, K. (1988) 'Paysage, langage et nationalisme au Japon du Meiji', 172–84 in G. Zanetto, (ed.) *Les langages des représentations géographiques. L'acte du colloque international tenu à Venise, 15 et 16 octobre 1987, vol.II*, Venice: Université de Venice.

Takeuchi, K. (1994a) 'The impact of the Japanese imperial tradition and Western imperialism on modern Japanese geography', 188–206 in A. Godlewska and N. Smith (eds) *Geography and Empire: Critical Studies in the History of Geography*, Oxford: Blackwell.

—— (1994b)'Nationalism and geography in modern Japan: with special attention to the period between the 1880s–1920s', in D. Hooson (ed.) *Geography and National Identity*, Oxford: Blackwell.

—— (in press) 'The formation of geographical images of the outside world in imperialist Japan (Mid-1880s to 1945)', in A. Buttimer, S. Brunn and U. Wardenga (eds) *Text and Image: Construction of Regional Knowledges*, Leipzig: Institut für Länderkunde.

Troll, C. (1947) 'Die Geographische Wissenschaft in Deutschland in den Jahren 1933 bis 1945. Eine Kritik und eine Rechtfertigung', *Erdkunde* 1: 3–47.

Tsujimura, T. (1925) 'Seiji chirigaku, Otto Maull', *Chirigaku Hyoron* 1: 814–23.

Uchimura, K. (1894) 'Chirigaku-ko in Uchimura Kanzo shinko-chosaku zenshu', under the title of *Chijinron* 4: 5–105.

Uno, S. (1981) '1930nendai ni okeru Nitchu no shinkinkan to sokoku; "Towa shinchitsujo" seimei zengo', in K. Miwa (ed.) *Saiko Taiheiyo senso zenya: Nihon no 1930nendai-ron to shite*, Tokyo: Soseiki.

Watanabe A. (1942) 'Chiseigaku no naiyo ni tsuite', *Chirigaku Kenkyu* 10: 1–14.

Watanuki, I. (1941)'Geopolitik no kaibo', *Chirigaku* 9 (8): 1–15.

—— (1942) 'Chiseigaku no tenkaisei', *Chirigaku Kenkyu*, 11: 1–8.

Wittfogel, K. A. (1929) 'Geopolitik, geographischen Materialismus und Marxismus', *Unter dem Banner des Marxismus*, 3: 26–64.

Yamaguchi, S. (1943) *Nihon o chushin to suru hankin chirigaku hattatsushi*, Tokyo: Seibido.

Yamamoto, N. and Ueda, Y. (1997) *Fukei no seiritsu:Shiga Shigetaka to Nihon fukeiron*, Osaka: Kaifusha.

Yonekura, J. (1947) *Towa chiseigaku josetsu*, Tokyo: Seikatsusha.

5

GEOPOLITICAL IMAGINATIONS IN MODERN ITALY

David Atkinson

Introduction

The geopolitics behind Italian history, and geopolitics in Italian history

Traditionally, many historians have commenced their interpretations of Italian history with discussions of geography; they frequently read the nation's development as "determined by geopolitics" (Bosworth 1996: 3; Serra, 1984). In his classic survey of Italian history for example, Denis Mack Smith (1959) opens with a section entitled 'A Geographical Expression' which claims that:

> Until 1860 the word Italy was used not so much for a nation as for a peninsula, and [Austrian Chancellor] Metternich wrote disparagingly of this 'geographical expression'. It is therefore with geography that Italian history must begin. Too often have poverty and political backwardness been blamed on mis-rule and foreign exploitation, instead of on climate and the lack of natural resources. We need not go so far as to believe that the destinies of a nation are altogether shaped by its wealth and position . . . but such characteristics are bound to define the scope of a nation within certain limits. It has always been historically important that the Apennines divide Italy from top to bottom and that the Alps cut her off from the rest of Europe; mountains may not be removed, even by faith.
>
> (Mack Smith 1959: 1)

Similarly, Christopher Duggan's recent *Concise History of Italy* (1994) begins with a chapter entitled 'The geographical determinants of disunity', and an opening line that states 'The history of Italy is tied up inseparably with its geographical position' (Duggan, 1994: 9). A basic geography of the peninsula follows: from coastline, topography and soils, to climate, deforestation and mineral resources, the factors that underpin political governance, agriculture, industrialization and demographics are described. Moreover, both Mack Smith and Duggan identify (but don't explain) what they call the 'geopolitical' significance of Italy's Mediterranean location. Even a recent text that acknowledges that ideas about Italy are socially produced nevertheless starts with a section entitled 'geographies' (Forgacs and Lumley 1996; Dickie 1996). If historians are to be believed, the 'geographical' and the 'geopolitical' provide crucial insight into the history of Italy.

By contrast, since the creation of the modern state in the 1860s, geography has remained a minority discipline in Italy. The formal geopolitical knowledges that this book considers have been accorded still less attention. Given this, my chapter briefly considers two occasions when the geopolitical traditions of the twentieth century *did* find expression in Italy, and Italians were encouraged to develop their geopolitical imaginations. In particular, I discuss the cases of *Geopolitica* (published between 1939 and 1942) and, more briefly, *Limes* (in print since 1993). Both publications promoted geopolitics as a way of interpreting the world, and both sought to develop an analytical space (Rose 1995) within academic, educational and more popular realms wherein Italians might learn to conceptualize the world 'geopolitically'. In addition, each journal also demonstrates some of the wider themes of this collection. Due to the constraints of space, it is a necessarily abbreviated version of the Italian geopolitical tradition that I discuss here (Antonsich 1997a; Atkinson 1995, 1996; Vinci 1990). However, in contrast to historians' accounts of the geopolitics *behind* Italian history, this chapter outlines something of the developments of geopolitics *in* Italian history.

Political geographies in Italy

This collection argues that geopolitical thought has found numerous different expressions in different places, yet also that there are some continuities through time as these ideas are re-worked at each different site. The Italian case demonstrates this usefully. For although the twentieth century witnessed various distinctive *Italian* strains of geopolitics, these were often connected to the broader, international debates that surrounded geopolitical ideas. Such patterns were evident from the emergence of modern political geography in Italy. In a 1903 discussion of Friedrich Ratzel's *Politische Geographie* (1897), Olinto Marinelli (1903), a leading figure in the development of Italian geography, identified the significance of *Politische Geographie* and its geographical analyses of the state. However, he also cautioned against the nationalism ingrained in Ratzel's work, whilst also positioning his critique in relation to that of the French geographer, Paul Vidal de la Blache (1898). This reflexive understanding of the political and cultural subjectivities entwined in the production of knowledges seems to have been readily accepted in a country that often drew upon the longer-established geographical traditions of other nations for intellectual innovation.

Certainly, as other Italian geographers began to discuss political geography, they often did so with an acute awareness of the national subjectivities of the authors. The First World War and its aftermath brought urgency to this process. Some Italian geographers recognized the extent to which geography had been mobilized by other combatant powers, and sought to provide an Italian analysis of the war and of Italian territorial claims (Revelli 1916, 1918, 1919). They lobbied for an Italian political geography that might argue their case at the peace conferences (Baratta 1918, 1919; Ricchieri 1920). One even spoke before the Royal Geographical Society in London to contest the maps of Serbian ethnicity that Johan Cvijic had created for the Versailles conference (Roncagli 1919; Wilkinson 1951). And while geographical knowledges throughout Europe became more overtly politicized through these years (Heffernan 1995, 1996; Sandner and Rössler 1994), political geography – promising analysis and comprehension of the shifting frontiers and new-born states of post-war Europe – found increasing favour in Italy (Gambi 1994).

Through the 1920s, Italian geographers increasingly engaged with broader European debates in their attempts to develop their *Italian* political geography (Ricchieri 1921; Filippo de Magistris 1923). The more established literatures of France and Germany were mined particularly, and again, Italian writers revealed an discerning sensitivity to questions of national subjectivities in academic knowl-edges (Toniolo 1923; Almagià 1926). One complained explicitly that German, French, British and American authors had all interpreted political geography from their own nationalist perspectives, especially at times of international tension (De Marchi 1929). Amongst the debates in Italy (Migliorini 1930; Toniolo 1930), the unqualified determinism of German geography was largely rejected (cf. De Marchi 1929). However, by the 1930s the state was broadly accepted as a topic of geographical analysis (Almagià 1923, 1936; De Marchi 1929; Toschi 1937). It is important to note that these developments drew upon international debates. Further, it was in this same fashion that geopolitical ideas were imported into Italy and renegotiated by Italians in their own distinctive contexts.

Geopolitica and geopolitics in inter-war Italy

Apart from Germany's *Zeitschrift für Geopolitik*, the Italian journal *Geopolitica*, pub-lished monthly between January 1939 and December 1942, was the largest and most significant collection of geopolitical writings in inter-war and wartime Europe. It had no monopoly over geopolitical thought in Italy. Indeed, it was viewed with suspicion by some geographers and contested by others (Atkinson 1996; Vinci 1990). However, *Geopolitica* did enjoy support from Mussolini's Fascist regime and was by far the most coherent and enduring source of geopolitics in

Fascist Italy. In the section that follows, I outline the origins and development of *Geopolitica* within Trieste and Italy more generally. I also consider examples of its geopolitical perspectives upon the wartime world. I aim to sketch the development of its approach from amidst the loose parameters of a European geopolitical debate, but equally to acknowledge the influential contexts and cultures of Fascist Italy. The sustained engagement with geopolitical thought, representations and analysis in *Geopolitica's* pages helps us towards a more comprehensive understanding of the histories of geopolitical thought in Europe.

The origins of Geopolitica in Trieste

The origins of *Geopolitica* are embedded in the Adriatic port-city of Trieste in north-eastern Italy. Although Trieste had entered the twentieth century as the main port of the Austro-Hungarian Empire, the city's predominantly ethnic Italian community ensured that it was the focus of long-standing Italian irredentist agitation (Millo 1987). Trieste was one of Italy's two main territorial claims at the peace conferences following the Great War, and the city was placed under the jurisdiction of Rome in 1919 (Burgwyn 1993; Goddi 1984). Despite widespread Italian celebrations at the 'redemption' of Trieste however, upon its incorporation into Italy the port lost virtually all of its established hinterland and economic functions, and throughout the 1920s the city entered a steady decline.

Consequently, academics at the newly founded University of Trieste were encouraged by local business and political elites to analyse the city's pressing economic problems. To this end, new chairs were appointed in international law and economic geography: subjects seen as offering analysis of, and informed solutions to, these problems (Vinci 1990). The appointment in economic geography was Giorgio Roletto. A prominent academic who had trained in the French geographical tradition in Grenoble and written extensively upon Alpine societies, Roletto was one of the founders of Italian geopolitics (Bonetti 1967; Valussi 1965). He shared this distinction with his student, supervisee, and then colleague at Trieste, Ernesto Massi (Lo Monaco 1987). By contrast to Roletto's background in the French-speaking Maritime Alps, Massi was born in Trieste in a household speaking both Italian and German. Although regarding himself as Italian, he was an Austrian citizen and was educated in German as demanded by the Viennese authorities (Vinci 1990). Being fluent in French and German, Roletto and Massi were well-equipped to engage with the dominant geographical literatures of inter-war Europe. But more especially, in Trieste they found themselves in a new Italian city that was a hotbed of Fascism, and working in a University with a remit to propose solutions to the growing economic problems of Trieste. It was a context where the intermingling of place, politics, individual perspectives and geopolitical traditions all informed the production of *Geopolitica*.

Compounding the situation, both Roletto and Massi were ardent nationalists and convinced Fascists, and each became prominent in local and national Fascist organizations. Finally, they were also committed geographers who, throughout their careers, proved fervent evangelists for a discipline that remained a minority concern in Italian academia (Atkinson 1996). Nevertheless, both men demonstrated an unwavering belief in the value of geographical perspectives and knowledges to the state and its governance. Consequently, in the face of the economic and political problems of Trieste, they developed their own, radical form of analysis that was an Italian strain of the geopolitical theorizing that circulated inter-war Europe.

It was Massi who first came across geopolitics in *Zeitshrift für Geopolitik* in 1930 (Massi 1939a). He was initially sceptical. Like other Italian academics, Massi was sensitive to the subjectivities of knowledges and he initially condemned *Geopolitik* as being overly skewed towards German nationalist agendas (Massi 1931). Moreover, he dismissed geopolitics as a branch of political science rather than an element of geography (Massi 1931; Roletto and Massi 1931). At the same time, however, his own interests in the broader category of 'political geography' were increasing exponentially as he recognized a potential window onto the material problems of geography and politics which faced Trieste. In the early 1930s, both Massi and Roletto read widely around the political geography emerging in the French, German and, to a lesser extent, English literatures (Massi 1930; Roletto and Massi 1931). And it was from amidst these debates that they developed their *own* strain of what they called a 'Dynamic Political Geography' (Roletto and Massi 1931). This distinctive form of political geography did not merely address the static facts of nation, state and territory, but also dealt with the ongoing, fluctuating patterns of global political affairs. It considered imperialism, trade-flows, nationalist and ethnic tensions, and other such geographical and political issues which increasingly convulsed the inter-war world (Roletto and Massi 1931). They claimed their 'Dynamic Political Geography' to be synthesizing and all encompassing in its range and grasp, and to offer a unique perspective upon the world.

With time, the Italians betrayed an ambivalent if increasing interest in the concepts that constituted geopolitics (Massi 1931). They began to acknowledge the similarities between their 'Dynamic Political Geography' and *Geopolitik* in a series of articles. In 1931, they admitted that: 'Geopolitics closely approaches, although can't be [directly] identified with, that which we have called a dynamic political geography' (Roletto and Massi 1931: 23). At about the same time, Massi argued that the genre of geopolitics, although unlikely to spread beyond Weimar Germany in his opinion, should be taken more seriously as a category of knowledge (Massi 1931). He cautiously admitted still more similarities between his own 'dynamic political geography' and German *Geopolitik*. By late 1932, Massi (1932) had become still more favourable towards the concept, and by 1933, in the light of

Haushofer's *Geopolitik*, both Massi and Roletto were describing their own work as 'geopolitical' (Atkinson 1996).

The strain of geopolitics which emerged in Trieste, however, was a self-consciously hybrid and negotiated *Italian* geopolitics that was developed from wider European literatures. While Massi had studied the geographical determinism of the German geopoliticians, Roletto contributed his reading of French critiques of *Geopolitik*, and brought an insistence upon the importance of human agency to the emergent Italian perspective. In 1931, Massi had written explicitly about the benefits that Italian geography, as a 'late developer' in his terms, could draw from the more established geographical 'schools' of Europe (Massi 1931). Unsurprisingly then, the geopolitics that emerged in Italy was influenced by both German and French precedents, but crucially, it was designed to be distinctive from both. It was in this manner that the infant geopolitical traditions of Europe spawned a further variant of geopolitics, although one that was equally informed by the specific contexts of Trieste and Fascist Italy.

Promoting geopolitics in Fascist Italy

Once the 'geopolitical' was accepted by the Triestine geographers, they began to promote their new perspective vigorously. Throughout the 1930s, a string of essays and articles espoused the geopolitical way of interpreting the world (Massi 1935, 1937a, 1937b, 1938; Roletto 1933, 1937, 1938). However, the ambitions of Roletto and Massi were not satisfied by academic audiences. They also sought wider influence throughout Italian society. In their attempts to promote the geographical and geopolitical awareness that they considered essential to the effective governance of the modern state, they took their message to the various quangos and institutions of the Fascist regime. Moreover, they made few distinctions between the regime's hierarchy, and other spheres of society such as business circles, educational policy, and the everyday popular cultures of Italy (Massi 1940; Roletto 1940a, 1940b). Their aim was to inscribe geography and a geopolitical imagination into all levels of society.

To these ends, one particularly useful organization was the Fascist Colonial Institute. Originally a colonial lobby whose membership included financiers, diplomats, politicians and many figures from Italian foreign policy circles, the institute had been 'Fascistized' in the 1930s to promote colonialism throughout society (Gambi 1994; Lando 1993). Both Roletto and Massi embedded themselves and their geopolitical project within this organization. They published geopolitical work in the organization's newspapers and pamphlet series (Massi 1937b). They also held key positions within the organization. Roletto was president of the influential Trieste branch and Massi was its leading activist: organizing evening classes, public exhibitions, and slide-shows to persuade Triestines of the critical importance of colonial territories to Italy. In 1935, when Massi moved to positions at the Universities of Pavia and Milan, he became still more involved with Fascist organizations in the heartland of the regime (Atkinson 1996). He was a leading figure in the local Fascist party and the 'cultural director' of the Colonial Institute. He also became involved with the extremist 'School of Mystical Fascism' (Marchesini 1976), and used these powerful connections to promote geography and to advocate geopolitics.

The ambitions of the Italians also extended to the growing international network of geopolitical thinkers. Massi began to forge links with Haushofer and other European 'geopoliticians'. He proposed Rudolf Hess for an honorary degree in Geopolitics from the University of Pavia. He also accompanied Fascist delegations to Nazi Germany: firstly to the centenary of the Frankfurt Geographical Society in 1935; and in 1936, to the *ReichsKolonial* ministry where he addressed the official function upon 'the concepts of geopolitics' (Atkinson 1996). Simultaneously, Massi sought further influence amongst the hierarchies of Fascist Italy. He used the European trips to badger the Fascist ministers he accompanied about the insights a geopolitical perspective might offer the state. And notwithstanding the hostility of some geographers, in 1938 Massi eventually managed to secure himself a meeting with Giuseppe Bottai, the Fascist minister of education (Atkinson 1996).

Bottai was the longest-serving minister in Mussolini's government. An intellectual and critic, his valuable original thinking ensured him a position near the heart of the regime. He nurtured a coherent and relatively sophisticated vision of a future utopian, Fascist society (De Grand 1978; Guerri 1976). And more than any other leading Fascist, he recognized the importance of culture in the re-casting of Italian society. Historians remember him primarily for his numerous cultural initiatives (Malgeri 1980), but amongst these was the encouragement of geographical knowledges within Italy. At the education ministry from 1936 to 1943, Bottai consistently patronized geography in universities, schools, and in everyday cultures. His self-appointed task was to raise the horizons of Italians from their traditional local and regional affiliations, and to nurture in them a broader *coscienza geografica*, or a geographical imagination (Atkinson 1995). This awareness of other places should, he argued, operate at an 'imperial level' appropriate to Italy's reborn imperial status upon the 1936 conquest of Abyssinia.

Aware of Bottai's interest in geography, Massi approached the minister with his ideas about geopolitics. After a further meeting with both Roletto and Massi, the minister was sufficiently impressed to arrange state funding and a publisher for the journal. By adding his own name to its masthead, Bottai installed *Geopolitica* amidst his portfolio of cultural initiatives, and situated it at the heart of the regime's support for geography. For their part, Massi and Roletto had established a regular forum through which to inform and educate the Italian public and policy-makers about their radical, new geopolitical perspective. Their analysis and explanations of Italy's place in the ever-changing world would appear monthly for four years.

Introducing Geopolitica

The first issue of *Geopolitica* appeared in January 1939. From the start, the journal's status as an explicitly *Italian* expression of the international geopolitical debate was evident. Roletto and Massi organized an editorial board for *Geopolitica* that by 1942 had included Karl Haushofer from Germany, Jaime Vicens Vives from Spain, representatives from Romania and Albania, as well as a range of Italian figures. The first issue also included a note from Haushofer that wished success to what he called *Geopolitik's* sister-journal (Haushofer 1939). *Geopolitica* was thus never conceptualized as a hermetic, Italian discourse, but as an Italian expression of the broader debates about geopolitical thought that circulated the inter-war world.

Geopolitica's official state sanction was evidenced by an introduction from Bottai that emphasized his vision of the centrality of geography to the modern state (Bottai 1939). The minister argued for a new conception of geopolitics that transcended orthodox, static political geography. Rather 'to avoid academic stagnation, [geography] has to raise itself to a political understanding of the world and the laws that direct and concern it' (Bottai 1939: 4), and geopolitics, he continued, should be attuned to political affairs and provide the nation with both a geographical and a political awareness: the *coscienza geografica* he envisaged (Bottai 1939). In turn, Roletto and Massi described the origins of their Italian geopolitics, its distinctiveness, and its proposed contributions to the Fascist State (Roletto and Massi 1939). '[Amidst] the new relationships that exist between science and politics in the Fascist state', they began, 'Italian geography has new duties to perform' (Roletto and Massi 1939: 5). And given the increasing complexity and interconnectedness of the modern world, they continued, geography was unique in its ability to synthesize its ever-more disparate elements. But rather than expose Italians to the subjective geopolitics of German or French writers, they argued, Italy deserved its own geopolitical perspective: one that consciously forged a path between these competing, foreign approaches (Roletto and Massi 1939).

The editors then outlined the new and encompassing vision they claimed for their journal. *Geopolitica's* subtitle announced it as a journal of 'Political, economic, social and colonial geography', and the editors made much of the breadth of their coverage (Roletto and Massi 1939). They argued that *Geopolitica* would encompass phenomena that fell beyond the scope of traditional political geography, like political alliances, colonialism, resistance and expansionism. None of these elements could be quantified or analysed before the emergence of geopolitics, they wrote. *Geopolitica* was:

> about studying the *geographical conditions* of the life and developments of states and the *geographical basis* of political problems that emerge

from their relations. In this manner one enters completely into the sphere of *Geopolitica*.

(Roletto and Massi 1939: 8 emphasis in original)

For while political geography measures the value and hierarchies of states . . . geopolitics extends this analysis to a wider basis that also considers cultural factors, spiritual factors, and the will to power and to empire.

(Roletto and Massi 1939: 10)

Finally, this radical, transcendent understanding of the world would render Italian geopolitics: 'the geographical doctrine of empire [that] should express in the most complete manner the geographical, political and imperial consciences of the Italian people' (Roletto and Massi 1939: 11). Such was the self-proclaimed nature of Italian geopolitics. A privileged comprehension of the politics, economies and geographies of the world system, it hoped to provide this insight to the Italian State, but also to educate the Italian nation about the geographies of the contemporary world. It was an attempt to define and prescribe a space wherein Italians might access this new way of conceptualizing the world. The next section examines some examples of this geopolitical vision.

Geopolitica: *representing the world*

In addition to their constant pleas that geography be incorporated into the fabric of Italian life, the editors of *Geopolitica* published an eclectic series of articles that revealed the global range of the journal's geographical coverage. Contributions were broadly connected by questions of space, territory, geography and politics, although, in line with the journal's conceptual agenda, these were usually studied as dynamic, fluid processes rather than fixed, static phenomena. There is not the space here to discuss the theoretical approaches of the journal more fully (Antonsich 1997a; Atkinson 1995, 1996; Vinci 1990). However, a taste of its themes, arguments, style and coverage can be acquired from the articles discussed below.

Many of the journal's main themes and preoccupations revolved around Italy's position in the international order of the late 1930s and early 1940s. At the regional scale, the Adriatic Sea that Italy claimed exclusively, was a focus of attention. Dalmatia and the Balkans were also sites of particular interest, given Italian ambitions for hegemony in the region, but anxieties about German intentions for the Danube basin also shaped Italian concerns (Vinci 1990). At a wider scale, the geopolitics of the Mediterranean, that Italy contested with the British, French and Spanish, were addressed repeatedly (Knox 1982). Meanwhile Italian colonies in North and East Africa and the Aegean were subject to speculation regarding their future development and their potential for further expansion (Atkinson 1995). The journal constantly assessed, debated and legitimated Italy's immediate strategic, economic and political concerns, while longer-term colonial ambitions were also evaluated regularly.

Geopolitica also directed its gaze at a broader, global scale, however. Its pages routinely contained analysis of diverse regions spread across the globe. In particular, there was much interest in the location of natural resources, their control and exploitation, and the politics and geographies of the trade routes that connected these resources to industries and markets. Similarly, the distribution of European colonial territories and their resources also concerned the Italians. These interests were derived from *Geopolitica*'s origins in the economically-stricken city of Trieste, but also from Italy's particular predicament in the late 1930s. Although the global recession of the inter-war years had not scarred Italy as badly as some more industrialized nations, nevertheless economic problems still caused much hardship and restructuring throughout society. In addition, Italian concerns about access to natural resources and the workings and control of the global economy were heightened by two factors. First, many believed that the Versailles peace conferences had failed to reward Italy adequately for its wartime sacrifices. Italy's fruitless claims for colonial territories in Africa and the Middle East were a particular grievance (Burgwyn 1993; Toscano 1937). Second, the 1935 invasion of Abyssinia had prompted unanticipated sanctions from the League of Nations. Already a resource-poor nation, Fascist rhetoric argued that the international community, driven by Britain and France, was deliberately excluding Italy from its rightful status as a major colonial power. This supposed conspiracy, and the Italian response of a policy of national self-sufficiency, were recurrent themes in *Geopolitica*. Indeed, such problems of geography and politics were deemed central to the journal's remit. In the following pages I discuss two examples of the global scope and synoptic vision that the journal claimed while it analysed and explained the world to its readers.

'Democracy, colonies and raw materials'

Italian concerns about the distribution of colonial territory and natural resources were demonstrated by a substantial article entitled 'Democracy, colonies and raw materials' by Ernesto Massi (Massi 1939b). To Massi, these long-standing problems were manifestations of dynamic phenomena (Massi 1937a, 1937b, 1938). As such they acquired 'an exquisitely geopolitical content [and] consequently, the problem entered into the field of [*Geopolitica's*] investigations' (Massi 1939b: 17). According to Massi,

all the post-war policies of the 'Great democracies' had been directed towards the consolidation of their supremacy, the impediment of every alteration to the balance of power and the obstruction of any change to boundaries. The League of Nations . . . perpetuated this hegemony, neutralizing every innovative movement and force. Humanity would therefore continue to be divided into rich peoples, abundantly supplied with raw materials, tropical territories, a higher potential for industry and for an elevated standard of living, and poor peoples, hard pressed demographically, scarcely supplied with raw materials, at a low standard of living.

(Massi 1939b: 18–19)

Massi described a binary division of the world. The 'rich peoples' of the 'Great Democracies' (particularly Britain and France) controlled resources, while the 'poor peoples', including Italy, were denied their share. The rich accumulated wealth while the poor were scantily rewarded for their labour. This uneven development was sustained and legitimated by the application of 'democratic principles' by these powerful states to the pliant world order (Massi 1939b). Such was the article's analysis of the 1939 global political economy.

In its place, Massi called for a new world order with a fairer distribution of global resources. Deserving peoples and the world's dispossessed nations would be measured by 'spiritual' criteria such as historical tradition, national consciousness, cultural or religious eminence or, he continued, the 'will to power and empire'. 'It was these factors, upon which one could base a criteria for a geopolitical differentiation between states' (Massi 1939b: 19). In this analysis, the 'deserving nations' were the revisionist powers dispossessed by the Versailles treaty, or other nationalist movements that opposed the established European-imperial world order. Instead of the global commodity control of Britain, France and the United States, Massi called for the redistribution of colonial resources and a collaborative approach towards their exploitation. Redistribution would avoid future conflicts between the revisionist nations and the 'democracies', and eliminate harmful and potentially destabilizing inequalities from the global economy, he added (Massi 1939b). Despite his complaints at the inequity of the world order, Massi did not challenge the colonial system itself, but simply argued for a 'fairer' share of colonial wealth for Italy (Massi 1937a, 1939b). Indeed, the colonial system was to be streamlined, improved and rendered more efficient in its exploitation of the colonial world once Italian territorial demands had been satisfied. This was an essentially modern vision of the colonial world as a resource-base to be exploited rationally.

Massi (1939b) concluded his analysis with the observation that it was political factors such as alliances and empires that were critical to understanding the global economy. This allowed him to emphasize that it was in the analysis of such dynamic phenomena that *Geopolitica* claimed special advantage. Here was the geopolitical imagination at work. Political geography might identify a given state's resources, population size, industrial capacity and other 'static' phenomena. *Geopolitica*, by contrast, could develop this empirical base but also accommodate the political contingencies of empire, treaties and mandated territories. This was the more ephemeral and dynamic currency in which *Geopolitica* dealt, and which enabled its geopolitical understanding.

'Self-sufficiency in the United States'

A further example of this global geopolitical vision is found in an article by Eliseo Bonetti entitled 'The geographical problem of self-sufficiency in the United States' (Bonetti 1940). Bonetti was a geographer based in Trieste. A regular and influential contributor to *Geopolitica*, his article appeared in the June-July issue of 1940. It is relevant for a number of reasons. First, it again demonstrates Italian concerns with the location, control and flows of the raw materials essential to modern industrial nations, and the collection of these concerns beneath the analytical umbrella of geopolitics. Second, the American theme also demonstrates the global scope of *Geopolitica's* gaze. Finally, the article was provoked by an essay in the 1940 volume of the *Geographical Review* entitled 'American raw material deficiencies and regional dependence' (Hull, 1940). Bonetti's re-working of these themes for an Italian readership demonstrates the circulation of ideas between 'geopoliticians' of different nations.[1] Likewise, the article also drew upon European geopolitical traditions as it contained two geopolitical maps (a technique adapted from German *Geopolitik*), one of which (Figure 5.1) referred to the Pan-regions concept that also originated in German geopolitics.

Bonetti's article surveyed what he called the 'geopolitical consequences' of the raw material requirements of the United States. Addressing the seventeen commodities declared of 'strategic' importance by the US government, Bonetti examined their geographies in peacetime and wartime. He demonstrated the global reach of American trade, but also claimed that given the resource potential of the Americas themselves, the only other resource-base that was indispensable to the US in times of crisis was that of Southeast Asia (Bonetti 1940). The corollary was that the US had to maintain Southeast Asia within its sphere of influence and this, concluded Bonetti, 'was extremely delicate from a geopolitical point of view, since it is here that the expansionist ambitions of [Japan] converge [with American interests]' (Bonetti 1940: 302).

Of the maps that accompanied the essay, Figure 5.1 demonstrates how *Geopolitica* adapted the 'suggestive cartography' and 'pan-region' concepts of *Zeitschrift für Geopolitik* (Herb, 1987). The pan-regions model posited a world divided into three vast self-contained longitudinal zones: Eurafrica, Pan-Asia and Pan-America (O'Loughlin and van der Wusten 1990). Figure 5.1 is an Italian representation of the 'Pan-America' concept. Using his sophisticated adaptation of the monochromatic, geopolitical cartography of *Zeitschrift für Geopolitik*, Mario

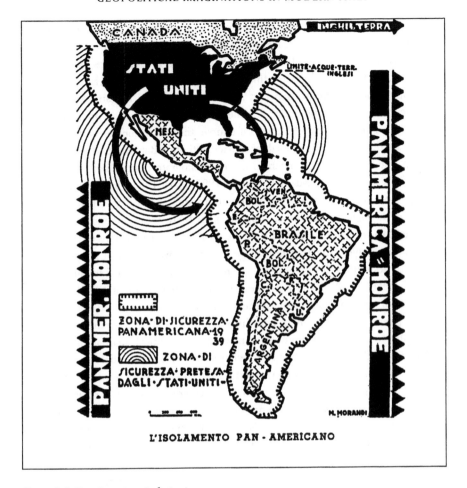

Figure 5.1 Pan-American isolationism

Source: Bonetti 1940: 299

Morandi (*Geopolitica's* main cartographer) reinforced the idea of the pan-region as largely self-sufficient through the two broad, black bars which delimit the western hemisphere longitudinally and are complemented by the serrated 'hostile front' symbol. The essential North–South relationship of this 'pan-region' is thus reinforced, as is the exclusion of other powers from the region.

Within the isolationism of the Pan-America region, the map also suggests the subservience of South America to the United States. The black bars which flank South America combine with the black mass of the United States to enclose Latin-American nations which, despite their political differences, are all shaded in a uniform, lighter tone. Graphically, America dominates these countries and the two black arrows that advance upon the continent indicate this influence symbolically.

DAVID ATKINSON

Their curvature suggests dynamic movement as their 'pincer-movement' encloses Central America. And whilst Canadian territory to the north is differentiated from the United States, the map fails to distinguish other British possessions in South America or indeed wealthy Argentina and its sizeable Italian emigrant population. This compounds the impression that American hegemony in Latin America was unchallenged, and that the entire continent was envisioned as an undifferentiated resource-base for the United States. Of course, this also resonates with the Italian plea for a redistribution of colonial territories. Although Bonetti's text was relatively detailed, the maps that accompanied it were deliberately simplified to translate a straightforward message visually (Atkinson 1995; Herb 1987). In combination, the messages of map and text were supposed to demonstrate the 'reality' of the Americas through the insight of *Geopolitica*. In turn, these understandings were supposed to advise, inform and educate the geopolitical imaginations of Fascist Italy.

In recent Anglo-American human geography, work in critical geopolitics has frequently identified the significance of the visual to geopolitical imagination (Ó Tuathail 1994, 1996). It was a privileged vision, whereby the initiated geopoliti-cian *observed* and thus *comprehended* the world in all its interwoven complexities, that the geopoliticians assumed. They imagined their shift from orthodox political geography to a more dynamic, encompassing geopolitical analysis had allowed them this additional perspective from where the world was rendered increasingly legible. There are clear masculinist and ethnocentric assumptions underpinning this all-knowing geopolitical gaze, with its unproblematized access to 'reality' and the casual reproduction of the racial hierarchies and categories of colonialism. Geography was also a casualty of this approach. The journal would often reduce complicated situations to simplistic arguments to communicate an unmistakable message. Whole regions might be homogenized, with differences obliterated and geographies flattened by the assumed authority of geopolitical representation (Atkinson 1995; Ó Tuathail 1993, 1996). While it is impossible to ascertain any impact these representations may have had in Italy, the geopolitical imaginations they hoped to catalyse would inevitably be partial.

The closure of Geopolitica

As the Italian war effort and the Fascist regime were collapsing in late December 1942, *Geopolitica* appeared for the final time (Vinci 1990). In responding to the need to understand the geographical and political factors that were destroying the econ-omy of Trieste, a form of analysis emerged that found wider application, and government patronage, against the backdrop of the instabilities of the inter-war period. At its peak, *Geopolitica's* print run was around 1,000 copies (Atkinson 1996). Again, it is not possible to quantify the political or cultural significance of *Geopolitica*. Nevertheless, the journal's story does increase our understanding of the nature of geopolitical ideas in inter-war Europe. *Geopolitica* is interesting for its geopolitical topography of the modern world, but perhaps its most significant legacy was its attempt to forge a geopolitical awareness in the Italian people and to carve out an intellectual space wherein *Italian* geopolitical analysis might flourish.

Remembering *Geopolitica* and remaking Italian geopolitics

The story of *Geopolitica* was seldom recounted in published form in post-war Italy. However tenuous the connections between German *Geopolitik* and Nazi imperialism (Bassin 1987; Heske 1987; Jacobsen 1979), the alleged associations cast a pall over geopolitics and political geography in Italy as elsewhere. Italian geographers distanced themselves from geopolitics via predictable strategies. In the immediate post-war years, practically no mention was made of geopolitics (Almagià 1946; Migliorini 1946; Nice 1943; Toschi 1949); historians of geography were often reluctant to address the Fascist era at all, never mind to touch upon the notoriety of geopolitics (Bonora 1987). The amnesia was near-total: in two almost encyclopaedic surveys of the history, development and orientations of political geography, Mario Ortolani – who had contributed to *Geopolitica* himself in the 1940s – neglected to mention the journal at all, despite writing at length about the geopolitics of several other nations (Ortolani 1956, 1976).

The situation only began to change in the 1980s. Geographers began to make brief mention of Italian geopolitics within wider accounts of political geography or the history of geography (Caraci 1987; Pagnini, 1987). On the rare occasions when *Geopolitica* was mentioned, the accounts stressed the Triestine origins of the journal, and the differences between *Geopolitica* and *Zeitschrift für Geopolitik*, especially the fact that *Geopolitica* largely eschewed environmental determinism and overt racial theorizing (Caraci 1987; Pagnini 1987). Even Ernesto Massi eventually wrote about *Geopolitica* and its place in the history of geopolitics. He too contrasted *Geopolitica* with German geopolitics, although his account was a largely unapologetic attempt to normalize *Geopolitica* within the history of geographical knowledges and their connections to politics (Massi 1986). The controversy attendant upon Massi's recollections demonstrated that geopolitics remained a problematic term within Italy.[2]

Perhaps it was partially because of the enduring notoriety of geopolitics that the first attempt to re-cast geopolitical thought in Italy was largely unsuccessful. *Hérodote-Italia* was published annually between 1978 and 1984. It was inspired by the French left-wing journal *Hérodote* founded by Yves Lacoste in 1976 (Claval and Hepple, both this volume). The Italian version published translations from *Hérodote* and original material from Italy (Antonsich 1997a). Its remit was to introduce *Hérodote's* perspectives into Italy and to radicalize Italian geography. The editor was

Massimo Quaini, a Marxist geographer whose interests in the connections between geography and Marxist theory underpinned the journal. Like *Hérodote*, the journal urged the recognition that geography was a politicized knowledge; it argued that geographers should pursue an active role in transforming society. And like the French original, *Hérodote-Italia* also reflected its concerns for interdisciplinary analysis and the importance of geographical education in schools (Desfarges 1998).

Yet despite a change of editor and name (to *Erodoto*) in 1982, the journal struggled for subscriptions and folded in September 1984 (Antonsich 1997a). Unlike *Hérodote*, the Italian journal did not use the term geopolitics in its sub-title: perhaps this label remained too sensitive for Marxists and geographers in 1970s Italy. However, the journal did attempt to re-introduce questions of geographies, power and politics into Italy, and it did refract something of *Hérodote's* efforts to re-work geopolitical analysis in France. In addition, like *Geopolitica* before it, *Hérodote-Italia* developed a self-consciously Italian version of a foreign form of political geography. Possibly, as Paul Claval comments in this volume, this attempt to revive geopolitical themes appeared when Italians were not yet ready to re-engage with the geopolitical tradition.

Limes: *a geopolitical Renaissance?*

The situation appeared to be different in the 1990s. As the certainties of the Cold War period dissolved and Italy's post-war political system fractured between 1992 and 1994 (Gundle and Parker 1996), a new journal that positioned itself as 'geopolitical' and claimed a special ability to 'understand the world like it is' began publication (Figure 5.2). Moreover, the journal immediately won a substantial popular audience for its declared insight into the foreign policy options of contemporary Italy.

The word *Limes* is the Latin term for 'frontiers', traditionally applied to the ancient borders of the Roman Empire. Since 1993 however, the term has also come to be associated with a quarterly publication with the self-appointed task of re-launching geopolitical thinking in Italy (Antonsich 1997a, Pfetsch 1993). Subtitled 'An Italian geopolitical review', *Limes* is a popular journal of current affairs and international relations. Its founding editors are Lucio Caracciolo, an Italian journalist and commentator, and Michel Korinman, the French historian of geopolitics and member of the *Hérodote* circle. The *Hérodote* connection is reinforced by the role of Yves Lacoste as 'Special editorial advisor'. The intention is to provide a forum for Italians to debate and develop a foreign policy appropriate for a changing world. This, they stated, was Italy's new responsibility in the post-Cold War world of the 1990s (Editoriale 1993). But to develop new policies, Italians first needed an accurate grasp upon the 'reality' of global issues. They had to understand the world as it *really* was, replete with elements such as nationalism,

Figure 5.2 Limes – an Italian geopolitical review, 'for understanding the world as it is'
Source: Bollettino della Società Geografica Italiana 1993: 176

ethnicity, frontiers and territoriality, that were not encompassed by the approaches of orthodox international relations. To enable this broader perspective, a *geopolitical* approach and 'geopolitical reasoning' were required (Editoriale 1993).

The inaugural editorial spelt out this understanding of geopolitics: 'More than a science, geopolitics is knowledge in Foucault's sense, or better still, it is a kind of reasoning' (Editoriale 1993: 9). As such, in addition to phenomena like nationalism and ethnicity, geopolitical reasoning would be sensitive to questions of scale and locality. It would give voice to all sides of a debate, and clarify problems with cartography (Editoriale 1993). *Limes* would establish a detached yet informed perspective, based upon a 'concrete reasoning' that would allow it an exclusive prospect upon foreign policy problems. It was this special vision and reasoning that *Limes* claimed as 'geopolitical', and that it applied to the 1990s world.

The first issue dealt with the Balkan crisis of the early 1990s. It rehearsed various national and ethnic perspectives on the conflicts from a series of different authors, including Croatian President Tudman (1993) and Bosnian President Izetbegovic (1993). However, alongside these voices, an entire section was devoted to the consideration of Italian interests in the region, and the future prospects of Trieste and Istria in particular (Segatti 1993; Pagnini and Galli 1993; Sema 1993; Ferraris 1993). It was in this manner that the journal planned to define Italian interests and priorities in the Balkans. An avowedly inter-disciplinary publication, contributors included 'security intellectuals', historians, diplomats, journalists, sociologists and a geographer. Prominent amongst these was Bruno Bottai (1993), the Permanent Secretary at the Italian Foreign Ministry and the son of Giuseppe Bottai, promoter of *Geopolitica*. Along with finance from Venetian and Triestine business interests (whose commercial prospects were revised by war in Yugoslavia), Bottai provided government funds for the 1993 Venice conference at which the journal was launched. He opened proceedings in Venice, and joined the journal's editorial board. The conference was concluded by the then Foreign Minister, Emilio Colombo, and a subsequent conference in Rome attracted the recently resigned Prime Minister Giulio Amato away from Italy's ongoing domestic political crisis. In its efforts to negotiate new foreign policies for Italy, *Limes* enjoyed influential support from the government.

In addition, *Limes* has a series of connections to the geopolitical traditions of Italy and France. Although the journal carefully denied any continuities from Friedrich Ratzel and 'nineteenth-century geopolitics' in its inaugural comments (Editoriale 1993), it is clear that in some respects *Limes* draws upon and renegotiates geopolitical traditions for its contemporary context. Most obvious are the adoption of the term 'geopolitics', and the links to *Hérodote*. But if the connections to the Bottai family and the city of Trieste are coincidental, the string of articles that discuss the theories and histories of geopolitics are not (Antonsich 1997b; Bonanate 1997; Bottai 1997; Ceretti 1997; Ciampi 1997). Similarly, the reprinted work of earlier geopoliticians also suggests that *Limes* has some sense of its own precedents (Haushofer, 1995). Finally, at a 1993 conference held in the Italian Geography Society in Rome to discuss geopolitical themes, Caracciolo spoke about *Limes* and its ambitions shortly before the aged Ernesto Massi reminisced about his journal, *Geopolitica* (Lucchesi 1993). Whether coincidental, implicit, covert or explicit, *Limes* has several connections to prior expressions of geopolitics in Italy, and thus it inevitably constitutes a further episode of Italy's haphazard geopolitical tradition.

Some critics have complained at this reinvigoration of geopolitics. Santoro (1996) worries that the 'pragmatic' or 'realist' methodological approach of *Limes* simplifies the complex histories of what he calls the geopolitical tradition. Raffestin, Lopreno and Pasteur (1995) criticize 'the nationalist project of *Limes*' and its 'insistence upon [an Italian] political-commercial living space beyond [Italian] frontiers' (Raffestin, Lopreno and Pasteur 1995: 303). Italian ambitions in the Balkans, for example, are stated very starkly in a recent English translation of a *Limes* edition (Caracciolo and Korinman 1998). If these more bullish statements are reminiscent of *Geopolitica*, so too is the assumption that the geopolitical perspective provides a unique and unproblematic window upon 'reality'.[3] In response to these criticisms, the editors of *Limes* and others point out that, while embedded within the 'western alliance' during the Cold war era, Italy was seldom permitted to develop foreign policies for itself. Consequently, an Italian debate upon foreign policies is well overdue (Editoriale 1993; Sfrecola 1997). Certainly, the vigorous debates now surrounding Italian foreign policy may be a response to the new instabilities of the 1990s, and particularly to the dilemmas and possibilities posed by the conflicts over space, territory and nationality in regions adjacent to Italy. But whatever explanation holds, it is through a self-proclaimed geopolitical framework that many Italians are starting to engage with these issues.

At first sight, *Limes* would seem to provide a neat conclusion to my essay. It uses the term geopolitics and sometimes expresses its vision through special maps. It is an attempt to interpret the instabilities of the modern world, and to access 'reality' by recourse to a geopolitical vision which encompasses phenomena that orthodox disciplines neglect. It sometimes claims an impartial perspective, but elsewhere promotes Italian foreign policy. Finally, it is a journal with government support arranged by a man called Bottai. This might all prove an interesting historical irony were it not for the astonishing and unexpected success of *Limes*. The first issue had a print-run of 14,000 copies which was increased to 24,000 for the second edition in June 1993. These sold out almost immediately. *Limes* now sells around 30,000 copies quarterly, although a recent edition on the Kosovo crisis sold 100,000. A French edition has appeared and a German one will follow. An English-language newsletter is also planned. Unlike *Geopolitica* and *Hérodote-Italia*, *Limes* has evidently located a vibrant constituency with a substantial taste for 'geopolitical reasoning'.

Conclusion

Historians use 'geopolitics' as shorthand for the geographies that underpin Italian history. By contrast, this chapter has considered the two most substantive examples of geopolitical thought in modern Italian history. They represent periodic attempts by Italians to comprehend the changing world through the application of geopolitical perspectives. Unfortunately, my accounts here have had to be brief; *Limes* in particular merits further attention. However, I have demonstrated that Italian geopolitics did not develop in isolation, but were consciously Italian negotiations of broader, international debates. In this respect both these journals drew upon, and contributed to, traditions of geopolitical thought. Furthermore, *Geopolitica* and *Limes* each claimed an exclusive grasp upon the 'reality' of world affairs, and both were eager to see their geopolitics of service to the state. Each also sought to promote their perspectives amongst the Italian public, thus encouraging the development of geopolitical imaginations in the population at large. And finally, both *Geopolitica* and *Limes* emerged in periods of international instability, when Italy was forced to re-negotiate its foreign policies in the face of, first the 1930s global slump and international sanctions, and second, the dissolution of Cold War frontiers and the crisis in the former Yugoslavia. In each instance, Italians developed forms of 'geopolitical reasoning' to help themselves understand these contexts, and it is perhaps just such periods of flux and anxiety that tend to catalyse geopolitical theorizing.

Acknowledgements

This chapter is drawn from a Ph.D. thesis funded by a University of Loughborough Scholarship, for which I appreciative. My gratitude also to Denis Cosgrove and Mike Heffernan for guidance along the way.

Notes

1 In 1941, Bonetti would introduce Hartshorne's 'The Nature of Geography' to the reader's of *Geopolitica* (Bonetti 1941); in the 1960s, it was Bonetti, still working in Trieste, who read Christaller's Central Place thesis in the original German, and introduced the theory and elements of quantification into Italian Geography. Once again, Trieste functioned as a gateway for foreign debates and ideas into Italian geography.

2 Massi had been invited to speak upon *Geopolitica* at the twenty-fourth Italian Geographical Congress in Turin, 1985. At the session organized by Giuseppe Dematteis, Claude Raffestin and Franco Farinelli provided critiques of geopolitics (Raffestin and Farinelli 1986), but Massi – now professor of geography at the University of Rome – sought to normalize *Geopolitica*. When Dematteis refused to publish Massi's account in the conference proceedings, Massi used his position as President of the *Società Geografica Italiana* to place the article in the organization's prestigious *Bollettino* (Massi 1986).

3 Although one commentator contrasts the ability of *Limes* to access 'geopolitical reality' with that of Italian geographers whose methodologies do not allow them such insight (Antonsich 1997a).

Bibliography

Almagià, R. (1923) 'La geografia politica. Considerazioni methodiche sul concetto e sul campo di studi di questa scienza', *L'Universo* 4: 751–68.

—— (1926) 'Una nuova opera di geografia politica', *L'Universo* 7: 353–60.

—— (1936) *Elementi di geografia economica e politica*, Milan: Giuffré.

—— (1946) 'I compiti attiali della Geografia e il Consiglio Nazionale delle Ricerche', *Ricerca Scientifica e Ricostruzione* 16: 3–11.

Antonsich, M. (1997a) 'La geopolitica Italiana nelle Rivista "Geopolitica", "Hérodote/ Italia" ("Erodoto"),"Limes"', *Bollettino della Società Geografica Italiana* 12, 2: 411–18.

——. (1997b) 'Eurafrica, dottrina Monroe del fascismo', *Limes* 1997, 3: 261–6.

Atkinson, D. (1995) 'Geopolitics, cartography and geographical knowledge: envisioning Africa from Fascist Italy', 265–97 in M. Bell, R. A. Butlin and M. Heffernan (eds) *Geography and Imperialism, 1820–1940*, Manchester: Manchester University Press.

——. (1996) *Geopolitics and the Geographical Imagination in Fascist Italy*, unpublished Ph.D. thesis, University of Loughborough.

Baratta, M. (1918) *Confine orientale d'Italia*, Novara: Agostino.

—— (1919) 'Giuseppe Mazzini e il confine orientale d'Italia', *Istituto Geografico de Agostini Quaderni Geografici* 7: 1–30.

Bassin, M. (1987) 'Race contra space: the conflict between German *Geopolitik* and National Socialism', *Political Geography Quarterly* 6: 115–34.

Bollettino della Società Geografica Italiana (1993) 11, 10: 176.

Bonetti, E. (1940) 'Il problema geografico dell'autosufficienza negli Stati Uniti d' America', *Geopolitica* 2: 296–9, 302.

—— (1941) 'Attraverso la storia della geografia', *Geopolitica* 3: 423–7 and 430–3.

—— (1967) 'Giorgio Roletto (1995–1967)', *Rivista Geografica Italiana* 74: 251–4.

Bononate, L. (1997) 'Qualche argomento contro l'interessa nazionale', *Limes* 1997, 2: 303–13.

Bonora, P. (1987) 'Umberto Toschi, 1897–1966', *Geographers Biobiliographical Studies* 11: 155–64.

Bosworth, R. B. (1996) *Italy and the Wider World, 1860–1960*, London: Routledge.

Bottai, G. (1939) 'Giuseppe Bottai alla Geopolitica', *Geopolitica* 1: 3–4.

Bottai, B. (1993) 'Vivere senza Jugoslavia', *Limes* 1993, 1: 143–50.

—— (1997) 'Jean Monnet visto da vicino', *Limes* 1997, 2: 149–56.

Burgwyn, H. J. (1993) *The Legend of the Mutilated Victory: Italy, the Great War and the Paris Peace Conference, 1915–19*, Westport: University of Connecticut Press.

Caraci, I. L. (1987) 'Storia delle geografia in Italia dall secolo scorso ad oggi', 47–94 in G. Corna-Pellegrini (ed.) *Aspetti e problemi di geografia*, vol. 1, Milano: Mondadori.

Caracciolo, L. and Korinman, M. (eds) (1998) *Italy and the Balkans*, Washington: Center for Strategic and International Studies.

Cerreti, C. (1997) 'San Giuliano e la non-geopolitica dei geografi', *Limes* 1997, 3: 249–60.

Ciampi, G. (1997) 'A che servono I geografi', *Limes* 1997, 2: 295–301.

De Grand, A. (1978) *Bottai e la cultura fascista*, Bari: Laterza.

De Marchi, L. (1929) *Fondamenti di geografia politica*, Padua: Cedam.

Desfarges, M. (1996) *Introduzione alla geopolitica*, Bologna, Il Mulino.

Dickie, J. (1996) 'Imagined Italies', in D. Forgacs and R. Lumley (eds) *Italian Cultural Studies. An Introduction*, Oxford: Oxford University Press.

Duggan, C. (1994) *A Concise History of Italy*, Cambridge: Cambridge University Press.

Editoriale (1993) 'La Responsibilità Italiana', *Limes* 1993, 1: 7–11.

Ferraris, L. V. (1993) 'Dal Tevere al Danubio: l'Italia scopre la geopolitica da tavolino', *Limes* 1993, 1: 213–25.

Filippo de Magistris, L. (1923) 'Geografia e politica', *Gerarchia* 2: 1033–9.

Forgacs, D. and Lumley, R. (eds) (1996) *Italian Cultural Studies. An Introduction*, Oxford: Oxford University Press.

Gambi, L. (1994) 'Geography and imperialism in Italy: from the unity of the nation to the "New" Roman Empire', 74–91 in A. Godlewska and N. Smith (eds) *Geography and Empire*, Oxford: Blackwells.

Goddi, E. (1984) *Trieste*, Bari: Laterza.

Guerri, G-B. (1976) *Giuseppe Bottai: un fascista critico*, Milan: Feltrinelli.

Gundle, S. and Parker, S. (1996) *The New Italian Republic. From the Fall of the Berlin Wall to Berlusconi*, London: Routledge.

Haushofer, A. (1995) 'Che cosa é un corridoio?', *Limes* 1995, 3: 177–96.

Haushofer, K. (1939) 'Der Italienischen "Geopolitik" als Dank und Gruss!', *Geopolitica* 1: 12–16.

Heffernan, M. J. (1995) 'The spoils of war: the Société de Géographie de Paris and the French empire, 1914–1919', **221–64** in M. Bell, R. A. Butlin and M. Heffernan (eds) *Geography and Imperialism, 1820–1940*, Manchester: Manchester University Press.

—— (1996) 'Geography, cartography and military intelligence: the Royal Geographical Society and the First World War', *Transactions of the Institute of British Geographers* 21: 504–33.

Herb, G. (1989) 'Persuasive cartography in *Geopolitik* and National Socialism', *Political Geography Quarterly* 8: 289–303.

Heske, H. (1987) 'Karl Haushofer: his role in German geopolitics and Nazi politics', *Political Geography Quarterly* 6: 135–44.

Hull, R. B. (1940) 'American raw material deficiencies and regional dependence', *The Geographical Review* 30: 147–62.

Izetbegovic, A. (1993) 'Dichiarazione islamica', *Limes* 1993, 1: 259–74.

Jacobsen, K-A.(1979) *Karl Haushofer: Leben und werk*, vols. I and II, Boppard am Rhein: Boldt.

Knox, M. (1982) *Mussolini Unleashed, 1940–1. Politics and Strategy in Fascist Italy's last war*, Cambridge: Cambridge University Press.

Lando, F. (1993) 'Geografi di casa altrui: l'Africa negli studi geografici italiani durante il ventennio fascista', *Terre d'Africa*: 73–124.

Lo Monaco, M. (1987) 'Ernesto Massi: mezzo secolo di analisi geografiche per la sintesi economica', in Dipartimento di Studi Geoeconomici, Statistici e Storici per l'analisi Regionale, Università di Roma, *Scritti in onore di Ernesto Massi*, Bologna: Zanicelli.

Lucchesi, F. (1993) 'Due Giorni si studio sul tema "Dalla Geografia Politica alla Geopolitica"', *Rivista Geografica Italiana* 100: 816–19.

Mack Smith, D. (1959) *Italy. A Modern History*, Ann Arbor: University of Michigan Press.

Malgeri, F. (1980) 'Giuseppe Bottai e "Critica Fascista"', 1–100 in G. De Rosa and F. Malgeri (eds) *Antologia di 'Critica Fascista', 1923–43*, Rome: Landi.

Marchesini, D. (1976) 'Romanità e Scuola di mistica fascista', *Quaderni di Storia* 4: 55–73.

Marinelli, O. (1903) 'Frederico Ratzel e la sua opera geografica', *Rivista Geografica Italiana* 10: 272–7.

Massi, E. (1930) 'Geografia politica e geopolitica', *La Coltura Geografica* 2: 137–45.

—— (1931) 'Geografia politica e geogiurisprudenza', *La Coltura Geografica* 2: 81–2.

—— (1932) 'Lo stato quale oggetto geografico', *Rivista di Geografia* 12: 169–76.

—— (1935) 'Aspetti geopolitici dell'Europa Danubiana', *Rassegna di politica internazionale* 6: 15–24.

—— (1937a) 'Il problema coloniale in Germania', *Vita e Pensiero* Febbraio: 3–12.

—— (1937b) *La participazione delle colonie alla produzione delle materie prime*, Milan: Istituto Coloniale Fascista.

—— (1938) 'Il valore economico dei "Mandati" Africani', *Africa* 56: 1–62.

—— (1939a) 'Römische und italienische Mittelmeer-Geopolitik', *Zeitschrift für Geopolitik* 16: 551–66.

—— (1939b) 'Democrazia, colonie e materie prime', *Geopolitica* 1: 17–35.

—— (1940) 'L'ora della geopolitica', *Critica Fascista* 18: 334–6.

—— (1986) 'Geopolitica: dalla teoria originaria ai nuovi orientamenti', *Bollettino della Società Geografica Italiana* 11, 3: 3–45.

Migliorini, E. (1930) 'Recensione e annunzi bibliografica', *Bollettino della Società Geografica Italiana* 6, 6: 620–2.

—— (1946) *La Terre e gli Stati, Geografia politica*, 2nd edition, Naples: Loffredo.

Millo, A. (1987) 'L'elite del potere a Trieste: dall'irredentismo al Fascismo', *Società e Storia* 36: 333–74.

Nice, B. (1943) 'Rassegna, Geografia politica', *Rivista Geografica Italiana*, 50: 48–162.

O'Loughlin, J. and van der Wusten, H. (1990) 'The political geography of pan-regions', *Geographical Review* 80: 1–20.

Ortolani, M. (1956) 'Orientamenti della geografia politica', *Il Politico, Rivista di Scienza Politiche* 21: 263–77.

—— (1976) 'Sviluppo storico della geografia politica', *Rivista Geografica Italiana* 83: 145–61.

Ó Tuathail, G. (1993) 'The effacement of Place: US Foreign Policy and the spatiality of the Gulf Crisis', *Antipode* 25: 4–11.

—— (1994) 'Problematising geopolitics: survey, statesmanship and strategy', *Transactions of the Institute of British Geographers* 19: 259–72.

—— (1996) *Critical Geopolitics*, London: Routledge.

Pagnini, M-P. (1987) 'La storia delle geografia politica', 409–42 in G. Corna-Pellegrini (ed.) *La storia della Geografia in Italia*, Milano: Mandadori.

Pagnini, M-P. and Galli, M. (1993) 'Contesa tra due patrie. I'Istria sceglie il regionalismo', *Limes* 1993, 1: 173–82.

Pfetsch, P. R. (1993) 'Sicherheit als begrift der internationalen Politik', *Geografische Zeitschrift* 81: 210–26.

Raffestin, C. and Farinelli, F. (1986) 'Seminario. "Geopolitica: dalla teoria originaria ai nuovi orientamenti"', in G. Dematteis (ed.), *Atti del 24 Congresso Geografica Italiana*, Turin: Einaudi.

Raffestin, C., Lopreno, D. and Pasteur, Y. (1995) *Géopolitique et Histoire*, Lausanne: Éditions Payot.

Ratzel, F. (1897) *Politische Geographie*, Munich and Leipzig: Oldenbourg.

Revelli, P. (1916) 'Una questione di geografia politica, L'Adriatico e il dominio del Mediterraneo Orientale', *Rivista Geografica Italiana* 23: 91–112.

—— (1918) 'Le origini italiane della geografia politica', *Bollettino della Società Geografica Italiana* 58: 394–416, 623–36 and 728–59. .

—— (1919) 'Le origini italiane della geografia politica', *Bollettino della Società Geografica Italiana* 59: 230–43, 279–308 and 395–422.

Ricchieri, R. (1920) 'La geografia alla conferenza per la pace a Parigi, nel 1919', *Rivista Geografica Italiana* 27: 103–9.

—— (1921) 'L'elemento geografico nella grandezza delle Nazioni secondo il Prof. Giuseppe Ricchierri', *Rivista Geografica Italiana* 28: 125–6.

Roletto, G. (1933) *Lezioni di geografia politica-economico*, Padua: Cedam.

—— (1937) *Le tendenze geopolitiche continentali e l'asse Eurafrica*, Milan: Mondadori.

—— (1938) 'Per una geopolitica italiana', *Politica Sociale* 11: 44–5.

—— (1940a) 'La geoeconomica al servizio dell'espansione commerciale', *Commercio* 1: 17–18;

—— (1940b) 'Funzione geopolitica di Roma', *Commercio* 5–6: 8–10.

Roletto, G. and Massi, E. (1931) *Lineamenti di Geografia Politica*, Trieste: Università di Trieste.

—— (1939) 'Per una geopolitica italiana', *Geopolitica* 1: 5–11.

Roncagli, G. (1919) 'Physical and strategic geography of the Adriatic', *The Geographical Journal* 53: 209–28.

Rose, G. (1995) 'Tradition and paternity: same difference?', *Transactions of the Institute of British Geographers* 20: 414–6.

Sandner, G. and Rössler, M. (1994) 'Geography and empire in Germany, 1871–1945', 115–27 in A. Godlewska and N. Smith (eds) *Geography and Empire*, Oxford: Blackwell.

Santoro, C. M. (1996) 'L'ambiguità di *Limes* e la vera geopolitica: elogio della teoria', *Limes* 1996, 4: 307–13.

Segatti, P. (1993) 'Trieste, un buco nero nella coscienza della nazione', *Limes* 1993, 1: 159–71.

Sema, A. (1993) 'Il triangolo strategico Trieste-Fiume-Capodistria', *Limes* 1993, 1: 183–95.

Serra, E. (1984) *La Diplomazia in Italia*, Milan: Mondadori.

Sfrecola, A. (1997) 'Il pensiero geopolitico italiano', 61–79 in P. Lorot (ed.) *Storia della geopolitica*, Trieste: Asterios.

Toniolo, A. R. (1923) 'I moderni concetti di geografia sociale e politica', *L'Universo* 4: 203–12.

—— (1930) 'Politica e Geografia', *Il giornale di politica e letteratura* 6: 303–24.

Toscano, M. (1937) 'Il problema coloniale italiano alla conferenza della pace', *Rivista di studi politici internazionale* 4: 263–96.

Toschi, U. (1937) *Appunti di geografia politica*, Bari: Macri.

—— (1949) 'L'oggetto centrale di studio della geografia politica', *Rivista Geografica Italiana* 56: 81–9.

Tudman, F. (1993) 'Deriva della verità storica', *Limes* 1993, 1: 247–57.

Valussi, G. (1965) 'L'Opera scientific di Giorgio Roletto', *Bollettino della Società Geografica Italiana* 9, 6: 313–26.

Vidal de la Blache, P. (1898) 'La Géographie politique, à propos des écrits de M. Frédéric Ratzel', *Annales de Géographie* 7: 97–111.

Vinci, A. (1990) '"Geopolitica" e balcani: l'esperienza di un gruppo d'intelletuali in un ateneo di confine', *Società e Storia* 47: 87–127.

Wilkinson, H. R. (1951) *Maps and Politics: A Review of the Ethnographic Cartography of Macedonia*, Liverpool: Liverpool University Press.

6

IBERIAN GEOPOLITICS

James Derrick Sidaway

Introduction

In spite of a passing reference in the writings of the Italian communist Antonio Gramsci – who had an eye for such things, despite, or perhaps because of, his long incarceration in a Fascist prison between 1927 and 1935 – to the rise of geopolitical discourse (Gramsci 1949: 221), and an interest in fashioning a 'Marxist or proletarian geopolitics' amongst a small number of German and Hungarian writers in the 1920s (see Bassin 1996), Anders Stephanson (1989: 137) could reliably claim that 'geopolitics has never found a congenial place within the Marxist tradition, let alone been properly theorized'.[1] That it belongs to the domain of ideology and to a broad realm of discourse, *signs* and cultures has however made it a fertile subject for recent post-structuralist analysis.[2] As the development of critical geopolitics has explained, once geopolitics is recognized as operating as a discourse, as a scripting of the world and at once a *pouvoir* (power) and a *savoir* (a practical knowledge), it is laid open to contextualization, deconstruction and reconstruction. It is in these senses that this chapter wishes to reconsider Iberian, that is Portuguese and Spanish (or, more exactly Castilian and Catalan), geopolitical traditions. The analysis here is preliminary, and a much longer, deeper and more thorough study would be required to do these traditions proper justice. This chapter therefore describes a selective cross-section of Portuguese and Spanish geopolitical discourses. Part of the selectivity arises from the way that they overlap with and form part of the wider rhetorics of the right-wing authoritarian regimes that dominated both countries for much of the twentieth century. However, although they merge into much broader colonial and nationalist politics, self-consciously geopolitical discourses, citing German, British, French and American traditions, are identifiable in both countries. Without claiming to be a wholly representative description of these, this chapter traces some key strands in each and identifies some of their key patterns.

The complex theme of relations between Portugal and Spain is largely left out here. Both countries saw calls for Iberian unity, particularly in the later years of the nineteenth and the first half of the twentieth century. Subsequently, Franco and the Falangist party had a predatory attitude towards Portugal in the early 1940s. Despite a 1939 Treaty of 'Friendship and Co-operation' between the two states, Franco took a dim view of what was perceived as a rather pro-allied stance by Portugal (both countries remained officially neutral for most of the Second World War, but in the Spanish case the reality was of a distinct pro-axis orientation). Later however, they grew closer, with Portugal making pleas to the allies for Spanish admission to NATO after the former was included but the latter excluded when the Atlantic Alliance was formally established in 1949.[3]

Historical accounts of geopolitical thought (at least those written in English) seem mostly unaware of the scale and significance of twentieth-century Portuguese and Spanish geopolitical discourse (let alone their nineteenth-century imperial antecedents). This is not to say that geopolitical discourse in Portuguese and Spanish has not been widely noticed. Partly as a result of the prominence of dictatorial 'geopolitically motivated' regimes in Brazil and the Southern Cone of Latin America in the 1970s, the profound historical (and ecological) significance of South American geopolitics has been widely commented upon, and many accessible accounts exist (Child 1985; Dodds 1993; Hecht and Cockburn 1989; Hepple 1992).

Turning however to the Portuguese and Spanish metropoles, there is no such prominent coverage of their own geopolitical writings. Yet the fact that they both suffered right-wing dictatorship into the mid-1970s and the imperial and some-times fascist pretences of Francisco Franco and António de Oliviera Salazar and their associates, provided (in both cases) relatively fertile ground for the circula-tion and elaboration of peculiar variants of geopolitics. The Portuguese case is both the least known and the most significant in terms of scale and the centrality of geopolitical discourse to the official ideology of the regime.[4] It will be the main subject of analysis here, but the account will also consider the main elements of Spanish geopolitics. Both studies are – it must be stated again – preliminary, and a great deal more work would be required to unpack the complexity of these tradi-tions, understand their interrelations, contrasts, contradictions, significance and, for what it is worth, restore them to their rightful (that is more central) place in the broader history of geopolitical traditions.

'Portugal is not a small country!' and 'Spain: an exceptional geopolitical entity'

Though always relatively peripheral to the main currents, both Spanish and Portuguese nationals participated in the broader circulation of geopolitical ideas that characterized inter-war Europe. In both cases, terms such as 'Pan-regions', 'vital spaces', 'world-island', 'heartlands' and so on became embedded in broader debates of Portuguese and Spanish politics and culture. In particular, references to 'geopolitics' became entwined with representations of colonialism, imperialism, regeneration and Spain's (and more particularly Portugal's) special 'missions' or pre-eminence amongst the colonizing nations (as the first trans-oceanic European powers), as well as notions of Christian mission in the supposed vanguard against the 'uncivilized', 'heathen' Orient. Later this would be articulated with Cold War rhetorics, in which 'Christian, civilizing' Spain and (particularly) Portugal and its global empire were (re)constructed as vital 'Western-frontlines' of opposition to 'atheistic, uncivilized, oriental' Communist powers and forces. In other words, to return to formal productions of 'geopolitics' in Portugal and Spain, we find complex local renegotiations of broader inter-national flows of ideas. And in Spain particularly, there were substantial engagements with the German geopolitical tradition produced by military (and some academic) writers and published in the early 1940s in the *Revista Ejército* (Army Review) and occasionally in *Estudios Geográficos*.[5] In addition a Spanish geopolitics was developed in the works of the Catalan historian Jaime Vicens Vives who in 1940 published a voluminous study (Vicens Vives 1940), replete with dozens of maps of *España: Geopolitica del Estado y del Imperio* (Spain: Geopolitics of the State and of the Empire) followed a decade later (Vicens Vives 1950) by his *Tratado de Geopolitica* (Treatise of Geopolitics). Vicens Vives' interest had been triggered in part by reading *Zeitschrift für Geopolitik* in the 1930s, and he would later publish an article therein on 'Spain and the Geopolitics of the New World Order' (Vicens Vives 1941). References to geopolitics and 'geopolitical mappings' also permeate his wider historical and journalistic writings, produced through the 1950s and 1960s. However, aside from Vicens Vives' and similar later historical elaborations on the classical tradition (Beneyto 1972), the Spanish military and academic geopolitical texts of the 1940s were not much reworked in subsequent decades. Instead geopolitical discourse migrated into officially sponsored secondary school texts on geography (e.g. Andrés Zapatero 1950) and found expression in some of Franco's notions of the geopolitical significance of Spain to the security of the 'west' following the 1953 bases agreement with the US and the reworking of an anti-Communist crusade mentality.[6] For Franco, the agreement with the USA provided the basis for a celebration of Spain's vital place in the (anti-communist) West. It was this anti-communism that allowed him to re-position himself as a Western ally after the defeat of the (Fascist or Nazi) Axis powers in 1945 (with which he had been closely identified during the Second World War).

However, a wider critical reading of notions of *Hispanidad* reveal a certain proto-geopolitical presence in Spanish literature and in debates on 'Spanish identities' following the loss in 1898 of most of what remained of the empire, notably Cuba and the Philippines. Questions of Spain's relationship to Europe and Iberia

(echoed in Portuguese, Gallego and Catalan literatures), to the old and now 'lost' empire in the Americas and Asia, the nature of Iberian cultural and infrastructural 'backwardness' and the prospects for 'national revival' were all key issues for the group of intellectuals who became known as 'the Generation of 1898'.[7] In addition to largely unreciprocated Spanish overtures to Latin America, backed up by a conservative and anti-communist notion of *Hispanidad*, one aspect of such notions of revival plugged into older Spanish designs on Africa for a new Spanish empire.[8]

In terms of a broad but active relationship between geographical knowledges and a geopolitics (of empire building), Capel (1994) has recently provided a basic account. However, the doctoral thesis by Hahs (1980) provides a fuller account of what became known as the *Africanista* lobby, its institutional expression in the Geographical Society of Madrid, and its subsidiary organizations and their intersection with the Generation of 1898 in the form of the complex personality of Joaquín Costa, a leading figure in both. Echoing other demands for Iberian unity, Costa also advocated a shortcut to Spanish empire through the annexation of Portugal writing in an 1882 booklet on 'Spanish Concerns and the Question of Africa' that:

> Portuguese soil, whether it is found in Europe or in Africa, is the soil of Spain, by the same right and for the same reason that Venetia and Rome were already provinces of Italy twenty years ago [i.e. long before they were incorporated into the Italian state]. I will not tire of repeating it: Portugal is Spain, unredeemed Spain.
>
> (cited in Hahs, 1980: 162)

Ultimately, as Hahs (1980: 303) concludes: 'the africanistas never transcended the limitations of a lobby to become the vanguard of a national movement. Although the africanistas were fully caught up in the Scramble [for African colonies], Spain was not.'

But a revisionist imperial lobby remained in Spain, evident in Franco's utterances, particularly after his victory in the late 1930's and in the early years of the Second World War (see Preston 1993) and in texts such as Areilza's and Castella's (1941) weighty *Reivindicaciones de España* (Spanish Reclamations).[9] After 1945, this was reduced to residual designs on Morocco and the Western Sahara. In respect of this, amidst the outpourings of geographical writings and analysis, Hahs (1980: 12) refers to the highest stage of *Africanismo*: 'its fullest elaboration africanismo asserted that Spain and Morocco constituted a single geographical area (bounded by the Atlas and the Pyrenees) which was distinct from both Africa and Europe.'

This was to be expressed in terms of a discourse of geopolitics which spoke of Spain's 'vital space' which included North Africa, of Spain as an 'exceptional geopolitical entity' beginning at the Pyrenees and stretching to the Atlas, or of the

scope for a space of 'Iberian-Maghreb' empire (Armado 1940; de Reparaz 1924). And it was in suppressing anti-colonial uprisings in Spain's 'new empire' of Morocco that the armed forces forged a new *esprit de corps* and style of campaign and rule (and recruited a mercenary force) that would later be unleashed by Franco on the republic. Franco's uprising against the republic meant that a new '*reconquista*' of the Iberian peninsula was pursued, replete with images from and references to the medieval wars against the Arab-Berber states and the Spanish conquest of the new world, and fortified with techniques of warfare and repression perfected in the colonial wars in Morocco. It was this historical-geographical mythology that, combined with anti-communism, bolstered Franco's sense of Spain's vital geopolitical significance for western defence.

Issues of empire also figure prominently in Portuguese geopolitics, which like that in Spain, overlaps with and takes many of its parameters from a broader colonial discourse.[10] Portugal was both the first and the last European country with a significant overseas empire, a fact that was celebrated within, and shaped, its geopolitical discourse. The scale of the twentieth-century Portuguese empire is graphically indicated in the map reproduced in Figure 6.1. The act of mapping represents a concentration of geopolitical discourse, as many other case studies of, for example, French, Latin American and Italian geopolitical/imperial cartographies (Basset 1994; Dodds 1993; Atkinson 1995) have shown. Such concentrated geopolitics takes on a particular significance when the maps are both bright and striking in appearance and claims, and are exhibited to mass audiences. In Portugal of the 1930s and 1940s, such 'dissemination' took place in an intense series of colonial expositions and also through display of maps in public buildings, schools and universities, and thereby continually (re)informed the national 'geographical imagination'.

The map reproduced in Figure 6.1 was originally produced as part of the *Primeira Exposição Colonial Portuguesa* (First Portuguese Colonial Exposition) held in Porto in 1934. Its sub-title '*Portugal não é um país pequeno!*' (Portugal is not a small country!) reinforces the cartographic expression of geopolitical will. Unlike revisionist maps which appeared from Germany (Herb 1996) and Japan at a similar time – or earlier Portuguese maps such as the *Mapa Cor de Rosa* (Rose Coloured Map) published in 1887 to present Portugal's claim for a territory in Southern Africa from the Atlantic to the Indian Ocean – such a map was, of course, not of any real threat or source of alarm to other European powers.[11] It showed, after all, simply that Portugal was already as great as these other imperial powers, and already as 'great' as any European continental power; and without the need for any new orders or *Anschluss*. Its compiler, the prolific (and ultimately enigmatic) Henrique Galvão, was certainly aware of a wider geopolitical discourse of Eur-Africa, i.e. the idea (significant in German geopolitics) that the world could be divided into three longitudinal imperial zones (Eur-Africa, Asia and the Americas) each having a share of resources and

Figure 6.1 'Portugal is not a small country'
Source: Henrique Galvão 1934: post-1945 reissue

being relatively self-sufficient (O Loughlin and van der Wusten, 1990).[12] Galvão refers to Eur-Africa in a 1936 publication on 'The Empire', where having told us that:

> the History of Portugal reflects successively : with the formation of the kingdom (Twelfth Century), a peninsular *finalidade* [which may be translated as purpose or finality]; with the struggle against the Moors – struggle that assured for Europe the defence against this anti-European element – a European *finalidade*; finally, with our overseas expansion, a world *finalidade*.
>
> (Galvão 1936: 5)

and that having rejected the 'moral and material ruin' of 'liberal politics', making Portugal 'with the best in the vanguard of the civilisations of the world' the country:

123

'offered to many others . . . a healthy example of the resurgence of order, of civilisation in equilibrium and of pure aptitude for the fulfilment of the Mission that our destiny gives to us' (Galvão 1936: 8). Furthermore, this 'agent of Order, of Civilisation and of Christian spirit' (the Portuguese Empire):

> constitutes in economically and politically sick Europe a case of health that is necessary *contar* [to tell, to rely upon and to sum up] . . . because it assures the order of things and persons . . . because it counts on a doctrine and because it is a *force*... because the bankruptcy of Europe before America, demands a spiritual, political and economic reorganization, that cannot possibly constitute itself – except through the formation of a new economic Eur-african continent to oppose the American economic continent. Our geographic position and the extent of our dominions in Africa, mark us a place of revelation in this work of eminently European range.
>
> (Galvão 1936: 8)

With such views echoed by Salazar, it is no surprise to learn that Galvão's map was re-issued in the 1940s and 'profusely displayed in schools and other public services',[13] or that other military and colonial officials noticed that other facets of German geopolitical reasoning might be usefully applicable systematizations for the Portuguese empire (Rita 1944). However, such Portuguese references to German language geopolitics are relatively insignificant when the sheer scale of codifications of Portuguese destiny and meditations on imperial mission and practice are taken into account.[14] In short, there was no real 'need' for Portugal to 'import' foreign geopolitical notions, in the way that, for example German geopolitical writings became influential in Italy or, soon afterwards in many Latin American states. Portugal had developed an original geopolitical discourse, which certainly belonged to the wider fascistic and imperial *Zeitgeist*, but one that registered considerable individuality, and which drew upon long-standing Portuguese ideas of mission and destiny.

For in the terms of Guimarães' excellent critical study of the ideology of Salazar's *Estado Novo*: 'The coherence of the system supported itself in a very particular geographic concept, that imagined that the world only existed through the means of the discovery by the Portuguese' (Guimarães 1987: 111).

So whilst the landed and merchant elites continued to reap most of the benefits, Portugal too could continue to play out notions of unique and heroic destiny on the world stage (see Clarence Smith 1985). It had developed a geopolitical discourse codifying this and a practical geopolitics of empire and official neutrality, steering (a sometimes precarious) course between the allies and the axis whilst selling vital raw materials to both.[15] This discourse and its peculiarities will be

described later. Suffice to say here that it persisted and was elaborated after the Second World War. For like that of Franco, Salazar's regime did not crumble (as many opposition figures had hoped) after the eclipse of European fascism in 1945. Indeed, a certain international recognition and legitimacy was extended to both regimes through participation in the American-led Atlantic alliance. Portugal was amongst the founder-members of NATO in 1949, and when the Cold War deepened in the 1950s, Spanish territory was formally integrated (through the US-Spanish Pact of Madrid in 1953) into the western military alliance system (although it should be added that owing to its close war-time association with the axis powers, Spain was still not formally included in NATO).[16]

In the new post-1945 world order, Portugal was rapidly able to locate itself quite securely (and with less initial difficulty than Spain, whose association with the defeated Axis powers was closer). In doing so, the imperial geopolitical formulations of the 1920s, 1930s and 1940s mutated into some new and rather fascinating forms. In the following sections of the chapter I shall attempt to unpack these and chart their increasingly contradictory elaborations up to and beyond the fracture generated by the leftist military coup of 25 April 1974. There follows some general remarks on the latter, including further comparisons with Spanish discourses, and the sketching of some further research agendas.

Integrar para não entregar
(Integrate so as not to hand over)

In terms of what might be termed an evolving 'geopolitical code, (after Taylor 1990, 1993), and at the risk of a certain schematization, a number of features became evident in post-1945 Portuguese geopolitical discourses.[17] Perhaps the most central was the preservation (indeed consolidation) of the empire. Portugal's long epoch as the nerve centre of a far-flung maritime empire already laid the basis for rigid centralism. However the long relative economic decline and the rise of other European powers meant that this was challenged. Brazil was lost in the 1820s when European politics, in the form of the Napoleonic wars, imposed upon the Portuguese project. What remained was the African empire, the (re)partition of Africa at the end of the nineteenth century having left Portugal with the territories of Angola and Mozambique in southern Africa, Guiné in West Africa, and the archipelagos of Cape Verde and São Tomé and Principe. In addition, Portugal retained some relatively small territories in Asia: Goa, Damão and Diu, plus Macau and East Timor.

Once the post-1945 geopolitical 'world order' had 'stabilized' (if that term can be used to describe the balance of nuclear terror and the continued dynamism of broadening formal decolonization and state formation in the periphery), a specific feature of Portugal's code included integration into the Atlanticist system of

NATO: of which – as has been noted – Portugal was a founder member. At the formation of the alliance Salazar unsuccessfully sought to extend the area included in the provisions of the NATO treaty to include the Portuguese empire; and NATO membership provoked a series of debates in the military concerning the mode of integration of Portugal into the new order, none of which however shifted the centrality of the empire to the geopolitics of the *Estado Novo*.[18] NATO membership also represented a certain continuity of Portuguese orientation, in terms of what Teixeira (1992: 124) refers to as 'a permanent and privileged alliance with maritime power in the Atlantic which was traditional in Portuguese foreign policy', whilst it simultaneously registered that the main pole of that power had now decisively shifted to the other side of the Atlantic. Furthermore NATO membership set in motion a rapid technical modernization of the armed forces, and when modern nationalist movements began to contest Portuguese rule (from the early 1960s) in Africa and East Timor, NATO arms and material (and a degree of diplomatic support) were vital in the Portuguese countermeasures (Crollen 1973; Minter 1972).

In turn, a significant feature of the geopolitical 'structure' put in place after 1945 was that the overseas colonies became overseas provinces of Portugal. This arrangement was formalized in 1951, with the publication of a revised Constitution which replaced the term colonies with that of *provinces*. In the terms of a book on the 'Overseas Politics of Portugal' published by the Lisbon Geographical Society a few years afterwards:

> we opted, in 1951, for the traditional designation of 'overseas provinces', which is considered better in confirmation with the principle of the unity and with the contracting of an intimate co-operation between all the populations that constitute the nation and between all the parcels of the Portuguese territory.

In this sense articles 134, 135 and 136 of the current constitution establish:

> Art. 134 The overseas territories of Portugal . . . are generically denominated as provinces and have been politically-administratively organized adequate to the geographic situation and the conditions of the social level.
> Art. 135 The overseas provinces as an integral part of the Portuguese State are joint between themselves and with the Metropole.
> Art. 136 The solidarity between the overseas provinces and the Metropole includes especially the obligation of contributing to an adequate form for assuring the integrity and defence of the whole Nation and the ends of national politics defined in the common interest by the organs of sovereignty.'
>
> (Bahia dos Santos 1955: 152–3)

In their own peculiar legalistic terminology, these articles express one of the core re-configurations of Portuguese imperialism after 1945. It can be summarized in the poetic line '*integrar para não entregar*' (integrate so as not to hand over). The 'colonies' were redesignated as 'overseas provinces' and so became legally rebound to metropolitan Portugal. Salazar became fond of saying that Angola or Mozambique were constitutionally just like other provinces of Portugal, such as the Algarve or Lisbon. The logic of this was an attempt to defuse calls for decolonization, for if the colonies were simply 'provinces', Portugal was not a colonial power, it was 'different' (from 'other' colonial powers), a 'pluri-continental power', with territories (provinces) throughout the world.

As de Figueiredo noted in an insightful critique:

> There followed another constitutional expedient, the success of which only the obscurity of Portugal and the 'cover' provided by Western diplomacy can explain. By an amendment introduced in 1951, the 'colonial empire' underwent a thorough verbal 'revolution' whereby all colonies were regarded as overseas 'provinces' of a single country. The words 'empire' and 'colonies' suddenly disappeared, which was no small task since they were printed on millions of items of stationery, coins and notes, government institutions, trademarks and the names of private concerns.
>
> (de Figueiredo 1975: 206–7)

De Figueiredo goes on to note how the 1951 revision was in fact a regression to the pre-Republican notion of a formally unitary empire, but one which had a distinctly contemporary function. For all this meant that Portugal could later claim that it was not a power that must (like other European empires) ultimately be forced to decolonize, since, as Salazar explained in a 1958 interview (Salazar 1960a): 'there are no Portuguese possessions, but [rather] footprints of Portugal disseminated in the world. In Lisbon, in Cape Verde, in Angola, in Guiné, in Timor or in Macau it is always the "*Pátria*".'[19]

Again Portugal seeks to separate itself as prior too, more noble than other imperialisms, whilst importing ideas from 'later' European imperial powers (both Britain in Ireland and France in Algeria had sought to redesignate colonies as 'provinces of the mother-country'). As contestation of Portuguese colonialism grew (from independence movements and the increasingly vocal independent African and Asian states), the redesignation of colonies as 'provinces' was to become more significant. In terms of the representation of broad geopolitical codes/orders' it also featured in an attempt to displace a potential geopolitical North–South fracture around issues of colonization/self determination–decolonization onto an East–West geopolitics of communism/anti-communism. Imperial policies and geopolitical discourse reinforced each other, in part through the figure and utterances of Salazar.

For example, Salazar had described the Atlantic pact in 1949 as 'the symbol and the expression of a new crusade: that of the defence of the Western and Christian civilisation' (quoted in Teixeira 1992: 121). Moreover, an article published in 1956 in the Proceedings of the Military Naval Club (a frequent forum for the codification of geopolitical discourse) on 'The Geopolitical Importance of Portugal for the strategy of the Free World' indicates how this was evolving in the mid-1950s. Its conclusions are as follows:

> In a tentative conclusive synthesis on the geopolitical factors of Portugal presented and of interest for the Strategy of the Free World, it is legitimate to affirm that:
>
> 1 The internal and external Portuguese politics constitute a stable element which integrates itself harmoniously in the high designs of Peace of the Free World and is an essential collaborative factor in the building of its security.
> 2 The spiritual power that irradiates the Portuguese universal history through the Lusitanian community represents a competitive current of major importance in the adhesion and co-operation that must exist between the people of the Free World.
> 3 The vast and promising Portuguese Euro-african resources represent a fundamental economic potential/power to the survival and sustenance of the defensive and attack capacity of the Free World, when integrated into its geoeconomy.
> 4 Portugal, by its heritage and historical continuity, and disposition of an impressive maritime geographical position in the Atlantic, Indian and Pacific [oceans] in the geostrategic maritime-world plan of the Free World, today more than ever valorized by the effect of the accentuated confrontation of the eastern political influence on the world, in particular in the Indian [ocean] and Asian Pacific.
>
> (Comprido 1956: 268)

Comprido's essay appeared with seven maps to demonstrate his themes (From which figures 6.2 and 6.3 are drawn). The rest of his article reworks some older ideas – the notion of 'Euro-Africa', and of a 'World Island', Portugal as 'Christian civilisation' – and anticipates other themes that were becoming more prominent within wider Western geopolitical discourses (Ó Tuathail, 1992), notably the idea of strategic choke points (especially the Cape of Good Hope) and of a coherent (expansionist) Soviet strategy for world domination emanating from the Asian heartland. These general points are unremarkable, except perhaps to note that Comprido saw that the combination of Portugal's 'political stability and order, as much in the metropole as in the overseas' (Comprido 1956: 235), and its strategic

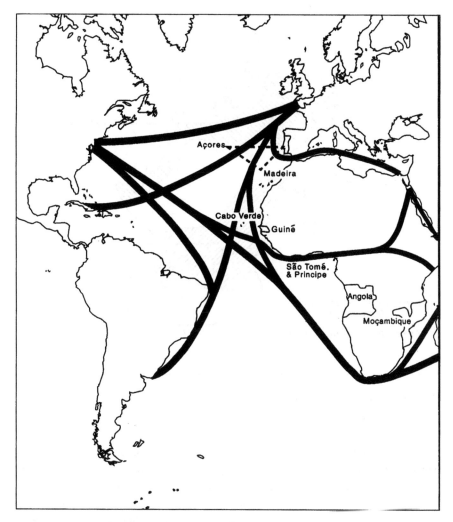

Figure 6.2 'Portugal and the Atlantic circulation'

Source: adapted from Comprido 1956: 258

location and the distribution of its overseas provinces, meant that it had to play a particularly vital ('a position of incontestable value') role in the defence of the 'Free World'.[20]

Four factors are held to be operating here. First the barrier (to Soviet invasion) of the Pyrenees, in some combination perhaps with the peculiarly 'Christian' and ordered (i.e. fascist) nature of Iberia. Second the historic Atlantic destiny of Portugal made it a core (perhaps the core), of the Atlantic Treaty (the 'Ocean of Western Civilisation' as Portuguese writers were fond of calling it). Third Portugal's

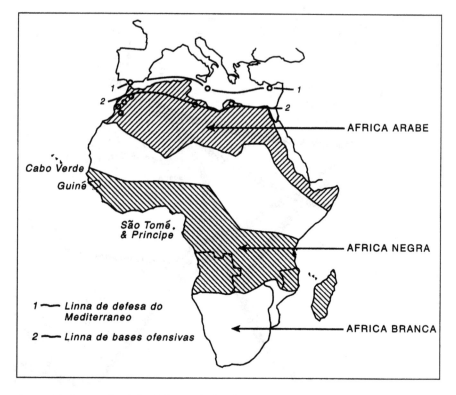

Cabo Verde
Guiné

São Tomé.
& Principe

AFRICA ARABE

AFRICA NEGRA

AFRICA BRANCA

1 — Linna de defesa do Mediterraneo
2 — Linna de bases ofensivas

Figure 6.3 'Portugal and the defence of Africa'
Source: adapted from Comprido 1956: 261

Euro-African dimension not only gave the West access to all kinds of strategic minerals (Comprido makes an extensive list) and scope for defence in depth, should Europe, in a worst case scenario, be occupied by the Soviets, but also 'the possibility to mobilize around 1.5 million blacks in our African territories, whom once organised by white officials and commanders constitute a human potential of high value for military ends' (Comprido 1956: 241–2). Finally, not only does Portugal control (through the Azores and Madeira) a vital strategic triangle in the Atlantic 'through which pass the five most important arteries that give life to Western Europe' (ibid.: 256) and the vital power of the 'luso-Brazilian sea' (ibid.: 232) of the South Atlantic between Brazil and Angola, but Portugal is within the World-Island:

> geostrategically a great *Arquipélago* dispersed through three oceans, with a territorial extent of 2,171,733 square kilometres, of which 4.2 per cent constitute the *Mother Island*, in the south-west extreme of Eurasia, 94 per cent flank southern Africa and the rest are disseminated through the Indian sub-continent, far-East and Pacific. We are then [in case the

reader should be in any doubt], an historically European country; culturally and spiritually western; politically, geographically and economically
maritime.[21]

(Comprido 1956: 252)

There is much more in this essay and in many other similar ones that appeared
– mostly it seems for internal consumption and/or written as part of official training programmes for military and naval officers – in Portuguese military
publications at around this time (Crespo 1956; Prior 1951, 1956; Sanches 1963)
than I can do justice to here.[22]

Instead I will focus on two themes, which will allow us to move into broader
comparative issues. The first is to briefly note that whilst the term geopolitics
tended to be avoided in much of Western Europe and North America in the
1950s and 1960s (while the discourse was displaced elsewhere, into expanding
disciplines such as international relations and strategic studies), in Portuguese
military publications utilization of the term, and explicit and mostly approving
reference to German geopolitics, was quite frequent after 1945. This does not
represent some uniquely Portuguese trait, as a glance at accounts of contemporary Latin American geopolitics mentioned earlier (such as Child 1985) would
indicate. It does however indicate something of how a pre-1945 order was
fossilized in Portugal, and how easily it adapted to the 'anti-communist' world
order of the Cold War. The variety of diplomatic tricks and re-inscriptions associated with this have been quite widely researched and commented upon by
Portuguese historians (Antunes 1986; Teixera 1992). Nor of course, was
Portugal (and later Spain) alone in being incorporated into the economic and
discursive space of the 'Free World' despite their impeccable anti-democratic
credentials. What has been less noticed is the way that in one of the Iberian states,
the process involved the re-production of certain pre-Cold War geopolitical
tropes: 'Heartland, Eur-Africa, World-Island'.

This relates to the second theme, which concerns the ways that the post-1945
geopolitical code of Portugal in particular articulated issues of north–south with
those of east–west. Already evident in essays from the mid-1950s, such as
Comprido, this took a rather more central place in many subsequent and much more
widely disseminated publications in the late 1950s and the 1960s, indeed up to (and
in a certain highly attenuated way, beyond) the reorientation set in place in 1974. We
can see a nascent version of this at work in Comprido's (1956) essay, when he links
the long-established historical myth of Portugal as a Christian civilizing and evangel-
izing power in Africa, with a notion that those territories which 'benefited' from this
mission had become of vital strategic significance to the defence of a wider
'Christian' western-civilization. In turn, this displaces the critique of imperialism
onto the (atheistic) East: it is the East which is the real imperial force. In this script,

it is only the Communist East which is expansionist, dangerous, aggressive and imperialist. Portugal and its 'overseas provinces' are, by contrast a force for 'civilization' and an expression of (divinely directed) historic 'destiny'. Franco and his apologists had made parallel claims in the 1940s and a variety of British, French and American commentators developed versions of this during the Cold War.[23] But it was more central to Portuguese discourses. At once we have operating here an attempted displacement of already growing condemnations of continued Portuguese colonial domination onto an 'other', and in such a way as to (at the same moment) embed Portuguese colonialism into the logic of global anti-communism.

Highly revealing too of what was to come, was Comprido's raising of the danger to the West of an emergent 'Afrasia' (1956: 227–9). In particular, the emergence of formally non-aligned states in what elsewhere was beginning to be called the 'Third World' was seen to pose a specific danger. Given 'the growing racial hostility of the movement against the West, it is possible that they [the non-aligned states] would approximate more to the communist bloc' (ibid.: 229). In the late 1950s and 1960s, this idea of anti-imperialism as simple 'racial hostility' (or rather its antithesis, the racially harmonious, Christian Portuguese overseas) would be mixed with the anti-communism and with the imperial mythologies of a civilizing, pluri-continental power to constitute a developed geopolitical discourse of Portuguese world-destiny.

The writings of military and colonial officials , such as Moreira (1955) and Neto (1963, 1968a, 1968b) register this. Moreira in particular in an article published in the bulletin of the Lisbon Geographical Society codifies a set of responses to the UN and emerging Third Worldism (the non-aligned movement and similar alliances between national liberation movements and newly post-colonial states of the South) on the lines already discussed. His response to the anti-imperialist challenge of the path-breaking conference of independent African and Asian states held at Bandung (Indonesia) in 1955, is to condemn it as 'racist' whilst differentiating Portuguese 'colonialism with a [civilizing] mission' from other colonialisms. Other colonial systems presumably lack such divine sanction and to contrast them with the avowedly 'benign' Portuguese missionary colonialism, Moreira calls the others 'vital space colonialism'. We are never really told (at least in Moreira's article) exactly what this mission of Portuguese colonialism is; and exactly why it is different from 'vital space' colonialism (or even who exactly practises the latter).

For those outside the inter-textual myths of the Portuguese empire, Moreira's article makes for a seemingly bizarre/eclectic mixture of contradictory propositions:

> the wholesale condemnation of colonialism at Bandung by Powers them
> selves members of the UN is, in a sense, a revolt against the UN itself. It is,
> in fact, a symptom of a new racialism in which the white man is the enemy.
>
> (Moreira, 1955: 15)

Given the 'chaos' that would follow any decolonization and given the fact that postcolonial states would inevitably find themselves mere 'satellites' of communist power, the only answer for Africa, for Europe, for what (once again) is called the Free World (*O Mundo Livre*), therefore lies in the 'rehabilitation of colonialism': under Portuguese expert guidance. This:

> is not a matter of invocation of heroic deeds, much as we are proud of them, but more simply of showing that we are saving for the free world an assemblage of peoples living together peacefully and voluntarily, assuring peace, work and prosperity, without recourse to any process of police or military repression. . . . The traditional vocation, of the essence of the Nation . . . indicates indisputably [that] the Portuguese [are] to initiate and take the ideological lead of the movement of regeneration of colonialism, in this way invigorating an ethics that will assure the ideological mobilisation of the west, that will eliminate the philosophy of defeat in which translates anticolonialism, that will realise the principle of the equality of humankind, that guarantees the solidarity of Africa with Europe. . . . I do not know people better indicated than the Portuguese, in the face of the standard of racial hatred that was raised in Bandung, to carry anew to the world the message of the equality of humankind. A task whose less important result will be one of creating a renaissance of faith and hope in the hearts of men, a dignified task of the Nation.
>
> (Moreira, 1955: 15)

Emerging here is the notion of a non-racist (indeed anti-racist) Portuguese overseas mission. This ideology was to take a more fully-fledged form in the 1960s, largely through state promotion of the works of the Brazilian sociologist Gilberto Freyre. Freyre codified his analysis of 'Portuguese Integration in the Tropics' (as his 1958 book was entitled) through the notion of 'Lusotropicalism'.[24] Having first introduced the reader to the geopolitical significance of the tropics, in particular in terms of determining which power would become the world leader, Freyre explains how the Portuguese have already developed a uniquely 'Lusotropical civilisation' (the highest stage of a wider set of 'Hispanotropical civilisations'). And particularly in the context of the 'tropical spaces as possible zones of expansion' (Freyre 1961: 30) of the imperial forces of the USSR and the 'Anglo-Americans':

> the organisation of the Hispanotropical civilisations into a trans-national system of culture, economy and politics appears as a necessity which I would not name geo-political, as if their organisation resulted mainly from situations called natural or geographical.
>
> (Freyre 1961: 31)

It is rather the result of 'an interpretation that exists in practice in a fashion that it is not exaggerated to call triumphal' (ibid.: 32).

It is important not to overemphasize the coherence or significance of these statements. More traditional anti-communist motifs remained as important as before; particularly after emerging resistance movements in Angola, Mozambique and Guiné looked to the USSR and its eastern European and Cuban allies and China for material support in the 1960s. The Soviet-made AK-47 rifle became a symbol of resistance and liberation for the nationalist movements in Portuguese Africa, and armed anti-colonial revolt was becoming widespread in parts of Angola, Guiné and Mozambique by the end of the decade. Portugal responded with counter-insurgency schemes which drew upon the American experience in Vietnam. Like the Americans, however, they could not defeat the guerrillas in the countryside.

Therefore amongst Portuguese condemnations of the African nationalist movements (which Portuguese government and military commentators referred to as 'terrorists, communists and tribalists'), it is possible to find more works on the strategic value of Portuguese Eur-Africa in particular, or of the 'ultramar' [overseas] in general, to the defence of the Christian West (Crespo 1956; Sanches 1963; Chassin 1961; Magalhães 1972, 1975). Nevertheless, Freyre's ideas provided another rationalization in the range available to the regime: and one that played a particular role in propagandizing amongst liberal circles abroad and (more importantly) at the UN, where Portugal fought a long and virulent battle against the emergent Third Worldist bloc. In total however, Freyre's eroticized 'Lusotropical' geopolitics constitute an extraordinary text, whether in its poetic Portuguese original (1957) or in the 1961 English translation, both produced under the auspices of Agência Geral do Ultramar.[25] For example:

> by the direct, living experiential knowledge of the tropics, both in the East and in Africa or America; and that the science to which they gave their contribution as pioneers, having been, in general, the science of space–time . . . in particular the science of tropical space–time . . . a new type of civilization was commenced for which a characterization as Lusotropical is suggested, in view of its singularly symbiotic character of union of European with Tropical – union that in no other European was ever so intense and symbiotic'.

> (Freyre 1961: 41)

Specifically:

> miscegenation should be considered inseparable, from the beginning, from any and all Portuguese effort in the tropics. Inevitably it appears sooner or later, with greater or lesser intensity, in all such efforts . . .

Lusotropical civilisation, when considered bio-socially, is no more than this: a common culture and social order to which men and groups of diverse ethnic and cultural origins contribute by interpenetration and by accommodation to a certain number of uniformities of behaviour of the European and his descendant in the tropics – uniformities established by the Lusitanian experience and experimentation.

(Ibid.: 47–8)

Freyre was able to draw upon and make an original adaptation of the mythologies of Brazilian nationalism. His referents, the socio-economic-racial formation that was Brazil by the middle of the twentieth century and the Portuguese colonial system, exhibited few of the ritualized taboos of racial contamination that had developed in certain formats of, for example, the US, British and South African settler-colonial formations. This allowed Freyre to project certain elements of Brazilian and Portuguese colonial society (notably the extent of black–white sexual relations) as a singular and historic reality of some kind of global significance. In so doing, he sidesteps vital questions of agency and power, and avoids any consideration of the hierarchical racialized class relations which from other (more critical or subaltern) vantage points were glaringly obvious.[26]

However, Freyre's work, as its English translation (of which 5,000 copies were produced initially) attests, became part of a wider propaganda effort of considerable scale. A set of texts for domestic consumption (Neto 1963), sometimes mixing Lusotropicology with other more 'classical' geopolitics as in the university extension course produced by Neto (1968b), and for military debate (Tavares 1964) were accompanied by a number of works published in English and designed for external consumption, particularly in diplomatic circles (de Andre 1961; Ministry of Foreign Affairs 1970) as well as numerous statements by Salazar himself (1960a, 1960b, 1961, 1962a, 1962b).

It is not possible to do full 'justice' to these writings and speeches here. However, many of the same themes keep cropping up, forming a kind of geopolitical-national-colonial intertext. First, 'lusotropicalism':

In all Portuguese territories, contrary to what has happened in most of those countries who regard themselves as paladins of the independence of peoples, racial or religious differences have never given rise to any discriminatory incident or measure.

(de Andre, 1961: 48)

Second, denial and projection of the self onto the other, at the same time as exploiting the many inconsistencies/contradictions of postcolonial states. For example in 1956, the Portuguese representative at the UN berated 'Indian colonialism' in

'Kashmir, the Andaman Islands, Bhutan and Sikkim', whilst at the same time stating that: 'there is no colonialism in Goa because politically as well as legally, Goa is an overseas province and is an integral part of the Portuguese nation: much as East Pakistan is an integral part of Pakistan' (statement made by the Representative of Portugal at the sixty-first Plenary meeting of the General Assembly on 6 December 1956; in Ministry of Foreign Affairs 1970: 277).[27]

Third, a re-working of Eur-Africa and the East–West/North–South displacement referred to earlier, evident in the classification on the part of Marcellano Caetano (Salazar's successor as prime minister, following the latter's incapacitation by a stroke in September 1968) of the Soviet Union as imperialist, and Portugal as the 'frontline' against Soviet imperialism/colonialism:

> The security of countries cannot today be defended in their frontiers. The nations are integrated in great spaces. . . . The liberty and independence of the Western European countries play themselves out not only in Europe itself but also in Africa. This is why we have to defend Guiné. In our interest it is certain, but also in the interest of Western Europe and of the Americans themselves.[28]
>
> (Caetano 1973: 12)

All this, expressing the singularity and unity of Portugal: 'we are a pluri-continental and pluri-racial country, with only one spirit, only one government and only one flag' (Caetano 1973: 71). And finally, the conspiracy against it:

> The various so called liberation movements that we combat in Guiné, in Angola and Moçambique were formed abroad with leaders that foreigners sustain and support and it is from foreign territories that they launch the attacks and send guerrillas. A vast organisation of African, Asian and socialist countries conspire against Portugal.
>
> (Caetano 1973: 41–2)

In short, much more could be said on how this geopolitical framework articulated with Portuguese culture, society and politics and how these have been reworked in the lead up to – and following on from – the military coup of 1974 which brought an end to the authoritarian order and the last significant European overseas empire. Lower-ranking officers in the military, who realized that the colonial wars were unwinnable and that the regime now led by Salazar's one-time deputy Caetano would neither voluntarily democratize nor decolonize, staged a coup in April 1974. The authority structures of the empire collapsed with the old regime and within eighteen months all of the empire (with the exception of Macao which China was unwilling to re-absorb) was gone. The African colonies gained independence and East Timor was occupied by neighbouring Indonesia. Despite the publication of crops of memoirs and interpretations over the last twenty-odd years, in terms of an analysis of geopolitical discourses and relations the scope for further critical analysis of the *Conjuntura Nacional* (National Conjuncture) – as one of its key texts termed the colonial question (Spínola, 1974) – remains open.[29]

Indeed in *both* Portugal and Spain, now established democratic polities and members of the European Union, there remain frequent direct references to the 'geopolitical' in a variety of publications. In part this is a reflection of the general *diffusion* or *dissemination* of the term. As O'Tuathail (1996: 16) notes, geopolitics sometimes stands in as 'an appealing and handy summary term for the spatiality of modernity as a whole'. It also gained an increased or recovered prominence in the 'statecraft' language of US 'security advisors' Henry Kissinger and Zbigniew Brzezinski in the 1970s (Hepple 1986; Sidaway 1998) which were noticed in Portugal and Spain. In Spain, one will still find the occasional newspaper article in the national daily *El Pais* drawing on the geopolitical cannon of Mackinder *et al.* These are mostly of a rather conservative genre and more often than not turn out to have been produced by functionaries in the military or foreign relations research institutes organically linked to the state.

In Portugal, less able to lose themselves in 'the labyrinth of nostalgic myths' of the empire, new book-length studies on Portuguese geopolitics by de Almeida (1994), Alves (1987) and Sacchetti (1987, 1989), and articles on geopolitics in newspapers and in the monthly journal *Estrategia*, lack something of the sparkle of pre-1974 texts.[30] Instead they tend to focus on more mundane themes such as the 'importance' of the 'Portuguese strategic triangle' (see for example, figures 6.4 and 6.5 from Alves, 1987), and recycle heartland–rimland discourses of conservative British-American writers such as Cohen and Mackinder and Spykman (Ferreira 1992).[31] Likewise the works published in the conservative Oporto-based journal '*Africana*' are mostly of interest for a backward looking nostalgia, mixed with references to the New World Order (e.g. the prolific de Carvalho, 1991, 1992, 1998).[32] But if we leave behind such (mostly) narrowly conceived and reactionary accounts to consider the presence of the 'geopolitical' in broader discursive terms, it is not difficult to recognize a reworking of the expressions of Portuguese identity. Today, as Portugal is re-imagined as a *European* society, this takes a number of formats. However as Almeida has pointed out, since 1974:

> the number of books on the subject of identity and related topics is staggering [Almeida reviews some of the most influential ones]. Dozens of volumes dealing with the theme have been published and countless times one encounters newspaper articles addressing the same topic.
>
> (Almeida 1994: 156)

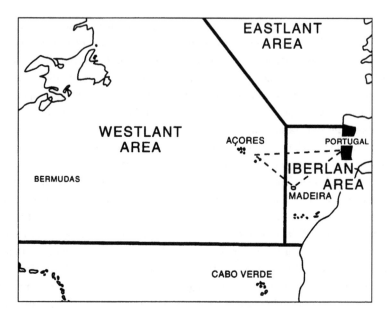

Figure 6.4 'Portugese strategic triangle'
Source: adapted from Alves 1987:128

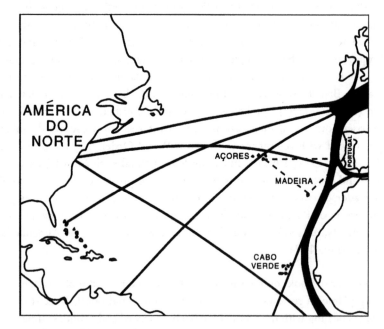

Figure 6.5 'Lines of navigation'
Source: adapted from Alves 1987: 128

Given the establishment of a formal Portuguese-speaking commonwealth-type association (see Cahen 1997; Wise 1995), ongoing European integration and the wider post-Cold War and *fin de siècle* era, such debate has not tended to diminish. There is no scope here to propose a general hypothesis concerning the mode of re-articulation of Portuguese and wider Iberian identities into a European orientation (Sidaway 1996). Suffice to say that in both Spain (Anon. 1996) and Portugal (Halliday 1992; Madureira 1995), this is happening through the (re)iterations and displacements of prior geopolitical moments and expressions. But this is another story. And the tale of new Iberian inscriptions, new differences, of another round of new orders and of the strange *geopolitics* of our own times must await other studies.

Acknowledgements

In addition to the general information offered by António de Figueiredo, I also wish to thank staff at the following Lisbon institutions who offered assistance and information in my ventures into the archives; *Biblioteca/Museu 'República e Resistência', Arquivo Histórico Militar, Biblioteca da Sociedade de Geografia de Lisboa, Instituto da Biblioteca Nacional e do Livro*. None of the people at these institutions are in any way responsible for my subsequent interpretation of the materials that they helped me retrieve. All translations from the Portuguese and Spanish originals are by the author, unless indicated otherwise.

Parts of this chapter are taken from a longer unpublished (and seemingly ever-lengthening) draft manuscript on Portuguese geopolitics. For comments on earlier versions of the latter I am grateful to David Atkinson, Jerry Brotton, Simon Dalby, Klaus Dodds, Les Hepple, Marcus Power, Richard Robinson, Gearóid Ó Tuathail, Jonathan Turton and Jo Sharp. Les Hepple also usefully supplied some references and copies of hard-to-find texts.

I would like to dedicate this chapter to the memory of all the British volunteers who died in Iberia fighting the forces that my chapter seeks to describe and understand.

Notes

1 In his own work on theories of aesthetics, ethics and ideology in the context of analysing the geopolitics of the influential American statesman George Kennan, Stephanson (1989) provides incisive observations on America's relationship with Portugal and the place of Portuguese geopolitics in the wider Cold War and Western context. This topic is developed in Sidaway (1999).

2 As noted in Dodds and Sidaway (1994) a critical 'geopolitical economy' did emerge in academic writing in the 1980s. But it remains underspecified, despite the suggestive text by Agnew and Corbridge (1995). On the material frame in which formal geopolitical discourse arose, Kearns (1984) is helpful.

3 Harrington (1994) prefaces his doctoral study of concepts of national identity in the Iberian Peninsula 1874–1925 with the observation that: 'in the last thirty years of the nineteenth century, and the first two decades of the present one, *el iberismo* flowed like a fresh stream under the frozen façade of the social institutions in Spain and Portugal. It existed as both a grand intellectual dream and a practical political alternative'. For

further details on these themes, see the references listed in note six and Preston (1993) on Spanish policies towards Portugal and the NATO issue.

4 Despite its promising title, the article by Roucek (1964) does not deal with Portuguese geopolitical discourses. Texts such as Bruce (1975) and the collection edited by Cahen (1994) are suggestive, but to my knowledge there is no comprehensive or systematic critical account of Portuguese geopolitical discourse. An otherwise fairly comprehensive account of '*Western* [European] *Geopolitical Thought in the Twentieth Century*' (Parker, 1985) makes no mention of Portuguese and Spanish traditions.

5 For details, see the review by Bosque-Maurel, Bosque-Sendra and Garcia-Ballesteros (1992); and the broader and more critical ones by Reguera (1990, 1991, 1992) and Raffestin, Lopreno and Pasteur (1995).

6 On the crusade mentality, see Preston (1993: 625–54) which is worth comparing with Southworth (1963). Marquina (1986) and Hughet (1997) are also useful analyses of discourses about Spain's key role for the defence of the Mediterranean and the West. On the 1953 bases agreement, see Viñas (1981). For a primer on subsequent US policies towards Spain, see Hadian (1978).

7 See Alexandre (1996), Lopes (1994), Ehrlich (1998), Harrington (1994), Lopes (1994) and Matos (1997) for studies of the literatures. For accessible accounts of proto-geopolitical discourses within the generation of 1898, see Cortada (1980), Mata Olmo (1995–96) and Pike (1971). On the historiography of 1898, see Moradiellos (1993).

8 Chapter Seven of Grugel and Rees (1997) contains a useful summary of and further bibliographic guide to, the record of *Hispanidad*. See too Cortada (1980) and Pike (1971). The regime of Primo de Rivera (1923–1931) which in many ways anticipated that of Franco also: 'adopted an ultranationalist, paternalistic variety of hispanamericanismo. A special Latin American department was created at the Foreign Ministry, new financial institutions were set up to encourage trade, and vast sums were spent on a grand Exposición Iberoamericana, held at Seville in 1929. The Latin American republics were assiduously courted at the League of Nations, a policy which resulted in the adoption of Spanish as one of its official languages' (Powell, 1995: 14).

9 José Maria de Areilza later became Spanish ambassador to Argentina and France, before splitting with Franco in the mid-1960's, advocating a conservative monarchism. Fernando María Castiella became Franco's Minister of Foreign Affairs from 1957 until 1969, during which time he pressed for British withdrawal from Gibraltar. Whilst he did manage to get the status of Gibraltar raised at the UN's decolonization Committee, this had no real impact on British policy.

10 On the colonial genealogies of much geopolitical discourse, see Agnew and Corbridge (1995) and Ó Tuathail (1996).

11 See Nowell's (1982) valuable study.

12 Galvão (1895–1970) was a remarkable character. Early in his life he exhibited a certain rebelliousness and was expelled from various schools. Later however, he supported the post-1926 military dictatorship and rose to positions of responsibility in the apparatus of the *Estado Novo*, including at various times governorships of cities in Angola, director of National Broadcasting (1935–1941) and organizer of the 1934 Porto Colonial Exposition and of the 'colonial section' of the 1940 Exposition of the Portuguese World. He also compiled several texts and monographs about the empire and a number of plays and novels set in Africa (one of which won the 'prize for colonial literature'). Later in the 1940s however, his rebelliousness resurfaced and he

became involved in a number of polemical accusations against other colonial officials, whom he (probably correctly) accused of corruption and incompetence. He presented a report to a 'secret' session of the National Assembly which condemned forced labour in the colonies. Eventually this dissent led him into conflict with the regime and in 1951 he was jailed for his part in plots against Salazar. He escaped from prison in 1959 and became a highly active figure of opposition involved in several anti-Salazar conspiracies during a long exile, variously in Argentina, Brazil, Venezuela and Western European capitals. Galvão died in Brazil in June 1970, just a short time before Salazar. Basic bibliographical details are provided in Ventura (1994). Galvão remained an advocate of some kind of Portuguese-colonial federation to the end; a theme elaborated (amongst other things) in his posthumously published book *A minha luta contra o Salazarismo e o Comunismo em Portugal* (My struggle against Salazarism and Communism in Portugal) (Galvão, 1975). For a biography, see Ventura (1994). See too de Figueiredo (1975) who speculates on what led Galvão into rebellion against the regime, and Raby (1988, 1994) for a wider study of exiled opponents.

13 Antonio de Figueiredo, personal communication 2/8/94.

14 It would be impossible for a preliminary study such as this to produce any kind of comprehensive survey of Portuguese colonial discourse. The catalogue of the *Sociedade de Geografia* contains thousands of articles, reports and monographs from the 1920s, 1930s and 1940s, which would define the category. A history of the Society is provided by Guimarães (1984). Her account is usefully read in context with Pereira's (1986) description of the development of anthropological expertise. I would like to cite just two articles from the library of the *Sociedade* here, which I believe (based admittedly on a more or less random survey) might be thought to typify the expression of the genre in the 1940s; Sampayo e Mello (1940–41) entitled 'Of the interventionism of the dominators in the social and economic organization of the dominated' and Pio (1944–45) entitled 'Socio-economic bases of a politics of racial contact'. The speeches and statements of Salazar from this period could also provide numerous examples. But since the collection of these runs to six volumes, and given that I shall draw upon them later in this chapter, I shall refrain from citing him in support of my argument here. However, aficionados might start with volume three of his discourses and political notes. His 11 March 1938 address to Legionnaires (Salazar 1943: 21) entitled 'We are a force destined to win' would be a suitable entry point. For some reflections on Salazar's life and deep conservatism, see Kay (1970), de Figueredo (1975) and Newitt (1981).

15 See de Figueiredo (1975) or Telo (1991) for a detailed analysis.

16 The signature of the Pact of Madrid was accompanied by an outpouring of official statements and media documentation of Spain and Franco's key role in the defence of the West. See Preston (1993) and Viñas (1981).

17 Following Gaddis (1982), Taylor (1993: 36) describes a 'geopolitical code' as: 'the output of practical geopolitical reasoning. These are the codes or geographical frameworks by which a government deals with the outside world. A national interest is defined and other states are evaluated in terms of whether they are real or potential aids or obstacles to that interest.' Although rather schematic (particularly in the World Systems Theory frame within which he places it), Taylor's designation remains suggestive.

18 Since Teixeira (1992) covers these, and given the limitations of space, they will not be considered here. The complex issue of Portugal in NATO is analysed in Crollen (1973), Telo (1996) and Sidaway (1999).

19 Interview with Serge Groussard originally published in *Le Figaro* 2 and 3 September 1958. Reprinted in Portuguese as 'Panorámica da Politica Mundial' in Salazar (1960a: 3–19). Portugal was not alone in this tactic. Integration of a colony into the metropolitan body politic was applied by France in Algeria and by Britain in Ireland. In all these cases, it means that the struggle for decolonization passes through the heart of metropolitan politics, as it has in Portugal, France and the UK.

20 Notably, for a Portuguese writer this 'Free World' certainly could not be defined in terms of its 'democratic' credentials (as it frequently was in US and British discourses of the time). Instead it was constituted by such things as a 'spirit of Western Civilisation', 'anti-communism' and 'a unity of origin and thought that resulted in a historic process that commenced in Greece of heroic times, was defined in its most important principles by Athens, took a juridical form in Rome, and was afterwards amalgamated into Christianity and finally created a Technology born of Science' (Comprido 1956: 229).

This 'Free World' was also defined against an 'Eastern Bloc' (Oriental Bloc): 'arising from the convergence of all the ideas created by the diverse imperialist Asian peoples and harbouring the living expression of an ideology impregnated of mysticism, that simultaneously constitutes the method and end of world hegemony practised by Russia' (Comprido 1956: 229). A clearer displacement of the 'self' onto the 'other' has rarely been written.

21 The notion of a Portuguese strategic triangle in the Atlantic between Lisbon, the Azores and Madeira is a frequent (and continuing) minor theme of twentieth-century Portuguese geopolitical discourse. Ferreira (1992) claims that the projection first cropped up in a 1906 article published in *Revista de Artilharia* (Covceira 1906). The projection of Portugal as an '*arquipélago*' also remains, though of course, post-1975, not on the world-scale envisaged by Comprido. For further details, see note thirty-one. It is also worth drawing attention to other subaltern scriptings of the Atlantic. Thus Paul Gilroy (1993: 15) notes how: 'a concern with the Atlantic as a cultural and political system has been forced on black historiography and intellectual history by the economic and historical matrix in which plantation slavery – "capitalism with its clothes off" – was one special moment'.

Likewise a different vision of common interest and historical linkage was expressed in the alliance between Marxist-nationalist movements in Guiné, Angola and Cuba.

22 Bordalio Lema (1992: 185) notes three institutional sites for Portuguese geopolitical discourses: the *Sociedade de Geografia de Lisboa* and its *Boletim*; the *Instituto Superior de Ciências Socias e Politicas da Universidade Técnica de Lisboa* and its review *Estudos Políticos e Socias*; and the *Instituto de Defesa Nacional do Ministério da Defesa Nacional* and its review *Nação e Defesa*. A fuller analysis would require a comprehensive review of the institutional roles and linkages of these sites. For a summary of the roles of key individuals (including Salazar) and their roles in the production of colonial/geopolitical discourse, see chapter two of Serapião and El-Khawas (1979). Human Geography in Portugal, epitomised in the main work of its leading twentieth-century figure Orlando Ribeiro (1987, first published 1945) on *Portugal: O Mediterrâneo e o Atlântico. Esboço de relações geográficas* (Portugal: the Mediterranean and the Atlantic. Sketch of geographical relations) with its stress on the organic unity-within-diversity of metropolitan Portugal heavily influenced by Vidal de la Blache's work, thereby bears traces of geopolitics evident in the latter (on which, see Gregory, 1994).

23 In a way that would be paralleled later in Portugal, some Spanish authors, such as the naval Captain Vazquez Sans (1941), contrasted 'Christian, civilizing' Hispanic colonialism with 'rapacious' British imperialism.

24 All citations here are from the English version (Freyre 1961). See Léonard (1997) for more on the regime's appropriation of Freyre's works.

25 Within critical writing on geopolitics and international relations a number of people have drawn attention to the gendering (specifically, the masculinity) of geopolitical discourses (Dalby 1994; Weber 1994). However, Freyre's writings are unusual in terms of the format that this takes. Madureira notes that:

> by effectively investing Portugal's semi-peripheral status within Europe (its cultural and economic 'backwardness') with a positive and dynamic civilizational value, lusotropicalism transcended Portugal's illusion of permanence in the tropics into a ribald tale of productive seduction. . . . Not only does this plausible narrative of Portuguese colonialism represent the spaces colonized by the Portuguese as so many Freudian 'gaps', so many 'sun-drenched wombs', but – concomitantly – and as a recent Brazilian critic of Freyre points out, the whole process hinges on the sexual availability of the 'native woman'.
>
> Madureira (1995: 23)

26 See, for example, Pina-Cabral (1989) or Nunes (1994). Just as it was in the Portuguese empire, racial discrimination is endemic in Brazilian society, but mystification and denial of race difference are widespread, sustained by the social construction of a 'superacial' Brazilian national identity.

27 Attacks on 'Indian colonialism' became more virulent following Indian occupation of Goa in 1961. See, for example, Salazar (1962b).

28 For a further sample of Caetano's after-dinner speeches and radio and television statements on geopolitical themes, see Caetano (1970, 1973). Along with many other invocations of Portuguese geopolitical destiny, further examples can be found in the journal *Ultramar*.

29 Coming from a figure associated with the regime, Spinola's book had tremendous impact. It is worth comparing with de Carvalho (1974) and Serrão (1976). For an accessible account of the transition, see MacQueen (1997). I attend to the 'geopolitics' of these issues in Sidaway (1996), drawing on the deconstructive analysis by Madureira (1995) of the spatiality of the imagination of national difference in Portugal.

30 The expression 'labyrinth of nostalgic myths' is from E. Lourenço (1978). *Saudade* signifies a poetic or romantic mixture of nostalgia, longing or yearning for that which is absent or distant.

31 Aurélio (1987–8) points out in a moderately critical article on Portuguese intellectuals and strategy that: 'the exact representation of the Portuguese territory is still not consensual in intellectual, military and political centres, oscillating between the image of a continental territory with adjacent islands and an *arquipélago* distributed through two sub-wholes of islands and anchored in the continent'.

32 The weekly newspaper column on things geopolitical by de Carvalho in *Diário de Notícias* ranges over similar themes. A collection of these newspaper articles has recently been published (de Carvalho 1995).

Bibliography

Agnew, J. and Corbridge, S. (1995) *Mastering Space: Hegemony, Territory and International Political Economy*, London: Routledge.

Alexandre, V. (1996) 'Questão nacional e questão colonial em Oliveira Martins', *Análise Social* 31: 183–201

Almeida, O. T. (1994) 'Portugal and the concern with national identity',*Bulletin Hispanic Studies* LXXI: 155–63.

Alves, T. L. (1987) *Geopolítica e geostratégia de Portugal, considerações sobre elementos históricos e actuais*, Lisbon: Gráfica Europa Lda.

Anderson, B. (1983) *Imagined Communities*, London: Verso.

Andrés Zapatero, S. (1950) *Los grandes paises de la tierra*, Barcelona: Liveria Elite.

Antunes, J. F. (1986) *Os Americanos e Portugal vl 1: Os anos de Richard Nixon (1969–74)*, Lisbon: Dom Quixote.

Anon. (1996) 'Moreover, resurrecting history', *Economist*, 6 July, 109–10.

Areilza, J. M. de and Castella, F. M. (1941) *Revindicaciones de España*, Madrid: Instituto de Estudios Politicos.

Armado, R. (1940) 'Así está escrito', *Ejercito* 7: no pagination.

Atkinson, D. (1995) 'Geopolitics, cartography and geographical knowledge: envisioning Africa from Fascist Italy', ???–??? in M. Bell, R. A. Butler and M. J. Heffernan (eds) *Geographical Knowledge and Imperial Power 1820–1940*, Manchester: Manchester University Press.

Aurélio, D. P. (19878) 'Os intellectuais e a estrategia – o caso Português', *Estrategia*, 4: 89–108.

Bahia dos Santos, F. (1955) *Política Ultramarina de Portugal*, Lisbon: Sociedade de Geografia de Lisboa.

Basset, T. (1994) 'Cartography and empire-building in nineteenth-century West-Africa', *Geographical Review* 84: 316–35.

Bassin, M. (1996) 'Nature, geopolitics and Marxism: ecological contestations in Weimer Germany', *Transactions of the Institute of British Geographers NS* 21 315–41.

Beneyto, J. (1972) *Historia geopolitica universal en el cuadro de las doctrinas políticas*, Madrid: Aguilar.

Bordalio Lema, P. (1992) 'Tendências da Geografia Politica em Portugal e Espanha', *V Coloquio Ibérico de Geografía León 21 al 24 de Noviembre de 1987: acta, ponencias y communicaciones*: 185–9.

Bruce, N. (1975) *Portugal:The Last Empire*, Newton Abbot and London: David and Charles.

Bosque-Maurel, J., Bosque-Sandra, J. and Garcia-Ballesteros, A. (1992) 'Academic geography in Spain and Franco's regime, 1936–55', *Political Geography* 11 (6): 550–62.

Cabral, A. (1973) *Return to the Sources: Selected Speeches*, New York and London: Monthly Review Press.

Caetano, M. (1970) 'Razões da Presença de Portugal no Ultramar', *Ultramar* 39: 115–34.

—— (1973) *Razões da Presença de Portugal no Ultramar*, Lisbon.

Cahen, M. (1994) *Géopolitiques des Mondes Lusophones, Lusopopie 1994.*

—— (1997) 'Des caravelles pour le futur? Discours politique et idéologie dans "l'institutionâlision" de la communauté de pays de langue portugaise', *Lusotopie* 1997: 391–433.

Capel, H. (1994) 'The imperial dream: geography and the Spanish Empire in the nineteenth century', 58–73 in A. Godlewska and N. Smith (eds) *Geography and Empire*, Blackwell.

Chassin, L. M. (1961) 'Geopolítica e a marinha do futoro', *Revista Marinha do Futoro* 455: 27–40.

Child, J. (1985) *Geopolitics and Conflict in Latin America*, New York: Praeger.

Clarence-Smith, G. (1985) *The Third Portuguese Empire, 1825–1975: A Study in Economic Imperialism*, Manchester: Manchester University Press.

Comprido, J. B. (1956) 'Importância geopolítica de Portugal para a estratégia do Mundo Livre', *Anais do Club Militar Naval* vol. LXXXVI 7–9: 223–68.

Cortada, J. W. (1980) 'Bibliographic essay on twentieth-century Spanish diplomacy', 261–73 in J. W. Cortada (ed.) *Spain in the Twentieth-Century World: Essays on Spanish Diplomacy 1898–1978*, London: Aldwych Press.

Covceira, P. (1906) 'O Triangula Estratégico e a Aliança Inglesa', *Revista de Artilharia* 26.

Crespo, M. P. (1956) 'Portugal na Política e na Estratégia Mundias', *Anais do Club Militar Naval* LXXXXVI 4–6: 135–66 and 275–98.

Crollen, L. (1973) *Portugal, the US and Nato*, Leuven: Leuven University Press.

Dalby, S. (1994) 'Gender and critical geopolitics: reading security discourse in the new world disorder', *Environment and Planning D: Society and Space* 12 (5): 513–634.

da Costa, F. S. (1994) 'The opposition to the "New State" and the British attitude and the end of the Second World War: Hope and disillusion', *Portuguese Studies* 10: 155–76.

de Almeida, P. F. A. V. (1994) *Ensaios de Geopolítica*, Lisbon: Instituto Superior de Ciências Sociais e Políticâs/Instituto de Investigação Cientifica Tropical.

de Andre, A. A. (1961) *Many Races – One Nation: Racial Non-Discrimination Always the Cornerstone of Portugal's Overseas Policy*, Lisbon: np.

de Carvalho, J. B. (1974) *Rumo do Portugal. A Europa ou o Atlântico*, Lisbon: Livros Horizonte.

—— (1991) 'Descobrimentos, Médio Oriente e países Lusófonos', *Africana* 9: 128–50.

—— (1992) 'Da racionalidade da Lusofonia', *Africana* 11: 189–202.

—— (1995) *O Mundo, a Europa e Portugal: artigos publicados em Diario de Noticias*, Lisbon: Sociedade Historica da Independencia de Portugal.

——. (1998) 'O regresso da geopolitica a África', *Africana* 19: 71–88.

de Figueiredo, A. (1975) *Portugal: Fifty Years of Dictatorship*, Harmondsworth: Penguin.

de Reparaz, G. (1924) *La política de España en África*, Madrid: Calpe.

Do, Ó. (1992) 'Salazarismo e Cultura', in F. Rosas (Coordinator) *Nova Historia de Portugal: Portugal e o Estado Novo (1930–1960)*, Lisbon: Editorial Presen_a, 391–454.

de Sousa Santos, M. I. R. (1992) 'An imperialism of poets: the modernism of Fernando Pessoa and Hart Crane', *Luso-Brazilian Review* 24,1: 84–95.

Dodds, K. J. (1993) 'Geopolitics, cartography and the state in Latin America', *Political Geography* 12: 361–81.

Dodds, K. J. and Sidaway, J. D. (1994) 'Locating critical geopolitics', *Environment and Planning D: Society and Space* 12: 515–24.

Ehrlich, C. E. (1998) 'Per Catalunya; l'Espanya gran: Catalan regionalism on the offensive', *European History Quarterly* 28, 2: 189–217.

Ferreira, V. M. (1987) 'Uma nova ordem urbana para a capital do império – a 'modernidade' da urbanização e o 'autoritarismo',do Plano Director de Lisboa, 1938–48', in *O Estado Novo das origens ao fim da autarcia 1926–59*, 359–75.

Ferreira, J. J. B. (1992) 'Portugal 'as Portas do Século XXI: Ensaio Geopolitico e Geostratégico', *Revista Militar* 44 (4): 243–68.

Freyre, G. (1958) *Integração Portuguesa nos Tropicos*, Junta de Investigações do Ultramar.

—— (1961) *Portuguese Integration in the Tropics*, Lisboa: Agencia Geral do Ultramar.

Gaddis, J. L. (1982) *Strategies of Containment*, New York: Oxford University Press.

Galvão, H. (1936) *O Império*, Lisbon: Edições SPN.

—— (1975) *A minha luta contra o Salazarismo e o Communismo*, Lisbon: Arcádia.

Gilroy, P. (1993) *The Black Atlantic, Modernity and Double Consciousness*, London: Verso.

Gramsci, A. (1949) *Note sul Machiavellia sulla politicia e sullo stato moderno*, Turin: Einaudi.

Gregory, D. (1994) *Geographical Imaginations*, Oxford: Blackwell.

Grugel, J. and Rees, T. (1997) *Franco's Spain*, London: Arnold.

Guimarães, A. (1984) *Uma corrente do colonialismo Português: a Sociedade de Geografia de Lisboa 1875–95*, Lisbon: Livros Horizonte.

—— (1987) 'O labirinto dos mitos', 107–21 in A.C. Pinto *et al.* (eds) *O Estado Novo ao fim da autarcia 1926–59*, Vol. 2, Lisbon: Editorial Fragmentos Lda.

Hadian, R. (1978) 'United States foreign policy towards Spain 1953–75', *Iberian Studies* 7, 1: 3–13.

Hahs, B. G. (1980) 'Spain and the scramble for Africa: the Africanistas and the Gulf of Guinea', Ph.D. thesis, University of New Mexico, Albuquerque.

Halliday, F. (1992) 'Atlantic connection', *New Statesman and Society*, 22 May, 25–6.

Harrington, T. S. (1994) 'The pedagogy of nationhood: concepts of national identity in the Iberian Peninsula 1874–1925', Ph.D. thesis, Department of Hispanic Studies, Brown University.

Herb, G. H. (1996) *Under the Map of Germany: Nationalism and Propaganda 1918–45*, London: Routledge.

Hecht, S. and Cockburn, A. (1989) *The Fate of the Forest: Developers, Destroyers and Defenders of the Amazon*, London: Verso

Hepple, L. (1986) 'The revival of geopolitics', *Political Geography Quarterly* 5: 21–36.

—— (1992) 'Metaphor, geopolitical discourse and the military in South America', 136–54 in T. J. Barnes and J. S. Duncan (eds) *Writing Worlds*, London: Routledge.

Hughet, M. (1997) 'Descubrir el Mediterráneo: una orientación recurrente en el ideario exterior franquista', *Cuadernos de Historia Contemporánea* 19: 89–115.

Kay, H. (1970) *Salazar and Modern Portugal*, London: Eyre and Spottiswoode.

Kearns, G. (1984) 'Closed space and political practice: Fredrick Jackson Turner and Halford Mackinder', *Environment and Planning D: Society and Space* 1: 23–34.

Léonard, Y. (1997) 'Salazarisme et Lusotropicalisme, historie d'une appropriation', *Lusotopie* 1997: 211–26.

Lopes, A. C. (1994) 'Dois projectos de geopolitica iberica, de matriz tradicionalista-Vazquex de Mella e Antonio Sardinha', *Revista da Faculdade de Letras* 16/17: 99–113.

Lourenço, E. (1978) *O labrinto da saudade, psicanálise mítica do destino português*, Lisbon: Dom Quixote.

MacQueen, N. (1997) *The Decolonization of Portuguese Africa: Metropolitan Revolution and the Dissolution of Empire*, London: Longman.

Madureira, L. (1995) 'The discreet seductiveness of the crumbling empire – sex, violence and colonialism in the fiction of António Lobo Antunes', *Luso-Brazilian Review* 32, 1: 17–29.

Magalhães, G. M. de (1972) 'A evolução do Oceano Índico', *Revista Militar 2-a epoca* ano 24, 7: 358–74.

—— (1975) 'A do Cabo na estratégia e na economia do Oriente', *Memorias do Centro de Estudos de Marinha* 5: 576–66.

Marquina, B. A. (1986) *España en la política de seguridad occidental*, 1939–86, Madrid: Ediciones Ejército.

Mata Olmo, R. (1995–6) 'Spain between Latin America and Europe: a geopolitical overview', *Boletín de la Associación de Geografos Españoles*, 21–2: 27–45.

Matos, S. C. (1997) 'Portugal: the nineteenth-century debate on the formation of the nation', *Portuguese Studies* 13: 66–94.

Ministry of Foreign Affairs (1970) *Portugal Replies to the United Nations*, Lisbon: Imprensa Nacional.

Minter, W. (1972) *Portuguese Africa and the West*, Harmondsworth: Penguin.

Moradiellos, E. (1993) '1898: a colonial disaster foretold', *Association for Contemporary Iberian Studies* 6 (2): 3–38.

Moreira, A. (1955) 'A conferência de Bandung e a missão de Portugal', *Separata do Boletim da Sociedade de Geografia de Lisboa* April–June.

Neto, J. P. (1963) 'O significado do Multirracialismo Portugués', *Separata da Revista Ultramar* no. 13–14.

——. (1968a) 'A evolução e tendencias recentes das hipóteses geopolíticas', *Separata de Estudos Politicas e Socias* 6 (1): 93–108.

——. (1968b) *As províncias portuguesas do Oriente pertante as hipótesas geopolíticas*, Lisbon: Instituto Superior de Ciencias Sociais e Política Ultramarina.

Newitt, M. (1981) *Portugal in Africa, The Last Hundred Years*, London.

Nowell, C. E. (1982) *The Rose-Colored Map: Portugal's Attempt to Build an African Empire from the Atlantic to the Indian Ocean*, Lisbon: Junta de Investigações Científica do Ultramar.

Nunes, Z. (1994) 'Anthropology and race in Brazilian modernism', 115–25 in F. Barker, P. Hulme and M. Iversen (eds) *Colonial Discourse/Poostcolonial Theory*, Manchester: Manchester University Press.

O'Loughlin, J. and van der Wusten, H. (1990) 'The political geography of pan-regions', *Geographical Review* 80, 1: 1–20.

Ó Tuathail, G. (1992) 'Foreign policy and the hyperreal: the Reagan administration and the scripting of "South Africa"', 155–76 in T. J. Barnes and J. S. Duncan (eds) *Writing Worlds: Discourse, Text and Metaphor in the Representation of Landscape*, London: Routledge.

—— (1996) *Critical Geopolitics*, London: Routledge.

Parker, G. (1985) *Western Geopolitical Thought in the Twentieth Century*, London: Croom Helm.

Pereira, R. (1986) *Antropologia aplicada na política colonial portuguesa: a missão de Fstudos das Minorias Étnicas do Ultramar Português (1956–61)*, Faculdade de Ciências Sociais e Humanas/UNL.

Pike, F. B. (1971) *Hispanismo, 1898–1936. Spanish Conservatives and Liberals and their Relations with Spanish America*, Notre Dame and London: University of Notre Dame Press.

Pina-Cabral, J. de (1989) 'Sociocultural differentiation and regional identity in Europe', 265–97 in R. Herr and J. H. R. Holt (eds) *Iberian Identity: Essays on the Nature of Identity in Portugal and Spain*, Berkeley: Institute of International Studies.

Pinto, J. E. R. (1956) 'Importância geopolitica de Portugal para a estratégia do Mundo Livre', *Anais do Club Militar Naval* 86, 10–12: 401–26.

Pio, S. M. (1944–5) 'Bases económico-sociais de uma Política de contacto de Raças', *Anuário da Escola Superior Colonial* years 23 and 24: 133–40.

Powell, C. T. (1995) 'Spain's external relations 1878–1975',11–29 in R. Gillespie, F. Rodrigo and J. Story (eds) *Democratic Spain: Reshaping Relations in a Changing World*, London: Routledge.

Preston, P. (1993) *Franco: A Biography*, London: Harper Collins.

Prior, G. (1951) 'A geopolitica e a estratégia mundial', *Anais de Marinha* ano 11, 16: 3–15.

—— (1956) 'A Africa na estratégia global', *Revista Militar* 8(II S'eculo), 5: 283–96.

Raffesin, C., Lopreno, D. and Pasteur, Y. (1995) *Géopolitique et Histoire*, Lausanne: Editions Payot.

Raby, D .L. (1988) *Fascism and Resistance in Portugal*, Manchester and New York: Manchester University Press.

—— (1994) *Portuguese Exile Politics: The 'Frente Patriótica de Libertação Nacional, 1962–73*.

Reguera, A. T. (1990) 'Origenes de pensamiento geopolitico en España. Una primera aproximación', *Documents d'anàlisi geogràfica* 17: 79–104.

—— (1991) 'Facismo y Geopolitica en España', *Geocritica* 94: 11–63.

—— (1992) 'Recepcion en España de la geopolitica Alemano. Desde los fundamentos ratzelianos hasta el radicalismo nazi', *V Coloquio iberico de geografiá León 21 al 24 de noviembre de 1989: acta, potencias y communicaciones*, Universidad de León: Departamento de Geografia: 221–33.

Ribeiro, O. (1987) (First edition 1945) *Portugal: O Mediterrâneo e o Atlântico. Esboço de relações geográficas*, Lisbon: Livaria Sá da Costa Editora.

Rita, J. G. da Costa Santa (1944) 'A geopolitica e as colónias de Angola e Moçambique', *Anuário da Escola Superior Colonial Anos* 23 & 24: 133–40.

Roucek, J. S. (1964) 'Portugal in geopolitics', *Contemporary Review* 205: 47688.

Sacchetti, A. F. (1987) *Geopolítica e Geostratégia do Atlântico*, Lisbon: Instituto Superior Naval de Guerra.

Sacchetti, A. F. et al. (1989) *Atlântico Norte e Atlântico Sul: Geopolítica e estratégia*, Lisbon: Instituto Superior Naval de Guerra.

Salazar, O. (1943) 'Nos somos uma força destinada a vencer', 21 in O. Salazar (ed.) *Discursos e notas politics Vol 3, 1938–43*, Coimbra: Coimbra Editora Lda.

—— (1960a) 'Panorama da Política Mundial', 3–48 in *Discursos e notas politicas, Vol 6 1959–66*, Coimbra: Coimbra Editora Lda.

—— (1960b) 'Portugal e a campanha anticolonista. Discurso pronunciada na sessão da Assembleia Nacional de 30/11/60', in *Discursos e notas politicas, vol. 6 1959–66*, Coimbra: Coimbra Editora Lda.

—— (1961) 'O Ultramar Português e o ONU. Discurso pronunciada na sessão extra-ordinária da Assembleia Nacional em 30/6/1961', in *Discursos e notas políticas vol. 6 1959–66*, Coimbra: Coimbra Editora Lda, 127–58.

—— (1962a) 'Defesa de Angola – Defesa da Europa. Discurso proferido na Cova da Moura em 4/12/62', 59–66 and 227–37 in *Discursos e notas politicas, vol. 6 1959–66*, Coimbra: Coimbra Editora Lda.

—— (1962b) 'Occupacão e Invasão de Goa', *Ultramar* 7/8: 179–91.

Sampayo e Mello, L.V. de (1940–41) 'Do intervencionalismo dos dominadores na organização social e economia dos dominados', *Anuário da Escola Superior Colonial* Anos 21 & 22: 189–221.

Sanches, M.C. (1963) 'Meditácões Geopoliticas-Geostratégicas', *Anais do Clube Militar Naval tomo* 93: 1–3 and 10–12; vol. 94: 1–3.

Serapião, L.B. and El-Khawas, M. (1979) *Mozambique in the Twentieth Century: From Colonialism to Independence,* Washington, D.C.: University Press of America.

Serrão, J. (1976) 'Repensar Portugal', *Nação e Defesa* April 1976.

Sidaway, J. D. (1996) *Geopolitical (Re)imaginations: The Portuguese Case,* unpublished manuscript, available from the author.

—— (1998) 'What is in a Gulf?: from the 'arc of crisis to the Gulf War', 224–39 in S. Dalby and G. Ó Tuathail (eds) *Rethinking Geopolitics,* London: Routledge.

—— (1999) 'American power and the fracture of the Portuguese empire', 195–209 in D. Slater and P. Taylor (eds) *The American Century: Consensus and Coercion in the Projection of American Power,* Oxford: Blackwell.

Southworth, H. (1963) *El mito de la cruzada de Franco,* Paris: Ruedo Ibérico.

Spínola, A. de (1974) *Portugal e o futuro: análise da conjuntura nacional,* Lisbon: Arcádia.

Stephanson, A. (1989) *Kennan and the Art of Foreign Policy,* Cambridge, Mass.: Harvard University Press.

—— (1996) 'Rethinking international relations', *New Left Review* 220: 137–42.

Taylor, P. J. (1990) *Britain and the Cold War: 1945 as Geopolitical Transition,* London: Pinter.

—— (1993) 'Geopolitical world orders', 31–61 in P .J. Taylor (ed.) *Political Geography of the Twentieth Century: A Global Analysis,* London: Belhaven.

Tavares, A. R. da Silva (1964) *Política Ultramarina Portuguesa, seus objectivos Históricos e Actuais, sua posição perante a conjuntura internacional* (Conferência Proferida no Instituto des Altos Estudos Militares em 28 de Fevereiro de 1964), Lisbon: Agência Geral do Ultramar.

Teixeira, N. S. (1992) 'From neutrality to alignment: Portugal in the foundation of the Atlantic Pact', *Luso Brazilian Review,* 29, 2: 113–26.

Telo, J A. (1996) *Portugal e a NATO: o reencontro da tradição Atlântica,* Lisbon: Ediçoes Cosmos.

Vazquez Sans, J. (1940) *España ante Inglaterra,* Barcelona: Angel Ortega.

Ventura, A. (1994) 'Resistências ao regime ditatorial: Henrique Galvão', in J. Medina (director) *História de Portugal; dos tempos pré-históricos aos nossos dias, Vol XXIII O Estado Novo opressão e resistência,* Lisbon: Ediclube, 235–8.

Vicens Vives, J. (1940) *España: Geopolitica del Estado y del Imperio,* Barcelona: Editorial Yunque.

—— (1941) 'Spanien und die geopolitische Neuordnung der Welt', *Zeitschrift für Geopolitik* 18, 5: 256–63.

—— (1950) *Tratado de geopolitica,* Barcelona: Editorial Vicens Vives.

Viñas, A. (1981) *Los pactos secretos de Franco con Estados Unidos: bases, ayuda económica, recortes de soberania,* Barcelona: Edicones Grijalbo.

Weber, C. (1994) 'Shoring up a sea of signs : how the Caribbean Basin Initiative framed the US invasion of Grenada', *Environment and Planning D: Society and Space* 12, 5: 513–634.

Wise, P. (1995) 'Portugal tries to preserve waning influence', *Financial Times* 26 July 1995: 3.

7

GEOPOLITICS AND THE GEOGRAPHICAL IMAGINATION OF ARGENTINA

Klaus Dodds

Introduction

In his popular text on geopolitics, *Que es la Geopolitica?* (What is Geopolitics?) the Argentine writer Jorge Atencio argued that his subject matter was concerned with:

> the influence of geographic factors in the life and evolution of states, with an objective of extracting conclusions of a political character . . . [Geopolitics] guides statesmen in the conduct of the state's domestic and foreign policy, and it orients the armed forces to prepare for national defence and in the conduct of strategy; it facilitates planning for future contingencies based on relatively permanent geographic features that permit certain calculations to be made between such physical realities and certain proposed national objectives, and consequently, the means for conducting suitable political or strategic responses.
>
> (Atencio 1986: 41)

Atencio's geopolitics was a concept for international affairs whereby the territorial requirements of states were established on the basis of a geographical evaluation of its location, resources and topography. As a former Army officer, Atencio's writings on Argentina and political space emphasized the organic state framework and a deterministic view of territory and populations. While space was not considered to be a mere backdrop to events, the social and cultural attributes of place were largely overlooked. Geographical features such as rivers, mountains and the seas were considered permanent factors that could be modified but never eradicated by human endeavour.

Atencio's monograph on the state of geopolitics was originally published in

1965 by the largest Argentine geopolitical publishers, Editorial Pleamar. Many strands of European and American geopolitical thought intermingle in the ensuing discussion of the organic analogies of Ratzel's writings to *Geopolitik* and Nicholas Spykman's observations on the role of geographical factors in determining foreign policy-making. Halford Mackinder's ideas concerning the 'Heartland' were frequently recycled by authors such as Atencio in order to produce sober analyses of Argentina's peripheral geographical position within global politics. Underlying these writings, however, was a sense of disappointment and frustration that Argentina was not involved in the grander machinations of global politics.

The importation of European and North American ideas concerning the nation-state, territory and international relations was not unusual for an intellectual culture which had been enriched by French art and literature, Italian opera, British commerce and financial thinking, German military strategy and Anglo-American models of patriotic education. This should not imply, however, that Argentine writers have only produced ideas and ideologies that are essentially derivative. Rather, it suggests that Argentine geopolitical thought is not only a complex ensemble of ideas and practices but also the result of a specific geographical location and the public culture of Argentina. Geopolitical writing in the southern cone as with other parts of the world has tended to 'exaggerate' the significance of space in terms of political conquest, territorial acquisition and the domination of place (Escudé 1984, 1992). This was most pronounced during the so-called 'Dirty War' of the 1970s when the Argentine military sought to impose a national security doctrine on the nation. Argentine officers and geopolitical writers frequently employed alarming metaphors of disease and infection to highlight the fragile health of the nation-state (Goyret 1980; Sibley 1995: 72–89). Argentine geopolitical writings (see below) have also drawn upon specific cultural frames in combination with bodies of geographical knowledge to produce particular representations of the Republic and its disputed international boundaries, territories, security zones and frontier regions.

While the literature on Argentine geopolitics has been carefully examined by a range of South American and North American writers, few have been attentive to the wider contextual settings which enabled this body of thought to take root in twentieth-century Argentina (Pittman 1981; Child 1985; Kelly and Child 1988). Moreover, there has been a reluctance to identify formal and popular strands of Argentine geopolitical thought. By adopting the theoretical platform of critical geopolitics, it will be illustrated how particular ideas on the state, territory and national identity were mobilized to the rich vein of geopolitical literature in the Argentine Republic (Ó Tuathail 1996; Ó Tuathail and Dalby 1998).

This chapter initially assesses the social and political conditions which existed at the time of some of the earliest citations of 'geopolitics' by authors such as Admiral S. R. Storni in his 1916 text, *Intereses Argentinos en el Mar* (Storni 1967). Thereafter, the discussion turns to the 1930s and the importation of German ideas and practices concerning the armed forces and geopolitical thought. This was a key period in the intellectual history of geopolitics, cementing a particular vision of Argentina as a territorially disadvantaged state. With the emergence of Colonel Juan Perón in the 1940s, Argentine political thinking began to embrace a 'third way' in international politics which ultimately ran counter to the concerns of Argentina's position in security organizations such as the Rio Pact by the United States. In the post-Perónist period, armed intervention by the military led to a significant transformation in geopolitical discourse, as internal security emerged as a priority in conjunction with long-standing aspirations concerning national integration and international strength. In the 1970s this concern for the internal security of the Argentine state coincided with the new external fears created by the strategic importance of the South Atlantic and the Antarctic. Finally, in the aftermath of the democratic revolution in 1983, Argentine geopolitics are considered in the light of various critiques of the subject within Argentine academic and political circles.

Geopolitics and the geographical construction of post-independence Argentina

The Republic of Argentina was officially declared on 25 May 1810, one of the earliest examples of a post-colonial state in Latin America. Over the next hundred years, various *caudillos* such as President Juan Rosas and General Juan Roca struggled to stitch together a nation-state from the various fragments of the disintegrating Spanish Empire and the administrative regions within the Viceroyalty of the River Plate (see Figure 7.1). The process of state formation and identity creation was violent, piecemeal and, as Nicholas Shumway has argued, subject to complex negotiation:

> Whereas in Europe and to some degree in the United States, myths of peoplehood on which nations could be built were available before the nations themselves were formed, in Spanish America, civil strife following independence forced nations to emerge in areas that had no guiding fictions for autonomous nationhood. Nation formation was further complicated by civil wars in post-Independence Spanish America which eventually broke four viceroyalties into eighteen separate Republics.
>
> (Shumway 1991: 2–3)

While the processes involved in Argentine state formation have received considerable attention, scant consideration has been devoted to the apparent lack of 'guiding fictions' and how this influenced geographical and geopolitical discourses concerning the ideological and physical boundaries of Argentina

Figure 7.1 Boundaries of the Viceroyalty of the River Plate

(Gerassi 1964; Walther 1976; Ozlak 1983; Rock 1987; Calvert and Calvert 1989; Escudé 1990; Shumway 1991).

Geographical and political construction of Argentina

The geographical creation of an Argentine Republic involved a series of bitter struggles between the various provinces and rival states such as Chile and indigenous Indian populations in Patagonia and the Pampa. Over fifty years, elites based in the Buenos Aires province sought to cajole and discipline the peoples and spaces that resisted the nationalist impulse. Nationalist writers such as Juan Alberdi wrote *Reconstruccion Geografico de America del Sur* (1879) with the explicit purpose of urging the Roca administration to develop the institutional power of the state in order to further the integration of the territories associated with the former Viceroyalty of the River Plate. Domingo Sarmimento's *Argiropolis: O la Capital de los Estados Confederadus del Rio del Plata* (1850) even called for the creation of a new capital city based on the Island of Martin Garcia in the River Plate so that Argentina could pursue the re-unification of the Viceroyalty in the last part of the nineteenth century.

Sarmiento's treatise on territorial integration was part of a wider project which assumed that Argentina had to seek inspiration from European liberalism in order to deal with the Spanish legacy of *Caudillismo* which encouraged long-standing dictatorships such as that of Juan Manuel de Rosas (1829–52). Under the Rosas regime, economic and political development was concentrated in the hands of a privileged minority and Sarmineto argued that Argentina's evolution as a state and society would be compromised by a lack of national feeling and integration (Dijkink 1996: 75–6). Later, countless intellectuals and commentators such as Juan Bautista Alberdi argued that Argentine citizens rather than embracing the notion of a participatory political and social system, displayed instead a considerable ambivalence towards citizenship and their relationship to the state.

The extending geographical boundaries of the Argentine Republic were facilitated by the development of cattle ranching in the Pampa and the creation of a railway system intended to transport crops and beef to the ports of Buenos Aires and the Plata basin, thereby developing a transport and commercial network and consolidating economic and institutional processes (Lewis 1983; Hodge 1984). However, in spite of the creation of an Argentine Ministry for Foreign Affairs and Worship in 1820, international relations and foreign affairs remained fraught with difficulties, as Argentine elites argued with their neighbours over common geographical boundaries such as the Andes and the Rio Plata. Despite imminent independence, frontiers and boundaries were poorly defined, partially mapped, and frontier regions generally unpopulated by European or Creole settlers.

More recent Argentine commentators such as Romero, Sanz and Ozlak have concluded that it is not possible, therefore, to speak of an Argentine state until the 1880s in the aftermath of the successful 'Conquest of the Desert' between 1879–1883. The final annihilation of the indigenous Indian population was the culmination of Argentina's state formation and the end of resistance from groups such as the Ranqueles situated in the Pampa and Patagonia. The institutionalization of property laws such as the Decree of 1867 encouraged farmers and colonists to occupy and demarcate the apparently 'empty spaces' of the Pampa and Patagonia. The diffusion of *estancierios* and cattle ranching co-existed with increasingly aggressive military campaigns designed to remove the Indian populations from southern territories. The most dramatic development occurred in the late 1870s when General Roca was instructed to remove the Indian peoples from the lands south of the Rio Negro. In 1879, an army of 8,000 men accompanied by geographers and surveyors swept through the Patagonian desert and sought to remove the Indian population from thousands of miles of territory. In his formal report to the War Ministry, Roca noted that: 'not a single place is left in the desert where the Indian can now threaten the colonists in the Pampas . . . Civilized populations will come and relieve our military forces of the simple indispensable services of police which is still required today, (Roca 1882 cited in Hashbrouck 1935: 223). Subsequently, twenty million hectares of Patagonian territory were turned over to Roca's 500 closest supporters in the form of private property. Pre-existing land Acts ensured, therefore, that this newly colonized territory in the South was administered and regulated by the Argentine state.

The 'Conquest of the Desert' was a landmark in the creation of the Argentine state as vast tracts of territory were seized from indigenous Indian communities at great financial cost (1.5 million pesos) to the state. At the same time, the newly created Argentine Geographical Institute (AGI) in conjunction with Colonel Manuel Olascoaga of the Army Office of Topography had completed and provided detailed surveys and maps of the region between the rivers Neuquen and Limay and the mountain landscapes of the Andes (Dodds 1993). This recently acquired geographical information of Patagonia was later to be deployed by Argentine negotiators in their discussions with the Chilean authorities over a common Andean boundary. General Roca argued in 1879 that the occupation of Patagonia confirmed that Argentina was a civilized and advanced state because 'the weaker race must perish in the face of the one favoured by nature' (cited in McLynn 1980: 28).

Geographical and racial imaginations of Argentina

The construction of 'Argentina' was not, therefore, just a process of physically delimiting a national space and particular territorial boundaries. Underwritten by appeals to Social Darwinism and racial theories, nineteenth-century soldiers and nationalists such as Juan Roca had trenchant views on Argentine citizenship. By the late nineteenth century and in the face of rising European immigration and economic growth, the construction of a national identity was based on two sets of processes which could be described as institutional and ideological (Solberg 1970; Vogel 1991). The former was assisted through the emergence of print capitalism and the expansion of transport and communication networks such as the railways and telegraph, and the latter via the institutionalization of public education.

State-funded education was a key feature in the creation of public culture and national identity. Armed with the discourses of Social Darwinism and race, Argentine Creoles began to identify those people who were worthy of patriotic forms of education (Szuchman 1990: 110–12). Racial codes based on genetic heritage and racial labels were used to differentiate the European citizen from the Afro-Argentine, the Indian and the gaucho communities (Andrews 1980). Creole Argentines were simply described as 'white' even though 'their claims to racial purity, were more often than not the product of some racial blending' (Shumway 1991: 5). The myth of a 'white' Argentina became a powerful element in the education programmes of the late nineteenth century as:

> Argentine writers and thinkers announced that Argentina had become a truly white society, racially superior to the other South American republics, and therefore, the premier nation of the continent. Non-European traces in the country's histories and geographies were ignored and eventually forgotten in the process of cultivating a myth of a white Argentina.
>
> (Andrews 1979: 39)

In the 1870s and 1880s, reform by the ruling elites in Buenos Aires extended the provision of educational training to a wider profile of the white Argentine and European immigrant population (Escudé 1990, 1992). The making of the Argentine citizen became a central theme in this pedagogic revolution as new forms of history and geographical lessons were introduced at primary and secondary levels. One of the most interesting dimensions of these forms of patriotic education was the depiction of Argentina as a 'vulnerable' state that was surrounded by unsteady neighbours such as Chile and Brazil and threatening others within the boundaries of the Republic. It was decided that Argentine children would have to be taught six hours of Argentine history and geography per week at all six of the primary school grades. In an 1888 decree issued by the Ministry of Justice and Instruction, teachers were urged to stress to children that Argentina faced many dangers from neighbouring countries. One of the great ironies of these depictions is that Argentine elites had been remarkably successful in colonizing territories to the south, to the west and to the north of Buenos Aires. The only 'loss' was the Falkland Islands to the British in 1833. The remaining part of the nineteenth century is one characterized by the Argentine state acquiring additional territory in virtually every direction from the national capital. As Carlos Escudé ironically noted, 'the perception of invasion was a cause of and a motivation for an extremist educational policy that was officially established . . . whose objective was to Argentinize (sic) the children and immigrants' (1992: 13). By 1902, it was decided that only Argentine citizens could teach history and geography at school and that children would have to start the day by singing the national anthem whilst standing by the national flag.

The ideological limits of Argentine citizenship were defined in opposition to a variety of threatening 'others' such as the Indian populations in the South and West of modern Argentina, Afro-Argentine communities and other social groups such as the British and the gauchos. These groups were liminal in the sense that they provided sources for the construction of difference (Norton 1988; Sibley 1995). In their occupation of territorial and ideological hinterlands, Argentine elites used these groups to differentiate the boundaries of citizenship. The massacres and the forced removal of Indian populations was one ironic element in this ideological transformation, for in the early part of the nineteenth century settlers had relied on the Indian populace as guides and farmers. However, by the 1880s, the interrelationships between the European Argentine and the Indian had become more stark as the latter was now represented as an obstacle to social progress and industrial development. The Indian had become a threatening 'other' to the Creoles and served as a point of differentiation. As David Campbell has noted, 'the demarcation of the self and the other, is, however, not a simple process that established a dividing line between the inside and the outside. It is a process that involves the grey area of liminal groups in a society . . . outsiders who exist on the inside' (Campbell 1992: 275).

By the first decade of the present century, Argentine education under the directorship of the head of the National Council for Education, Maria Ramos Mejia, ensured that Argentine citizens were inculcated with a geographical and historical sense of their nation (Spalding 1972). The 1908 Patriotic Education Plan was intended to ensure that: 'In conversation, in all grades, issues of a patriotic character must be frequently included: the flag, the coat of arms, the monuments, the national anthem, the national heroes. . . . Scrap books of a patriotic character should be made by all school children' (cited in Escudé 1990: 33). Geographical education was central to this patriotic programme because it was seen to provide the academic and practical resources for generating a sense of regional awareness and spiritual affiliation with the Argentine Republic. The depiction of the state as 'vulnerable' to outside aggression contributed to narratives about territorial losses and legitimized the role of a strong state in safeguarding the rights of its citizens. As the Ministry for Public Education noted in 1932 via a decree:

> We, with an independent life of less than a hundred years old and with a population of that has doubled in less than the span of this time . . . cannot yet entrust the family that intense and noble task [of education and citizenship]. Here the nationalist bulwark must inevitably be the school. It is the school that must create in the soul of children and grand children of foreigners, a clear and firm national feeling.
>
> (cited in Escudé 1990: 67)

This legacy of Argentine education and national identity is complex and controversial. Revisionists such as Carlos Escudé have argued that the education system perpetuated pernicious narratives of national identity which alienated internal groups such as the Afro-Argentines and generated exaggerated fears of neighbouring states such as Chile and Brazil (see also Escolar *et al.* 1994). However, it is possible to distinguish a number of issues that have had a profound effect on the subsequent geopolitical and political discourses within the Republic. These included: a widely held sentiment that Argentina had been a great economic power and that her subsequent decline in the 1920s and 1930s was caused by western powers such as the USA and the UK; a belief that Argentina's greatness could be secured through the exploitation of underdeveloped spaces such as Patagonia, the oceans and Antarctica; a sense that Argentina had been the victim of territorial losses such as the Falklands/Malvinas in 1833; a conviction that the country had been marginalized in the dramas of world politics; and a strong sense that Argentina lacked a common purpose or 'national project which would bind all its citizens together'. As one North American author noted, Argentina remains 'a headless chicken, scurrying about in quixotic search for an identifiable future. . . . [Given] the impossibility of getting the Argentines to agree about anything, creating a new sense of Argentinity (sic) .. seems utopian' (Shaw 1985 cited in Calvert and Calvert 1989: 211). In contrast, Argentine geopolitical writers have believed throughout this century that geopolitical theories could be developed for the purpose of creating a sense not only of common purpose but also for restoring national pride (Dijkink 1996).

German geopolitik and the 'Prussification' of the Argentine armed forces

One of the earliest texts explicitly concerned with European and North American theories of geopolitics was written by Admiral Roberto Storni in

1916. Using the naval writings of the American Admiral A T Mahan, Storni turned his attention to the seas surrounding the Argentine Republic. In the *Intereses Argentinos en la Mar*, the sea is divided up into a series of geographical regions and traffic patterns. The Atlantic and Pacific Oceans are conceived as empty spaces which await the attention of the economically successful Argentine state. Storni argues that it is unfortunate that the oceanic routes and the resources of the sea have been neglected by a country which depends upon global networks of maritime traffic and trade. The seas and oceans around Argentina could not only assist the economic development of the country but also act as a source of unity and purpose: 'in this common attractive force resides one of the real and permanent causes of national union; it makes geographical unity and in the end political unity' (Storni 1967: 22, originally published in 1916).

While Storni's analysis of the sea draws upon the writings of Mahan and Mackinder, his concern for 'unity' reflects the wider belief that the country's economic and political position had declined since the late nineteenth century when it led world production of maize, linseed, beef and wheat (Rock 1987). Storni's attraction to the ideas of Mackinder and Mahan was in part related to their concern for the role of seapower and the geopolitical significance of the oceans. As a nation dependent on maritime trade, the Argentine armed forces appraised the ideas of these Anglo-American writers once earlier optimism on the export-led economy had given way to fears that decline and decadence had invaded the public imagination of its citizens. The subsequent economic crash during the 1929 Great Depression precipitated an ideological and cultural crisis in Argentina as the old equation of economic development, immigrant labour and limited political participation gave way to concerns that liberal capitalism and democracy could produce economic and social chaos. The final act in this process of disillusionment appeared to be the removal of President Yrigoyen in September 1930 by the Argentine military (Rock 1993: 17–25).

The rising profile of the Argentine armed forces was facilitated in part by the patterns of professional interaction between German, Italian and Argentine military officers in the 1930s. In 1900, the German officer Alfred Arent had been appointed as head of the Superior War College. By the first decade of the present century, 50 per cent of the staff were of German origin and contributed to a two-year training programme for military officers. When Juan Perón was sent by the army to Italy to study Italian mountaineering strategy and organizational methods in 1938, many German officers continued to provide instruction on military strategy and geopolitics. His visit to Rome was also combined with a series of trips to Germany where Perón held meetings with some of the German officers who had earlier taught at the Argentine Army's staff college. He was later to return as Commander of the Mendoza Mountain

Detachment and wrote numerous and widely circulated books and papers on the First World War, nineteenth-century history and strategic studies (Crawley 1984: 63–5). German officers such as Johannes Kretzchmar remained involved in the training of Argentine officers until 1940, and were responsible for cementing a resolutely hierarchical chain of command. Admission to the Argentine military academies, however, was not open to all citizens as the armed forces insisted that their trainee officers were Catholic. The interaction of militarism and religion was later to emerge as a significant theme in the regimes of terror of the 1960s and 1970s.

In terms of shaping the overall politics and identity of the Argentine army, German military training played a key role in moulding military ideologies, and Ratzelian theories on the state were frequently combined with references to Social Darwinism, religion and national security (Potash 1980; Waisbord 1991). However, there is little evidence to suggest that German geopolitical thinking played a seminal role in military training in the 1920s despite of references to Ratzel in Argentine and Chilean military texts (Hepple 1992). Robert Potash has noted, for example, that the ideology of the Argentine military was based on a variety of sources including anti-communism (particularly after the 1930s and 1940s), Catholicism and virulent nationalism initially in the face of difficult trading relations with Britain (Potash 1969). From the 1930s onwards, the role of the military in public life increased markedly as leaders of the 1930 coup such as General Juan Bautista Molina began to see themselves as the defenders of the nation against the evils posed either by communist totalitarianism or democratic governance. The latter was considered with a great deal of scepticism because of a belief that democracy encouraged weak government and corrupt political parties. Admiration for German governance and military politics in the 1930s was combined with an organic view of the Argentine state, and this corresponded with a belief that the Republic's needs were more significant than the individual rights of citizens (Rock 1993: 178–80).

By the late 1930s, Richard Henning and Leo Korholz's account, *Einführung in die Geopolitik*, was required reading in most of the Argentine military colleges (cited in Pittman 1981). As a consequence, a Spanish translation of the book appeared in 1941 under the title *Introduccion a la Geopolitica*. The latter was the first book on German *Geopolitik* to appear in Argentine military circles, but in contrast to the Spanish military it did not generate a formal exchange of ideas between Argentine and German military intellectuals (see Sidaway in this volume). It was apparent that the central themes of the book concerning the organic state and the need for strong armed forces in an uncertain world had been thoroughly digested by students, just as they had been in the Brazilian military (Hepple 1986). Within Argentina, Mario Travassos's work on the Brazilian frontier was also translated into Spanish and used in the Superior War College because it apparently offered helpful guidance on the need to protect and preserve distant national frontiers such as those proximate to the Amazonian basin and Patagonia (Pittman 1981).

The most sustained interrogation of North American and European theories of geopolitics was carried out by Jasson and Perlinger's *Geopolitica* in 1948 (cited in Pittman 1981), and Isola and Berra in their *Introduccion a la Geopolitica Argentina*, first published in 1950. The latter was the more significant in terms of coverage and circulation within the military academies because it was written by an Army Major (Emilio Isola) and Colonel (Angel Barra) who had both taught at various military academies and universities in Argentina. In their foreword to the book, Isola and Berra warn that foreign theories of geopolitics may not be appropriate and hence there was an:

> absence, in our environment, of a text in which is summarized the geopolitical influences in the formation of our state. This has brought to us the task of uniting the principle antecedents that permit an evalua- tion of the influence of the geographical environment in the political development of our country.
>
> (Isola and Berra 1950: 1)

One of the striking features of this dense regional geography of Argentina is the use of cartographic images to locate Argentina within the wider world. Their widespread employment of polar-centred projections of the world was intended to highlight the importance of Argentina in the defence of southern hemispheric space against threatening communist sources such as the Soviet Union (see Figure 7.2). As with many maps found within Italian and German geopolitical writings, threatening black arrows and flows became a defining feature of Argentine geopo- litical cartography as maps were used to chart new dangers against the Argentine state. In the light of President Perón's endorsement for the Argentine Antarctic sector, the map also highlights tri-continental Argentina (the mainland, the Argentine Antarctic and islands such as the Malvinas).

The use of propaganda cartography combined with sympathetic assessments of inter-war German geopolitical writings meant that South American geopolitical writing went in a different intellectual direction to that of Anglo-American polit- ical geography (Hepple 1986; Henrik-Herb 1989). Geopolitics was not shunned as an academic term because it was felt by many South American military writers to demarcate a valuable intellectual terrain concerning the inter-relationship between territory, state and governance. Under the influence of German military training and later Brazilian national security thinking, Argentine writers stressed the importance of space and territory in shaping foreign and security politics. Maps and charts were used to highlight the pressing needs of the Argentine state, including the imperative to develop frontier regions and ensure that the Republic

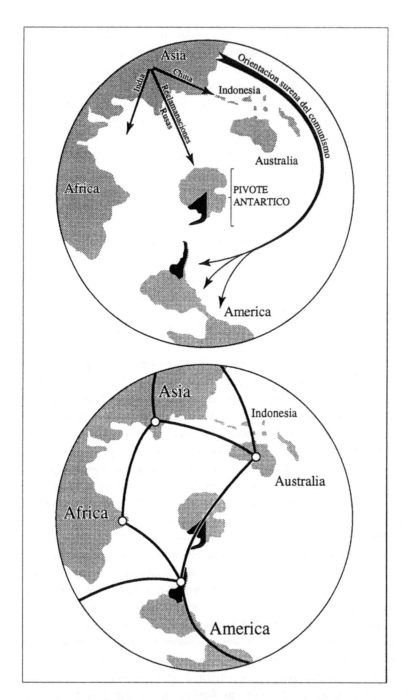

Figure 7.2 Geopolitical projections of Argentine air traffic and flows of communism
Source: Adapted from Isola and Berra (1950)

was not overwhelmed by imperial powers such as the UK and the USA and/or regional rivals such as Brazil.

During the inter-war period, Argentine interest in geopolitical theories was mainly confined to the military colleges and institutions such as the Circulo Militar. Admiration for German ideas on statecraft, however, also fuelled unease amongst senior officers such as General Juan Bautista Molina that Italian and German forms of fascism spawned potentially unstable political leaders (Rock 1993). The politics of the Argentine military was, as Robert Potash has noted, politically sympathetic towards the German armed forces without necessarily endorsing extreme forms of fascism and Nazism (Potash 1980).

Perón and Argentine geopolitical thought

The emergence of Colonel Perón and the Grupo Unido Officiales in 1943 signalled a major change in Argentine political and cultural life. Whilst few North American or European commentators have identified Perón as a geopolitical writer, his approach to international affairs and geographical education revealed a profound sense of Argentine national space and Argentina's place in world affairs. Military writers such as Juan Briano have sought, however, to locate Perón within an Argentine geopolitical tradition of confederalism and national integration stretching back into the nineteenth century through the writings of Alberdi and Sarmineto. Perónismo represented a partial break from earlier geopolitical writings through its concern with geopolitical non-alignment and Argentine economic self-sufficiency (Pittman 1981; McLynn 1982; Rock 1987). Within Perón's writings and speeches it is possible to develop, therefore, a number of distinct themes which have had some bearing on the evolution of Argentine geopolitical thought and these include: a belief that Argentina had been the victim of colonial aggression in the Falklands/Malvinas and the Antarctic; a concern for Argentina's economic and financial self-sufficiency; and a desire to avoid the rigid bi-polar politics of the Cold War (Pittman 1981).

The formal election of Perón in 1946 further cemented his grip on the Argentine political system. The decade and a half after the armed forces coup in 1930 had seen a further decline in Argentine economic and political fortunes. In the aftermath of the inter-war depression, Argentina had attempted to develop new trading links with the United States but the Second World War had interrupted this growing exchange of primary and secondary products. Argentina's ambivalence towards German fascism during the Second World War did not endear the regime to the American or British governments. In the face of apparent economic decline and political disinterest from the Great Powers, Perón attempted to develop a new sense of mission for the Republic with the aid of public education and the newspaper media, to promote a vision of Argentina as a politically self-sufficient and territorially confident nation.

A striking aspects of Perón's rule in Argentina (1946–1955) was a belief that Argentina had to ensure that its sovereign rights in the Argentine Antarctic sector and the Malvinas were secured not only through the political and legal means but also through the education of Argentine citizens on the geographical realities of the Republic. In 1948, Perón ordered the *Instituto Geografico Militar* (IGM) to produce new maps and charts of the Argentine Republic which displayed Antartida Argentina and the Malvinas. Henceforth, it was an offence to produce any maps which did not depict these polar and insular claims. Every map or chart of Argentina regardless of its actual geographical coverage had to depict the Argentine Antarctic sector in a corner of the map (see, for example, the cover of the Argentine journal *Revista Cruz del Sur* 1983, Figure 7.3). British and Chilean claims to the Antarctic were considered illegal or even irrelevant. Public education was used to promote geographical awareness of tri-continental Argentina (main-land, insular and polar Argentina) and in 1946 children were being instructed to memorize the territorial area of the country. By 1947, in the minds of its citizens, Argentina had expanded from 2.8 million to 4 million square kilometres and the Ministry of Foreign Affairs and Worship created a separate ministry for the Malvinas and the Argentine Antarctic sector (Dodds 1997: 52–3).

The construction of tri-continental Argentina reflected the considerable power of the map and survey. State-sanctioned histories and geographies of Argentina played an important role in explaining and legitimating past events and future aspirations. The Argentine historian Carlos Escudé has contributed a number of important studies which demonstrate how geography textbooks were altered to reflect the Perónist values concerning Antartida Argentina:

> In 1939, he [L. Dagnino Pastore. a popular textbook writer] wrote that Britain 'possess' more than eight million square kilometres in Antarctica (to which he applied the British term, the Falkland Islands Dependencies); in 1940 he changed the word 'possess' for the expression 'attributes to itself', adding that Argentina might get part of this if the criterion for the distribution of territory used in the Arctic were (*sic*) applied; in 1944 he stated that Argentina had 'unquestionable rights' . . . in 1946 he reports that Argentina has made it known to the world that it claims the Antarctic sector over which 'it has rights'; and finally in 1947 he writes matter-of-factually (*sic*) of an Argentine sector over which Argentina 'exercises authority.'
>
> (Escudé 1992:10)

Escudé concluded that Argentina under President Perón effectively perpetuated

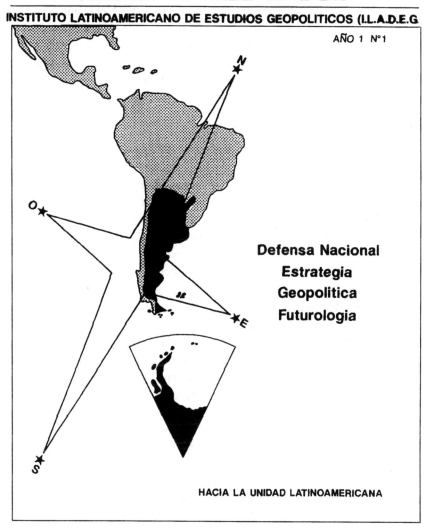

REVISTA

CRUZ DEL SUR

INSTITUTO LATINOAMERICANO DE ESTUDIOS GEOPOLITICOS (I.L.A.D.E.G

AÑO 1 N°1

N

O

E

S

Defensa Nacional
Estrategia
Geopolitica
Futurologia

HACIA LA UNIDAD LATINOAMERICANA

Figure 7.3 Cartographic image of tri-continental Argentina

Source: Revista Cruz del Sur, Argentine geopolitical journal

a particular narrative of territorial loss which stipulated that lost territories such as the Islas Malvinas, the Sandwich del Sur islands and the Argentine Antarctic sector had to be recovered. The 'intentional silences' of the maps concerning other territorial claims (e.g. Chile and the UK) were then reproduced on postage stamps, surveys, atlases and on public mural maps. In turn, these kinds of popular geographical motifs contributed to a public culture which concentrated on the recurrent idea that Argentina had been the victim of the grand designs of other powers.

In a speech to the Argentine War College in 1953, Perón developed his belief that Argentina also had to be cautious not only about her external relations but also to ensure that her primary resources were used to encourage the industrial development of the country:

> What I want to say, clearly, is that we are threatened that one day the over-populated and superindustrialized countries, that do not have food and raw materials, but do have extraordinary powers, will wield this power to despoil us of those elements that we possess a surplus with relation to our population and our necessities. Here is the problem set forth in its fundamental bases, but also the most objective and realistic.
>
> (Perón 1953 cited in Pittman 1981: 745)

As Edward Milensky has argued, Perónism encouraged and sustained a long-running debate within Argentina as to the role the country should play in international affairs. Perón's policy of Justicialism, non-alignment and self-sufficiency continued a nineteenth-century tradition of statist-nationalism in contrast to the more participatory forms of liberal internationalism (Milensky 1978). In the immediate post-war period, therefore, Argentina's foreign economic policies were predicated on establishing controls on foreign investment, diversifying markets and challenging the industrial privileges of the Northern states.

Ironically, therefore, the 1950s were often considered to have been relatively unimportant in the development of the Argentine geopolitical tradition because few explicit texts were produced in the country. While North American and British geographers were reluctant to use the term 'geopolitics' in their writings, geopolitical ideas about the state, development and territorial security circulated around the Argentine military academies and the public arena (Hepple 1986). A host of popular geopolitical sources comprising cartography, stamps and dictionaries such as the *Diccionario Historico Argentino* reproduced Perónist thought concerning Argentina's Antarctic destiny and politico-economic security (see, for example, Piccirilli *et al.* 1954). Specific Argentine texts such as Isola and Berra's 'Introduction to Argentine Geopolitics' were reprinted and continued to circulate around the military academies. When

additional books such as *Tratado General de Geopolitica* by the Spanish writer
Vicens Vives became available in Argentina in 1950, followed by Albert Escalona
Ramos' *Geopolitica Mundial y Geoeconomica* published in 1959 (after initial expo-
sure in Mexico), Vicens Vives' work in particular enjoyed many citations by
Argentine geopolitical writers in the 1960s and 1970s, not least because of his
concern for robust governance, territorial security and historical awareness:

> Geopolitics is the science of vital space. It summarizes the results from
> historical geography and political geography in an explanatory synthesis
> that it seeks to apply to the consideration of political and diplomatic
> events. It does not concern itself only with the science of geography.
>
> (Vicens Vives 1950: 76)

Whilst it would be fair to conclude that there was little published material
available on Argentine geopolitical affairs until the return of major geopolitical
monographs in the 1960s and 1970s, it would underestimate the thorough infu-
sion of geopolitical ideas and the complex politics of geographical knowledge
within post-war Argentina.

Internal security and external dangers: Argentine geopolitics and the military

In the aftermath of the Second World War, Argentine officers participated with
other Latin American officers in Brazil, Chile, Peru and Venezuela in various intra-
military debates over geopolitics, national security and strategy (Pittman 1981;
Ewell 1982; Gorman 1982; Child 1985). In Brazil, for example, leading geopolitical figures such as General Golbery do Couto de Silva were re-working Ratzelian
themes on the organic state in conjunction with European and North American
writings on national security and development. In contrast to some of the French
literature on counter-insurgency and security, Brazilian geopolitical writers
emphasized a more comprehensive vision of 'societal security' rather than a
limited operational view about military tactics in limited encounters. However, it
became apparent that Argentine military officers in the 1950s were also being
exposed to articles on French counter-insurgency theories through the Argentine
military journal, *Revista de la Escuela Superior de Guerra*. The development and
evolution of national security doctrines in the Southern Cone became a defining
feature of the military regimes that dominated political life between the mid-
1960s and the 1980s (Hepple 1992: 146–7). As David Pion-Berlin noted:

> Counter-insurgency principles, for example, first made their entry into

Argentina during the late 1950s with the visit of French military missions. Officers who would later rule during the Proceso [1976–1982] had their formative training and indoctrination into military life at about this time. . . . A flurry of articles (many of them authored by French officers) soon appeared in Argentine military journals . . . warning of Argentina's vulnerability to international communism and preparing the country for countersubversive struggle.

Pion-Berlin 1989: 99

Ironically, the Perón and post-Perónist governments of Frondizi and Illia were never dominated by revolutionary socialism even when Perónist followers were anxious to protect the social and labour rights of the working classes.

By the mid 1960s, military governance or forms of 'bureaucratic-authoritarianism' began to emerge which witnessed the armed forces seeking to re-organize the nation armed with the doctrines of national security and developmentalism. Inspired by Brazilian geopolitical writings on security and development and American counter-insurgent training in institutions such as the School for the Americas (in Georgia, USA) and the Inter-American Defense (sic) College in Washington DC, Argentine and other Southern Cone National Security Doctrines (NSD) sought to contain the danger of internal wars in the wake of the global threat of communism and the emergence of Castro's socialist Cuba in 1959. Later presidents of Argentina, such as General Galtieri studied at the School for the Americas and was instructed in counter-subversion training. According to Colonel Mario Horacio Orsolini, hard-line French and American views on counter-insurgency had been widely accepted by the Argentine armed forces in the mid-1960s (cited in Pion-Berlin 1989: 99).

In contrast, European geopolitical authors such as Michel Foucher and Leslie Hepple argued that the genealogy of the national security doctrine lies not only in France and the United States but also in South American states such as Brazil (Foucher 1986; Hepple 1986, 1992). The Brazilian geopolitical writer and leading military figure, Golbery do Couto e Silva began to revise traditional geopolitics, including the organic theories of the state. In doing so, Golbery suggested that the dangers facing the polity were not only located in the international realm but were also to be found inside the corpus of the state. The Cold War climate of anti-communism played a significant role in shaping these new-found fears concerning the internal health of the nation-state. Events such as the 1959 socialist revolution in Cuba and the activities of high-profile figures such as Che Guevara in combination with the failure of US operations such as the Bay of Pigs fiasco in 1961 cemented a widely held view within the South American military that communism was escalating in the region (Barton 1997: 62–4). In contrast to Northern hemispheric formulations, Golbery's writings were not limited to a concern for limited counter-insurgency strategies but were also concerned to demonstrate that national security was tied to the overall economic, political and even cultural development of the state (Hepple 1992: 147).

Within Argentina, the military borrowed Golbery's ideas about national security in the 1970s and then argued (as it had done since the 1930s) that civil society and the national economy were threatened by subversive forces which sought to under-mine the Christian and occidental values (and spaces) of the nation-state. By generating significant links with industry and commercial interests, the armed forces sought to identify and then destroy those elements in society which were considered 'subversive' and/or 'dangerous'. Underwritten by a strident form of anti-communism and a hostility to socialist economics, military leaders such as Ongania, Videla and Viola sought to use these apparent 'challenges' as a means of revitalizing the state and its citizens. Political instability, turmoil and labour strikes in Argentina were interpreted in the 1960s and 1970s as growing evidence that the country was threatened by ideological subversion that touched the entire corpus of society from the family to the professions (Timmerman 1981; Pion-Berlin 1989).

During this period, the military intervened in Argentine politics in 1963, 1966 and 1976. On each occasion, military leaders argued that the threat of communist aggression combined with economic failure meant that the armed forces had to defend the state against murky dangers. Armed with the doctrine of national secu-rity and the organic metaphor of the state, military figures employed geopolitical discourses to geographically locate these dangers and thereafter to 'purify' these contaminated spaces. New institutions such as the *Instituto Argentino de Estudios Estrategicos y de las Relaciones Internacionales* (INSAC) and the *Instituto de Estudios Geopoliticos* (IDEG) were created in Buenos Aires for the purpose of providing expert commentary on the challenges facing Argentina. Headed by General Juan Enrique Gugliamelli, the chief of the National Development Agency during the Ongania military regime (1966–1971), INSAC published the influential journal *Estrategia* between 1969 and 1983. The style of *Estrategia* was highly nationalistic in the sense that most of the contributors adopted an approach to their work which stressed the need for the Argentine state to develop areas such as Patagonia and to ensure that strategic places such as the River Plate, the Malvinas, the Antarctic and the Beagle Channel were not dominated by rival states such as Brazil, Chile and or the UK (Gugliamelli 1974, 1983).

Three major geopolitical works appeared in the mid-1960s: Justo Briano's *Geopolitics y Geostrategica Americana*, Atencio's *Que es la Geopolitica*, and Milia's *Estrategia y Poder Militar*. There are a number of similarities within these three texts not least because all were written by serving officers of the Argentine armed forces. Justo Briano, a former colleague of Emilio Isola at the Army War College in the early 1940s, who had then been promoted to the Chief of Staff of the General Command of the Military Regions in the late 1940s, wrote an account of Geopolitics and American Geostrategy which was a robust defence of the South

American engagement with German geopolitical writings, and argued that Argentina should borrow American and Brazilian ideas about world strategy (including Golbery's about national security) in order to prepare for the next century as a world leader in international affairs. Jorge Atencio, a former Colonel in the Argentine army, argued that geopolitics should seek to provide guidance for the statesman (sic) and help to identify the territorial and resource needs of the state. He was also bitterly critical of the condemnation of German geopolitics in the 1940s by writers such as Isaiah Bowman and Robert Strausz-Hupé. Atencio believed that the association of German fascism should not be allowed to 'contaminate' the basic geopolitical significance of territory and politics. Fernando Milia was an Admiral in the Argentine Navy and head of the Argentine Institute for Strategic Studies when his book appeared in 1965. In contrast to the alleged expansionist strands of German *Geopolitik*, all these writers sought to promote the development of existing Argentine territories (even the disputed Malvinas and the Antarctic peninsula) rather than advocate the annexation of further territories in South America and beyond.

In charting the rising interest in geopolitics within military academies and the wider public in the late 1960s, General Guglialmelli argued that two basic reasons explained this transformation: first, there was a growing dissatisfaction with the current civilian governance of Argentina and second, there were widespread fears that Argentina's territories were threatened by Chile, Brazil and the United Kingdom. Moreover, the ever-present danger posed by alleged Soviet-backed revolutionaries in Latin America contributed to a climate of fear and uncertainty. In the late 1960s and early 1970s, however, Gugliamelli's geopolitical writings were strongly infused with concerns over Brazilian expansionism in the River Plate region:

> The frontier [so Brazilian theorists claimed] is a force in the service of political contingencies, an 'isobar' which establishes the equilibrium between two pressures. This notion clearly indicates the extreme sensitivity and significance which Brazil attaches to its traditional border policy, which has been in evidence since the violation of the Treaty of Tordesillas.
>
> (cited in Child 1979: 96)

This obsession with Brazil's activities north of the River Plate was rooted in a strong perception of territorial injustice in the post-colonial period. The imperial division of South America in the fifteenth century was frequently interpreted by Gugliamelli as a quasi-divine intervention (rather than an imperial imposition) which had been unfavourably altered against Argentina in favour of Brazil, Chile and Paraguay. As Carlos Escudé has noted, this form of extreme territorial nationalism was predicated on a bizarre reading of history and geography far removed from the realities of a situation characterized by the extermination of indigenous populations, incomplete imperial power and post-colonial battles for territory and resources. However, leading figures within Argentine politics such as Italo Luder, the former head of the Argentine Senate's Foreign Relations Committee, often drew upon these analyses by Gugliamelli and Nicholas Boscovich to draw attention to Brazilian territorial expansion in the Rio Plata basin (Boscovich 1974–5).

General Gugliamelli's geopolitical writings occupied a position of consider-able significance within military-dominated Argentina (1976–1982). As with other high profile writers such as General Osiris Villegas, a former Interior Minister and key negotiator during the Beagle Channel crisis with Chile, Gugliamelli was head of the Fifth Army Corps, Head of the War College and a member of the Ongania military government between 1966–1971. As an army officer with responsibilities for the frontier regions such as Patagonia and the South Atlantic, Gugliamelli articulated an agenda which emphasized that Argentina's geographical disposition was peninsular: the country was composed of distant frontier regions both North and South (Dodds 1997: 64–8). Using distinct cartographic images, he argued that Argentina had to secure and develop these frontier regions in order to prevent other powers from compromising national sovereignty. Whilst his analysis rarely engaged with the writings of Kjellén and others, Gugliamelli's geopolitics was a complex mixture of organic views of the states combined with quasi *dependista* arguments concerning Argentina's economic vulnerability. As he noted in *Geopolitica del Cono Sur*:

> The agro-exporting role of the country, that has led inevitably to [our] vulnerability that condemns us to external dependency, reduces the capacity for national decision-making and does not permit the satisfac-tion of [fulfilling] necessities of well-being and prosperity. And in the attainment of national security, it restricts the freedom of strategic action, it creates grave vulnerabilities in the power relations among the member countries of the Southern Cone, and internally it confronts us with a picture of permanent dissatisfaction and social agitation.
>
> (Gugliamelli 1979: 255)

Argentine geopolitics, therefore, re-emerged in the 1970s in a wider setting which linked the organic state to security and development, disputed territories, regional development and particular regions such as Patagonia and the Falklands/Malvinas (Hepple 1992). His colleagues at the INSAC and other officers such as General Goyret were equally adamant that the dangers facing Argentina lay inside and outside the boundaries of the state. Goyret, for instance, later established a journal called *Armas y Geoestrategia* which sought to translate Golbery's ideas on security and development for an Argentine context which stressed the development of remote regions and the need to counter threats inside and outside the state (Goyret 1980). More generally, these ideas about national security were used by the ruling military regimes of the late 1960s and mid 1970s to transform concerns over limited terrorist activities by groups such as the Montoneros into a 'dirty war' against labour, student and church groups who were deemed to be a threat to the internal security of the state. The activities of paramilitary groups such as the Alianza Anticommunista Argentina (created in 1974) in combination with vicious national security policies identified so-called 'subversives' who were then kidnapped, tortured, murdered and imprisoned by the military and secret police (Staub 1992: 217).

The geographical coverage of the national security doctrine in Argentina was also substantial:

> It was geographically comprehensive, reaching into the most remote corners of the nation. From the tropical provinces of Misiones, to the windswept and sparsely populated expanses of Patagonia, no part of the country was left unaffected. The armed forces established a set of security zones, subzones, and areas that effectively parcelled the territory into increasingly smaller units.
>
> (Pion-Berlin 1989 cited in Scarpezi and Frazier 1993: 7)

While the military were fighting a 'dirty war' against their own citizens, military planners and geopolitical writers such as Gugliamelli and Carlos Moneta were either advocating the re-capture of the Falklands and/or the need to defend the South Atlantic from communist or even Brazilian expansionism. Professor Moneta, for example, warned in 1975 that Brazil would attempt to occupy Argentine Antarctica by 1990 because the Brazilian military recognized the strategic significance of the polar continent and the Drake's passage (Moneta 1975: 29–30). The South Atlantic and the Malvinas also offered the military an opportunity to articulate a new sense of mission for the country, and military authors such as Vicente Palermo, Fernando Milia and Pablo Sanz argued that the oceanic spaces surrounding Argentina offered unprecedented chances for economic development and for demonstrating the country's commitment to the western community of Christian nations. Discourses of danger and threat were inserted into the maritime spaces surrounding Argentina, as the armed forces sought to create a South Atlantic Treaty Organization (SATO) with the assistance of the United States (Dodds 1997: 64–8). In the process these representations of the South Atlantic presented an opportunity for securing particular political identities. Argentina was described as a Christian and western state, threatened by Soviet-inspired subversive forces. As Patricio Siliva noted,

In this new foreign policy the concept of *Occidentalidad* or the western nature and orientation of the country was reformulated. In the recent past the military governments had utilised this principle to legitimise their active anti-communism on the international scene.

(Siliva 1989: 89)

In contrast to the more extremely nationalist currents of geopolitical thoughts in the journal *Estrategia*, the journal *Geopolitica* advocated internal integration projects rather than annexation. Under the leadership of Andres Bravo and Augusto Rattenbach, *Geopolitica* was published by the IDEG, and the earliest editions of the journals concentrated on regional integration, national development and international co-operation. One of the original board members, Cirigliano was a strong source for Perónist ideals about continental integration, non-alignment and the need to develop regional areas such as Patagonia. In his major text, *Argentina Triangular: Geopolitica y Projecto Nacional*, Cirigliano outlined his geopolitical vision of an Argentina which had overcome two great weaknesses: the under-exploitation of its 'open spaces' in Patagonia and Antarctica and the over-concentration of people and industries in the Buenos Aires province.

In the midst of the terror and the introduction of secret detention centres, Cirigliano's geopolitical writings on territorial integration appear at odds with the devastating impact of the 'dirty war' on Argentine society. This obsession with integration was, however, in keeping with the military's concerns that Argentina was a divided and vulnerable country. As Cirigliano noted with reference to the integration of the La Plata basin: 'for a country with a destiny – as we continue to believe Argentina has – whose geography has been dismantled and her history adulterated, the primary objective is to repair her geography and her history' (Cirigliano 1975: 36–7).

The formal geopolitical reasoning of *Estrategia* and *Geopolitica* had been supplemented by a range of institutions and sources which ensured that by the late 1970s school-level courses on geopolitics were coinciding with formal training offered in military colleges and the Argentine Foreign Service Institute (Pittman 1981: 858). Most of the formal teaching was unquestionably reactionary and devoted to endorsing the security policies of the military regimes. This position was without doubt problematic given that the military later concluded that teaching was a site for subversion. In 1980, for example, the military issued a publication on terrorism in Argentina which concluded that the introduction of subversive thought had been facilitated by teachers:

Subversive operations were carried out by biased teachers who, because of their pupils' age, easily influenced their minds' sensibility. The instruction was direct, using informal talks and readings of

173

prejudiced books published to that effect. Using children's literature, terrorism tried to convey the kind of message which would stimulate children, and make room for self-education, based on freedom and the search for 'alternatives'.

(cited in Staub 1992: 216)

The military blamed educated professionals such as teachers, university lecturers and even priests for corrupting the young. In a parallel with Argentina in the 1890s, military elites identified public education as a critical mechanism for the infiltration of subversive, immoral and secular thought (Comblin 1980).

In the aftermath of the regime of terror, a number of Argentine scholars critically engaged with the problematic relationship between geopolitics and national security. In their Marxist-inspired analysis of the 1982 Malvinas War, Dabat and Lorenzano argue that the issues raised by geopolitical writers such as Villegas and Gugliamelli were used by the military regimes to divert attention from pressing social and economic reform within Argentina at the expense of chasing territorial ambitions in the South Atlantic and Antarctica. While these authors' express support for the Argentine claim to the Malvinas, they refused to accept the intellectual terrain of the Argentine geopolitical tradition: 'certainly the Malvinas are important. But of much greater importance is the recovery of the national territory for the people's use' (Dabat and Lorenzano 1984: 223).

In a similar if distinct vein, Carlos Escudé published a series of important articles and books (in English and Spanish) in the 1980s which sought to contest the orthodox territorial histories of Argentina. He argued, amongst other things, that Argentine geopolitical and historical writers had perpetuated a serious fallacy regarding the territorial evolution of the Republic. In contrast to traditional writings, Escudé presents a detailed evaluation of Argentina's post-independence history which acknowledges the fact that the Republic was extremely successful in claiming and colonizing territory in the North (Chaco) and the South (Patagonia). Moreover, Escudé has emphasized the rhetorical and political significance of Argentine geopolitics in terms of being able to mobilize popular support (and associated geographical imaginations) for territorial adventures such as the Falklands/Malvinas (called the Little Sisters in Argentine mythology) invasion rather than to concentrate on fundamental socio-economic reform (Escudé 1984, 1987, 1988, 1992).

Carlos Reboratti of the University of Buenos Aires (UBA) published a powerful critique of Argentine geopolitical thinking (Reboratti 1983). Under the title of 'The Charm of Darkness', Reboratti identified a number of facets of geopolitical reasoning which had been employed during the period of military rule between 1976 and 1982. He argued that military writers and officials had used geopolitics in a crude fashion in order to justify particular national security strategies and development programmes which in turn had attracted very little critical evaluation from social science scholars. At the heart of geopolitics lies a tremendous degree of vagueness about the defining theories and concepts which enabled the military to think geopolitically. Writers such as Briano, Fraga and Guglialmelli often reduced geopolitics to a crude form of geographical deter- minism which failed to capture the complexities of the late modern world. While few critical scholars in Argentina may not have been attracted to the legacies of geopolitical thinking, Reboratti argued that there was an important need to inter- rogate these ideas and concepts because of their significance in determining national policies such as the Falklands invasion in 1982 and the near war with Chile in 1978 over the ownership of the Beagle Channel.

Reboratti's critique of geopolitics linked forms of geographical reasoning to aggressive forms of nationalism and militarism. His line of reasoning was strongly influenced by Lacoste's stinging critique of the role of geographical knowledge and academic geography's links to military planning, war and violent nationalism (Lacoste 1976). Given the immediate aftermath of the Dirty War, it is not surprising that Reboratti's critique of Argentine geopolitics appears uncompro- mising. The violence of the military regimes in the 1970s and early 1980s generated considerable bitterness, anxiety and a culture of fear. As Carina Perelli reminds us:

> From 1974 onward, innocent bystanders disappeared from the scene: everybody – rich and poor, men, women and children, workers and busi- nessmen, students and homemakers – was at least potentially at risk of being caught in the grinding machine of destruction.
>
> (Perelli 1992: 418)

The organic framework of the state had played its part in the violence of the state and helped to sustain a suspicious and untrusting public culture which only a few groups such as the Mothers of the Plaza del Mayo dared to resist (Radcliffe 1993).

South American geopolitics and Anglo-American engagement

In the late 1970s and early 1980s, Professor Jack Child of the American University was the foremost authority on South American geopolitical writings. As an American citizen born in Buenos Aires, Child had an intimate and sympathetic knowledge of Argentine nationalism and geopolitical writing. In later life, as an army officer who specialized in US-Latin American relations, he encountered the geopolitical writings of Armed Forces officers in Argentina, Brazil and Chile. In alliance with other colleagues such as Howard Pittman and Lewis Tambs, Child produced some of the earliest English-language papers and later books on South American geopolitical thought (Child 1985, 1988). His 1979 paper on 'Geopolitical thinking in Latin America' was his mainstream appraisal of geopolitical thought in the so-called ABC (Argentina, Brazil and Chile) countries (see also Tambs 1979). Subsequently, his student and fellow US army officer, Colonel Howard Pittman, produced a monumental doctoral study on South American geopolitics published in 1981 which included interviews with the then President of Chile, General Pinochet and many of the leading authors of geopolitics (Pittman 1981). As former US army officers, Pittman and Child enjoyed access to these South American geopolitical authors/officers at a time of violent military dictatorship in Argentina and other South American countries. Both authors subsequently published their work in Spanish in journals such as *Geopolitica* (Argentina) and *Geosur* (Uruguay).

There were two elements in this North American analysis of South American geopolitical thought. The first was based on a desire to re-negotiate the traditional histories of geopolitics which had been located in the experiences of the Anglo-American world. South American geopolitics presented a stark contrast to the intellectual decline in English-speaking geopolitical exchanges in the 1950s and 1960s. The contested legacy of *Geopolitik* and Nazism was not evident in Argentina, Brazil and Chile in the sense that writers such as Jorge Atencio and General Golbery of Brazil presented significant surveys of European and American writers including Karl Haushofer. In doing so, Child and Pittman drew Anglophone scholars' attention to the fact that the so-called 'decline of geopolitics' was not occurring in South America, where numerous military and civilian authors were constructing analyses of South America's position within the international system, and engaging with French, German and Anglophone geopolitical literature.

The second element of this assessment process was predicated on an analysis of this material in terms of the historical and geographical circumstances of these Southern Cone states. Jack Child, for example, argued that the Argentine geopolitical journal *Estrategia* was 'Latin America's (and possibly the world's) most sophisticated and penetrating journal of geopolitics' (Child 1979: 95). Generally speaking, their assessments were sympathetic (there was very little intellectual competition from other geopolitical journals *vis-à-vis Estrategia*) to the geopolitical agendas of these South American writers, even if Child and a later co-author Phil Kelly were to complain that 'geopolitics is conceptually and theoretically vague. An adequate geopolitical model does not exist' (Kelly and Child 1988: 9). This was a common complaint about geopolitics in South America because it was inferred that this body of writing was incapable of producing vigorous models of foreign policy and strategies of national development. Most of the South American authors who contributed to important English-language collections such as the *Geopolitics of the Southern Cone and Antarctica* were more concerned with developmental priorities for particular countries rather than articulating a robust theorization of the 'geopolitical'. In part, this was a reflection of the fact that many South American geopolitical writers possessed a strong commitment to developing applied policies and programmes rather than enter the ephemeral world of geopolitical traditions.

In his recent account of *Checkerboards and Shatterbelts*, Phil Kelly returns to South American geopolitical thinking by providing a wide-ranging overview which remains pessimistic about the capacity for intellectual exchange between Anglo-American political geography and South American variants:

> I doubt that the new geopolitical theories described in the British journal *Political Geography* and like sources will receive much acceptance among South American academicians, who will likely continue to follow traditional geopolitical thinking.
>
> (Kelly 1997: 211)

He believes that geopolitics will remain an important facet of national policy-making in South America even if many of the authors pursue traditional themes of national development, territorial integration and continental co-operation. The reasons for this reluctance to embrace the Anglophone literature of critical geopolitics, for example, lies in the fact that many South American geopolitical writers are retired military officers and civil servants with a very different cultural, geographical and linguistic world-view. In that sense, the issue of 'acceptance' does not enter the equation because many geopolitical writers in Argentina would be distinctly uneasy with recent critical geopolitical debates over the declining significance of territory and state sovereignty (Dodds 1998). However, it is also apparent that Kelly's appraisal of those who worked on South American geopolitical themes excludes a range of critical scholars such as Carlos Reboratti and Carlos Escudé who took a more robust view of the intellectual culture that has sustained these ideas about territory, nationalism and international relations.

In contrast, British and French authors such as Leslie Hepple and Michel Foucher have sought to problematize the intellectual heritage of Iberian and South American geopolitical thought. Hepple's critique of the organic metaphor of the state is a significant interrogation of geopolitical thinking not least because it is situated in the unstable and sometimes violent political and cultural conditions of South American regimes and governments (Hepple 1992). However, as Hepple has noted, this does not imply that all currents of South American geopolitical thinking are intimately connected to militarism and power politics; rather it highlights how ideas about space and territory are negotiated through particular national cultures and intellectual traditions (see, more generally, Livingstone 1992; Driver 1994). Moreover, it can also be used to illustrate how German ideas on the nation-state travelled to South America and were adapted and negotiated. The evolution and subsequent adoption of the national security doctrine also provides another fascinating example of how ideas developed in France and the USA were received in Brazil and then further developed and negotiated in Argentina and Chile.

Conclusions

Recent historians of geopolitics have drawn attention to the role geographical knowledge has played in shaping public cultures and specific geopolitical ideas such as the organic metaphor of the state (Livingstone 1992; Escolar et al. 1994). The South American geopolitical literature has invited much commentary in terms of exploring the interconnections between territorial growth, political programmes, geographical education and international boundaries. Some of these linkages were unpleasant and violent, as Argentine geopolitical thinking was unquestionably linked to the ideologies and policies of the military regimes of the 1960s and 1970s. While the horrors of these years have been recorded (in part) in formal government reports such as *Nunca Mas*, the connections between Argentine geopolitics and the military remain to be thoroughly documented even if there is evidence that geopolitical writers contributed to military programmes on education and policy development (Pittman 1981, Child 1985).

The geopolitical journal *Estrategia* ceased to exist with the death of General Juan Gugliamelli in 1983. In general, formal geopolitical writing in Argentina suffered a temporary crisis of confidence during the immediate years of the democratically-elected Alfonsin administration (1983–1989). However, the journal *Geopolitica* continues to attract a number of long-standing geopolitical writers and editorial board members such as Nicholas Boscovich, Carlos Moneta and Hugh Gaston Sarno, to address the developmental and security needs of Argentina (Sarno 1997). Under the editorship of Andres Alfonsin Bravo, *Geopolitica* produces a bi-annual magazine which is sold in the leading bookshops of Buenos Aires and other major towns. It has succeeded in attracting international contributions, largely through Bravo's daughter's ability to translate English manuscripts into Spanish. Moreover, the publishers Editorial Pleamar continue to produce Spanish-language editions of North American and European geopolitical theorists such as Karl Haushofer, Halford Mackinder and Saul Cohen.

For some academics in universities and institutions such as the University of Buenos Aires and Cordoba, Argentine geopolitics continues to be seen as a reactionary doctrine. More critical forms of reflection have not developed within the writing circles of *Geopolitica*, possibly because many of the contributors remain former military officers and officials who are unfamiliar with the more recent debates concerning the politics of geographical knowledge, let alone critical geopolitics (cf. Sarno 1997; Ó Tuathail 1996). However, some geographers attached to the Institute of Geography at UBA (created in Perónist Argentina) such as Marcelo Escolar and Carlos Reboratti have led the way in terms of critically evaluating geographical discourses through their careful analyses of the institutional development of Argentine geography and contemporary proposals for the development of Patagonia and the relocation of the capital to Viedma (Escolar 1991; Reboratti 1982, 1987). As Carlos Reboratti noted, the then President Raul Alfonsin, employed the term 'geopolitics' to justify this initial proposal of relocating the capital southwards: 'its geopolitical value for access to the Pacific and defending national sovereignty, its symbolism as the southernmost capital in the world, its visionary model on the threshold of a twenty-first century' (Alfonsin 1986 cited in Reboratti 1987: 76).

In his trenchant critique of such proposals, Reboratti demonstrates how long-standing Argentine geopolitical ideas have sustained particular geographical imaginations based on a belief that Argentina's 'under-populated' southern regions need to be developed and populated, to ward off Chilean incursions. It also serves to remind observers of Argentine affairs that particular geographical imaginations (such as a concern over Argentina's 'empty spaces') can remain powerfully ingrained even if there is a shift away from the intellectual terrain of geopolitics.

The belief in a 'Southern project' such as the recovery of the Malvinas remains a powerful feature of Argentine political culture. President Carlos Menem (1989–1999) has been a staunch advocate of such a policy in spite of the country's commitment to democratic government and peaceful diplomacy. He has declared on numerous occasions, in Argentina and elsewhere, his commitment to the retrieval of the Falklands/Malvinas by the year 2000. Public education in Argentina continues to teach children from Grade I onwards that the Malvinas remain Argentine and promotes the teaching of patriotic songs such as the 'March of the Malvinas'. This is not to imply, however, that Argentina is hell-bent on the violent recovery of the Falklands; rather it is to illustrate that particular geographical imaginations can retain considerable symbolic appeal. Presidents Alfonsin and Menem have, in their different ways, sought to exploit long-standing interest in specific territorial spaces at the expense of more fundamental social and economic reform in the context of welfare, health and education.

Within institutions such as the *Instituto Torcuato di Tella* and FLASCO in Buenos Aires, political scientists such as Roberto Russell and Carlos Escudé have also interrogated the foreign and security policies of post-democratic Argentina. In particular, Carlos Escudé has illustrated how Argentine public culture and constitutional governance has tended to stress the territorial rights of the state above and beyond the rights of the populace (Escudé 1984, 1992). In his first critical examination of Argentine politics and society, *La Argentina: Paria Internacional?* the publishers Editorial del Belgrano insisted that Escudé include a question mark in front of the title in order to add a degree of uncertainty concerning Argentina's standing in the world during the military regimes of the 1970s. During the early 1980s, contemporary Argentine geopolitical writing was investigated and critiqued in a manner which had not previously been possible in the restrictive and often suffocating public culture of the 1970s.

While much has changed since the dangerous and violent days of the military regimes, geopolitical discourse remains a significant force within wider Argentine public culture. The integrationist strand of *Geopolitica* touches upon the major debates in Argentina during the 1980s and 1990s; regional economic integration and MERCOSUR, democratic governance and non-alignment, globalization and US-Latin American relations. Geopolitics continues to provide a pathway for the discussion of territorial, resource and developmental questions with particular reference to a democratic Argentina. However, well-established geopolitical agendas such as territorial security and boundary negotiation also remain prominent not least because the recovery of the Malvinas dominated President Menem's priorities. Yet, it is likely that conservative strains of Argentine geopolitical thinking can enter the new millennium with some confidence. As Hans Weigert once noted, 'Each nation has the geopolitics it deserves' (Weigert 1942: 23).

Acknowledgements

I owe a debt of gratitude to David Atkinson and Leslie Hepple for their very helpful comments on earlier drafts of this paper. Thanks to Aloisius for his editorial advice. I also owe thanks to the Arts and Humanities Research Board for a 1998–9 Research Award which allowed this chapter to be completed. The usual disclaimers apply, however.

Bibliography

Alberdi, J. (1879) *Reconstruccion geografico de America del Sur*, Buenos Aires: La Facultad.

Andrews, G. (1979) 'Race versus class association: Afro-Argentines in Buenos Aires 1850–1900', *Journal of Latin American Studies* 11: 19–39.

Andrews, G. (1980) *The Afro-Argentines of Buenos Aires 1800–1900*, Madison: University of Wisconsin Press.

—— (1986[1965]) *Que es la Geopolitica?*, Buenos Aires: Editorial Pleamar.

Barton, J. (1997) *A Political Geography of Latin America*, London: Routledge.

Boscovich, N. (1974–5) 'Analisis comparativo: Argentina y Brasil en el espacio geoeconomico del Cono Sur', *Estrategia* 31–2: 34–60.

Briano, J. (1965) *Geopolitica y Geoestrategia Americana*, Buenos Aires: Circulo Militar.

Calvert, P. (1989) 'The primacy of geopolitics', The World Today 45: 33–6.

Calvert, P. and Calvert, S. (1989) Argentina: Political Culture and Instability, London: Macmillan.

Campbell, D. (1992) Writing Security, Manchester: Manchester University Press.

Campo Wilson, J. (1920) Geografia Politica de America, Buenos Aires: Editorial Americale.

Casellas, A. (1974) El Territorio Olvidado, Buenos Aires: Centrol Naval, Instituto de Publicaciones Navales.

Celerier, P. (1953) Geopolitica y Geoestrategia, Buenos Aires: Editorial Pleamar.

Child, J. (1979) 'Geopolitical thinking in Latin America', Latin American Research Review 14: 89–111.

—— (1985) Geopolitics and Conflict in South America, New York: Praeger.

—— (1988) Frozen Lebensraum: South American Geopolitics and Antarctica, New York: Praeger.

Cirigliano, G. (1975) Argentina Triangular: Geopolitica y Projecto Nacional, Buenos Aires: Humanitas.

Comblin, J. (1980) The Church and the National Security State, New York: Orbis.

Crawley, E. (1984) A House Divided 1880–1980, London: C. Hurst.

Dabat, A. and Lorenzano, L. (1984) Argentina: The Malvinas and the End of Military Rule, London: Verso. Expanded and revised translation of 'Conflicto Malvinese y Crisis Nacional', Mexico City: Teoria y Politico, 1982.

Daus, F. (1950) Geografia y Unidad Argentina, Buenos Aires: Centro Naval, Instituto de Publicaciones Navales.

Dijkink, G. (1996) National Identity and Geopolitical Visions, London: Routledge.

Dodds, K. (1993) 'Geography, identity and the nineteenth century Argentine state', Bulletin of Latin American Research 12: 361–81.

——. (1997) Geopolitics in Antarctica: Views from the Southern Oceanic Rim, Chichester: Wiley.

—— (1998) 'Political geography I: The globalization of world politics', Progress in Human Geography 11: 595–606.

Driver, F. (1994) 'New perspectives on the history and philosophy of geography', Progress in Human Geography 18: 92–100.

Escolar, M. (1991) 'A harmonia ideal de um territorio fictio', in Actas de Conferencia A Cuestao Regional e os Movimento Sociais no Terceiro Mundo, Sao Pablo: USP-UNESP.

Escolar, M., Quintero, S. and Reboratti, C. (1994) 'Geographical identity and patriotic representation in Argentina', 346–66 in D. Hooson (ed.) Geography and National Identity, Oxford: Blackwell.

Escudé, C. (1984) Argentina: El Paria Internacionale?, Buenos Aires: Editorial del Belgrano.

—— (1987) Patologia del Nacionalismo: el caso Argentino, Buenos Aires: Instituto de Torcuato Di Tella/Editorial Tesis.

—— (1988) 'Argentine territorial nationalism', Journal of Latin American Studies, 20:139–65.

—— (1990) 'El Fracaso del Proyecto Argentino', Buenos Aires: Instituto de Torcuato Di Tella/editorial tesis.

—— (1992) 'Education, public culture and foreign policy: the case of Argentina', Working Paper Series of Duke-UNC Program of Latin American Studies.

—— (1997) Foreign Policy Theory in Menem's Argentina, Gainsville: University of Florida Press.

Ewell, J. (1982) 'The development of Venezuelan geopolitical analysis since World War II', *Journal of Interamerican and World Affairs* 24: 295–320.

Foucher, M. (1986) *L'invention des Frontières*, Paris: Foundation pour les Etudes de Défense Nationale.

Fraga, J. (1979) *Introduccion a la Geopolitica Antarctica*, Buenos Aires: Direccion Nacional del Antarctica.

Gerassi, M. (1964) 'Argentine Nationalism of the Right', unpublished Ph.D. thesis, Department of History, Columbia University.

Gorman, S. (1982) 'Geopolitics and Peruvian foreign policy', *Inter-American Economic Affairs* 36: 65–88.

Goyret, J. (1980) *Geopolitica y Subversion*, Buenos Aires: Ediciones Depalma.

Gugliamelli, J. (1974) *La Cuenca del Plata*, Buenos Aires: Tierra Nueva.

—— (1979) *Geopolitica del Cono Sur*, Buenos Aires: El Cid Editor.

—— (1983) 'La Crisis Argentina: una perspectiva geopolitica', *Estrategia* 73/74: 9–30.

Hashbrouck, A. (1935) 'The conquest of the desert', *Hispanic American Historical Review* 15: 198–228.

Henrik-Herb, G. (1989) 'Persuasive cartography in Geopolitik and national socialism', *Political Geography Quarterly* 8: 289–303.

Hepple, L. (1986) 'The revival of geopolitics', *Political geography Quarterly* 5: 21–36.

—— (1992) 'Metaphor, discourse and the military in South America', 136–54 in T. Barnes and J. Duncan (eds) *Writing Worlds*, London: Routledge.

Hodge, J. (1984) 'The role of the telegraph in the consolidation and expansion of the Argentine Republic', *The Americas* 41: 59–80.

Isola, E. and Berra, A. (1950) Introduction a la Geopolitica Argentina, Buenos Aires: Editorial Pleamar.

Kelly, P. (1997) *Checkerboards and Shatterbelts: The Geopolitics of South America*, Austin: University of Texas Press.

Kelly, P. and Child, J. (eds) (1988) *Geopolitics of South America and Antarctica*, Boulder: Lynne Rienner.

Lewis, C. (1983) *British Railways in Argentina 1857–1914*, London: Athlone Press.

Livingstone, D. (1992) *The Geographical Tradition*, Oxford: Blackwell.

McLynn, F. (1980) 'The political thought of Juan Domingo Perón', *Journal of Latin American and Caribbean Studies* 32: 15–23.

Milia, F. (1965) *Estrategia y Poder Militar*, Buenos Aires: Editorial Circular.

Milensky, E. (1978) *Argentina's Foreign Policies*, Boulder: Westview.

Moneta, C. (1975) 'Antartida Argentina: los problemas de 1975–90', *Estrategia* 31–2: 23–36.

Morris, A. (1996) 'Geopolitics in South America', 272–93 in D. Preston (ed.) *Latin American Development: Geographical Perspectives*, Harlow: Longman.

Newton, R. (1992) *The Nazi Menace in Argentina 1931–1947*, Stanford: Stanford University Press.

Norton, A. (1988) *Reflections on Political Identity*, Baltimore: Johns Hopkins University Press.

Ozlak, O. (1983) *La Formacion del Estado Argentino*, Buenos Aires: Editorial del Belgrano.

Ó Tuathail, G. (1996) *Critical Geopolitics*, London: Routledge.

Ó Tuathail, G. and S. Dalby (eds) (1998) *Rethinking Geopolitics*, London: Routledge.

Perelli, C. (1992) 'Settling accounts with blood memory: the case of Argentina', *Social Research* 59: 415–51.

Piccirilli, R., Romay, F. and Gianello, L. (1954) *Diccionario Historico Argentino*, Buenos Aires, Ediciones Historias Argentinas.

Pion-Berlin, D. (1989) *The Ideology of State Terror*, Boulder: Lynne Rienner.

Pittman, H. (1981) 'Geopolitics of the ABC Countries', unpublished Ph.D. thesis, Department of Latin American Studies, The American University.

Potash, R. (1969) *The Army and Politics in Argentina 1928–1945*, Stanford: Stanford University Press.

—— (1980) *The Army and Politics in Argentina*, Stanford: Stanford University Press.

Radcliffe, S. (1993) 'Women's place/Lugar de mujeres: Latin America and the politics of gender identity', 102–16 in M. Keith and S. Pile (eds) *Place and the Politics of Identity*, London: Routledge.

Reboratti, C. (1982) 'Human geography in Latin America', *Progress in Human Geography* 6: 397–407.

—— (1983) 'El encanto de la oscuridad: notas acerca de la geopolitica en el Argentina', *Desarollo Economico* 23: 137–44.

—— (1987) *Nueva Capital, Viejos Mitos*, Buenos Aires: Sudamericana Planeta.

Rock, D. (1987) *Argentina 1516–1987*, London: I. B. Taurius.

—— (1993) 'Argentina 1930–1946', 173–242 in L. Bethell (ed.) *Argentina: Since Independence*, Cambridge: Cambridge University Press.

Sanz, P. (1976) *El Espacio Argentino*, Buenos Aires: Editorial Pleamar.

Sarno, H. (1997) 'El pensamiento Geopolitico y sus escuelas', *Geopolitica* 61: 43–52.

Scarpaci, J. and Frazier, L. (1993) 'State terror: ideology, protest and the gendering of landscapes', *Progress in Human Geography* 17: 1–21.

Shumway, N. (1991) *The Invention of Argentina*, Berkeley: University of California Press.

Sibley, D. (1995) *Geographies of Exclusion*, London: Routledge.

Siliva, P. (1989) 'Democratisation and foreign policy: the cases of Argentina and Brazil', 86–10 in B. Galjart and P. Silvia (eds) *Democratisation and the State in the Southern Cone*, Amsterdam: CEDLA2.

Solberg, C. (1970) *Immigration and Nationalism in Argentina and Chile 1890–1914*, Austin: University of Texas Press.

Spalding, H. (1972) 'Education in Argentina 1890–1914', *Journal of Interdisciplinary History* 3: 31–61.

Staub, E. (1992) *The Roots of Evil*, Cambridge: Cambridge University Press.

Storni, S. (1967[1916]) *Interes Argentinos en el Mar*, Buenos Aires: Instituto de Publicaciones Navales.

Szuchman, M. (1990) 'Childhood education and politics in nineteenth century Argentina: the case of Buenos Aires', *Hispanic American Historical Review* 70: 109–38.

Tambs, L. (1979) 'The changing geopolitical balance of South America', *Journal of Social and Political Studies* 4: 17–35.

Timmerman, J. (1981) *Prisoner Without a Name, Cell Without a Number*, New York: Alfred Knopf.

Vega Valencia, A. (1965) *Geopolitica en Bolivia*, La Paz: Juventud.

Vicens Vives, J. (1950) *Tratado General de Geopolitica*, Barcelona: Editorial Circular.

Vogel, H. (1991) 'New citizens for a new nation: naturalization in early independent Argentina', *Hispanic American Research Review* 71: 107–31.

Waisbord, S. (1991) 'Politics and identity in the Argentine army', *Latin American Research Review* 26: 157–71.

Walther, J. (1976) *La Conquista del Desierto*, Buenos Aires: Circulo Militar.

Weigert, H. (1942) *Generals and Geographers: the Twilight of Geopolitics,* London: Methuen.

Part 2

GEOPOLITICS, NATION AND SPIRITUALITY

8

SPIRITUAL GEOPOLITICS

Fr. Edmund Walsh and Jesuit anti-communism

Gearóid Ó Tuathail (Gerard Toal)

Introduction

Considerations of the histories and traditions of geopolitics have rarely engaged the relationship of these traditions to the problematic of religion. The reasons for this neglect seem straightforward for geopolitics appears as a thoroughly modern and secular set of discursive practices. As a state-centric territorial imagination, it supposedly triumphed at the expense of a religious cosmography in the sixteenth and seventeenth centuries with the adoption of the *cuius regio, eius religio* formula at the Peace of Augsburg in 1555 and its consolidation after the Thirty Years War in the 1648 Treaty of Westphalia. Westphalia has long been a watershed event for orthodox international relations, the founding moment of the modern state system. It is held to mark a decisive displacement of a medieval imaginative geography, which organized space as a vertical hierarchy in relationship to a Christian God, by a modern geopolitical imagination, which organized space as a horizontal set of competing territorial orders (Huxley 1944; Shapiro 1992: 109; Agnew and Corbridge 1995: 18). This political displacement was anticipated by intellectual and scientific shifts in the notion of space within Europe as Nicholas Copernicus, Giordano Bruno and others called the hierarchical conception into question and argued for a notion of space as infinite, homogeneous and measurable (A. Crosby 1997: 95–108). The modern geopolitical imagination, it seems, begins where the medieval geo-religious imagination falls away.

The notion of a decisive break between medieval religious space and modern geopolitical space at Westphalia, however, is questionable. Rather than a clear and clean rupture, the already existing relationship between the secular and spiritual, the territorial and the ecclesiastical was re-organized and re-conceptualized at Augsburg, Westphalia and numerous other historical moments since. Medieval religious notions were re-cycled into the emergent mythology of a diversity of European states, each of which claimed variations on heavenly inspiration, providential blessing and/or divine leadership. Rather than geopolitical traditions and religious traditions being at odds, they are more often than not deeply interwoven and mutually constitutive. Normative and spiritual vertical hierarchies of sacred space justified imperialistic and worldly horizontal hierarchies of geopolitical space. The historical development of the modern state system in Europe and its violent imposition across the globe saw multiple and complex (con)fusions of geopolitical and religious discourses (Cosgrove 1999). The overseas expansionism of the European empires into the Americas, Asia and later Africa was in significant part driven by religious motivations and sanctioned by the Church. In numerous cases the pioneers of imperialist encounter and conquest were men belonging to religious orders such as the Jesuits. Puritan jeremiads and religious zeal helped establish the meaning of America (Campbell 1992). Notions of providential will and divine destiny were vital elements in the nineteenth-century conquest of the American West (Stephanson 1995).

The emergence of a self-conscious 'geopolitics' as a tradition of theorizing about geographical relationships, state territoriality and world power in the United States in the first half of the twentieth century is also characterized by a (con)fusion of geopolitical and religious discourses. One of the leading figures in the dissemination of 'geopolitics' as a domain of knowledge in the United States was the Jesuit priest Father Edmund Walsh (1885–1956), founder of the Georgetown School of Foreign Service in 1919. Today faculty from Georgetown University, a university founded by the Jesuits in 1789, are regularly appointed to leadership positions in the US foreign policy bureaucracy while the School of Foreign Service has for decades produced students who have staffed and run that bureaucracy. As the first Regent of the School, Father Walsh taught courses on geopolitics to future diplomats and military leaders, functioning also as an 'expert' on European geopolitics to government institutions and to the American public in his many public addresses and writings. Walsh's life was not only an eventful one that spanned the major upheavals of the first half of the twentieth century but his explicit interest in geopolitics, especially German *Geopolitik* and Soviet geopolitics, and his restless political advocacy make him a particularly interesting figure in the history of American geopolitics. This chapter provides an introduction to Walsh's geopolitical philosophy as found in his major books and speeches. A detailed consideration of Walsh's political activities, diplomatic endeavours and army service (which lead him to have a key role in the US army's interrogation of Karl Haushofer after the Second World War for example) is not attempted here.

Walsh's career can be divided into five different phases.

First, born into an Irish-American family in Boston, he was ordained a Jesuit priest in 1916 and appointed Dean of the College of Arts and Sciences at Georgetown University in 1918. The same year he was appointed as a member of the Special Commission of the War Department to administer the Student Army

Training Corps. The following year, after demobilization, he helped found the School of Foreign Service as a department of Georgetown University, becoming its first Regent.

Second, in 1922 Walsh was appointed by Pope Pius XI as Director-General of the Papal Relief Mission to Soviet Russia and Vatican representative concerning church interests in the Soviet Union. He served in Russia for a year and a half, developing a lifelong antipathy for Bolshevism. While in Russia, Walsh had mixed success defending certain Church officials from being persecuted by the Bolsheviks. In one instance, a Church prelate was murdered.

Third, from 1923 to 1945 Walsh was a leading anti-communist campaigner in the United States. In 1929 he published *The Fall of the Russian Empire*, a narrative history of the fall of the Romanov dynasty and the Bolshevik revolution woven together with his own experiences in Russia. He was a vigorous opponent of US recognition of the Bolshevik government. His gave public lectures on a regular basis in Constitution Hall in Washington DC and preached an anticommunist message to audiences of US military officers and FBI agents. When President Roosevelt decided to recognize the Soviet Union in 1933, he called Walsh to the Oval office to personally explain his decision and reassure him. Walsh, however, was not reassured and Catholic leaders organized public demonstrations, mass meetings and petition drives to protest FDR's action (Crosby 1978: 6). After war broke out between Russia and Finland in 1939 Walsh organized and directed the Finnish Relief Fund in Washington D.C.

Fourth, as a consequence of his studies and lectures on German *Geopolitik*, Walsh was asked to serve as a consultant to Justice Robert H. Jackson at the International Military Tribunal at Nuremberg. Walsh helped interrogate Karl Haushofer in October 1945 and gave a morality and ethics test to Rudolph Hoess, the SS commander of Auschwitz, amongst others. In 1946 Walsh was appointed a member of the US President's Advisory Commission on Universal Military Training. Walsh travelled to Japan in 1947 to study educational and religious issues for the Jesuits where he toured Hiroshima and met with General Douglas MacArthur.

Finally, from 1948 Walsh resumed his anti-communist activities in Washington DC. Walsh reportedly influenced the thinking of the Jesuit-educated Senator Joseph McCarthy at an informal dinner in Georgetown, encouraging him to undertake a crusade against 'known communists' in the US government (Halberstam 1972: 146–7). This contention, however, is disputed (Crosby 1978 47–50). Walsh published *Total Power: A Footnote to History* in 1948 on German geopolitics and the new threat from Soviet geopolitics. In 1951 Walsh published *Total Empire: The Roots and Progress of World Communism* which was devoted solely to examining Soviet geopolitics and its threat to Western civilization. In 1952 he celebrated his Golden Jubilee in the Society of Jesus, suffering a stroke soon afterwards. He died in 1956.

Walsh's life and career were not typical for an American Jesuit. First, Walsh's interest in geopolitics was somewhat unusual for a Catholic priest. Walter Giles, his personal secretary from 1944 to 1950, recalls that 'his professional lifestyle and his prominence as a political commentator made him an unconventional Jesuit for his period, when it was virtually unheard of for a member of the Catholic clergy to be a political activist in the public forum' (Giles in Watkins 1990: 6). One exception was Father Charles Coughlin whose populist radio broadcasts between the Great Depression and the Second World War were influential for a while (Kovel 1997). Unlike Coughlin, Walsh was neither a populist nor an isolationist but an internationalist who dined in elite Washington society circles. Giles remembers that some Jesuits criticized Walsh for having interests and concerns which were too worldly (a charge often levelled at Jesuits in the past). It was thought that 'he lacked the kind of spirituality, and commitment to strictly religious interests and activities, deemed appropriate at that time for a Catholic priest' (Watkins 1990: 6; interview 1999).

Second, Walsh largely eschewed the dominant social mission of American Jesuits in the first half of the twentieth century for an international geopolitical one. Walsh was almost certainly a product of that social mission which was stimulated in the late nineteenth century by Pope Leo XIII's directives about the perils of capitalism and the dangerous attractions of socialism to the urban poor and working class. Jesuit priests undertook social missions to the rough and tumble working class immigrant enclaves of the United States, in overcrowded neighbourhoods in Boston, Chicago and New York (McDonough 1992). Teeming with poor Irish, German and Italian Catholics, the Jesuits established high schools, social work programmes and universities, facilitating upward mobility while solidifying the faith, checking the growth of socialist sentiment and recruiting talented boys for the priesthood. This social mission, of course, was not entirely separate from the larger geopolitical one of the Catholic Church. But its later articulation in Pope Pius XI's *Quadragesimo Anno* (1931) was cautiously approached by Walsh who showed an awareness of the inappropriateness of its corporatist rhetoric – shaped by an admiration for Mussolini's Italy – in the American context (McDonough 1992: 65–75). Walsh's 'social mission' was to help educate the future cadre of American diplomats and to advise and inform America's governing elites, those that made up 'Georgetown society,' about the threat posed by communism. In this, he could well claim to be following Ignatius de Loyola's charge to cultivate the favour of the powerful.

Third, the School of Foreign Service at Georgetown University was perceived as a break from traditional Jesuit educational institutions and practices. It was a professional school organized as a national, non-sectarian institution of higher learning. Initially, it was located in the Law School building in downtown Washington and physically separate from the Georgetown University campus. It was only in 1932 that the School moved onto the 'Georgetown Heights' campus (Tillman 1994: vii). According to Giles, some influential Jesuits on the Georgetown campus during Walsh's lifetime tolerated but never really accepted Walsh's 'foreign school' because of its worldly focus and non-religious character (Watkins 1990: 7). Walsh deliberately cultivated a professional and ecumenical image for the School to enhance its effectiveness and enrolments. Countering suspicion of the Jesuits and the Catholic religion more generally – a concern that lead Scottish Rite Masons in 1928 to fund a rival 'non-sectarian' School of Government at George Washington University, subsequently to become the Elliott School of International Affairs – was also a concern (Tillman 1994: 17). Yet, the majority of the students were Catholic and they were obliged to take one religion course, which was perceived as congenial especially when discussions often concerned baseball (Giles interview, 1999)! Only male students were admitted until the 1940s, after which a few female students were admitted, 'rather grudgingly and in small numbers' (Tillman 1994: 5).

While Walsh's position and profile were somewhat unusual, his activities and writings fit squarely within the history and tradition of the Jesuit order. That history and tradition is a diverse and eclectic one. It is not a monolithic history of subterranean influence, confessorly intrigue and pernicious manipulations. This 'black legend' was largely an invention of the many enemies the Society of Jesus acquired from its establishment under the leadership of Ignatius de Loyola in 1540 (Lacouture 1995: 348–75). A universalist order pledging absolute obedience to the papacy, the Jesuits were made up of men from many different cultures and backgrounds. Its organization was hierarchical and militaristic, Loyola having been a soldier before his remarkable journey to the Church and Rome (Lacouture 1995; Mitchell 1980). He deliberately adopted military metaphors to describe the 'Society of Jesus' – the name itself was a bold statement – as an elite unit with a general for a head and members who were to think of themselves as soldiers for Jesus and the one true Roman Catholic faith. The geographical and institutional setting they worked within varied tremendously.

Yet, though heterogeneous and diverse, Jesuit history and tradition is given coherence by certain transcendent preoccupations and concerns. The Jesuits became the shock troops of the Counter-Reformation. They represented the Catholic Church militant and organized, a counter-offensive force against Protestantism and heresy. Their operational environment was one defined by a clear and present enemy that needs to be firstly discerned and then confronted and vanquished. Their goal was the defence and propagation of the true faith of Christianity, the spreading and consolidation of the Kingdom of Christ on earth. They were thus *propagandists* in an original sense of the word. To this end, they employed a wide variety of tactical methods of conversion, the most successful of which was their establishment of institutions of education and learning throughout the world and their staffing of these institutions with well trained Jesuit educators. The Jesuits were innovators in methods of communication and strategies of pedagogy. They were encouraged by Loyola to become eminent in some field of endeavour. They stressed discipline, regular spiritual exercises, and thorough preparation. They sought to cultivate the powerful and convert foreign rulers, to impress them with their science and great learning. Their ultimate purpose was to illuminate the mind and move the will of their subjects in the spiritually correct direction.

It is within this tradition that we can locate the geopolitical writings of Father Edmund Walsh. To introduce Walsh's world-view, I have organized his writings and activities around four key themes which are central not only to Walsh's geopolitical philosophy but to Jesuit history and tradition. The analysis is necessarily succinct and does not do justice to the larger intellectual and political context of Walsh's work .Nevertheless it provides an introduction to Walsh's 'spiritual geopolitics' and hopefully demonstrates why the problematic of geopolitics and religion is an important one deserving further research.

The modern fall of man

The Catholic Church, like many other religions, has had a fraught and difficult relationship to modernity. The Jesuits, for McDonough (1992: xii), incorporate more accurately than any other group in Catholicism the tension between modernity and tradition. By the early twentieth century, the leadership of the Catholic Church was at war with the modern world. The tone was established by the papal declarations of Pius X condemning 'Americanism' in 1899 and repudiating the heresy of 'modernism' in 1907. Within the Jesuits, the anti-modernist creed was lead by Father Wlodimir Ledochowski who governed the society as its Superior General from 1915 to his death in 1942. The son of Polish nobility serving the Habsburg court, Ledochowski embodied the reactionary sentiments of old regime Catholicism (McDonough 1992: 65–8). Writing after the collapse of the Habsburg empire in 1919 he argued that 'all is tottering in modern society . . . modern society resembles in its wretchedness the poor paralytic of Bethsaida in that it has no strength within itself to rise from its sick bed' (Ledochowski in Schmidt 1945: 380). 'Freethinkers', those with a contempt for God and religion, 'promised that science would solve all the problems of life, becoming the provident and generous dispenser of happiness to the generations of mankind so thirsty for perfect happiness. However, science and the state have proved to be false gods. Now the souls of men are feeling ever more and more the terrible void created by the conspiracy of governments to tear away modern generations from Christ and His Church' (Ledochowski in Schmidt 1945: 382).

Father Edmund Walsh's anti-modernist sentiments permeate his writings and practices. Like many of the famous geopoliticians of the first half of the twentieth century – Halford Mackinder, Karl Haushofer and later George Kennan (Stephanson 1989) – Walsh was an organic conservative whose meta-narrative was that of the 'fall' of Man (the language is, as might be expected, unreflectively patriarchal) in the modern world from a previous state of organic harmony and unity (an idealized and élitist conception of the past that ignores the exploitation, structural violence and class repression that underpinned so-called 'harmonious' communities). Alienated from many aspects of twentieth-century modernity, these figures interpreted the political events of their times as a decline from a previous golden age, usually an idealized medievalism and/or antiquity. Unlike the others, Walsh gave this story of the fall of Man a distinctively religious cast. The first fall of Man was in the Garden of Eden. The second *modern* fall of Man was from the sixteenth century onwards (Walsh 1947: 27). The meta-narrative of the modern fall, expressed particularly in revolts and revolutions, is the story of Man unified, balanced and whole during the medieval age of Christendom stumbling into a condition of disunity, imbalance and division as he negotiated the trials and temptations of modern times. Christianity was not solely a religion for Walsh but a universal civilization with a normative code of moral conduct that has its origins in Greek and Roman antiquity (Walsh 1948: 190–5).

Walsh's reasoning was grounded in a distinctive Catholic ontology. The condition of Man is one of struggle between nature and civilization, between man's animal nature and his better nature as represented by Christianity. In Catholic dogma Man is endowed with God's grace but nevertheless has free will and must struggle to construct a moral life and find his way towards God. 'Men', Walsh argues:

> are born helpless infants dependent on adult authority; they progress to maturity and to the grave under the unpredictable preferences of free will, and they act responsive to the inequalities of intelligence, talents, the pressure of diverse interests, group outlook, secret prejudice, and moral imperatives, all of which require regulation if society is not to become a planned chaos.
>
> (Walsh 1948: 78)

The regulation and guidance provided by the Church is fundamental to Man realizing himself.

Man is a composite of matter, spirit and will, a body of flesh with instincts and a mind with an intellect that requires proper discipline and training. 'The material elements of our nature are so integrated with the spiritual, their functioning so interdependent, that a laboratory separation is unreal psychology and dangerous empiricism' (Walsh 1948: 177). Recognizing the indissoluble interdependence of matter and spirit, body and mind is the starting point of any truly humanist pedagogy. Fixation on one of these elements to the neglect of the whole produces an unbalanced Man. Reason, Walsh (1948: 178) argues, certainly 'should steer in human voyaging and intellect set the course for all navigators; but intuition, interest, love, hatred, spiritual perception, toleration, faith, and tradition enter into the table of wind and tide to an extent that may irritate the perfectionist but will not surprise the realist or your true humanist.' The observation reveals an inherited tradition of Catholic ontology and a developed Jesuit philosophy of pedagogy. Not surprisingly, Walsh believed that a Jesuit education is a truly humanist education. It recognizes and caters to the innate dualism of Man, providing him with not only a solid intellectual training but also a clear moral compass.

There are at least four different stages or 'scenes' in the modern fall of Man for Walsh. The first is associated with the Reformation and the splintering of Christendom. The Protestant revolt instigated by Luther, he claimed, 'must rank as the most unfortunate domestic tragedy in the family of Christian nations' (Walsh 1948: 148). The break-up of Christendom was the beginning of the break-up of Man. 'As the unity of Christendom was shattered in its ecclesiastical organization by the religious schisms of the sixteenth century, so the intrinsic unity of man himself was breached by the miscalled rationalism of succeeding generations' (Walsh 1948: 177). As a Jesuit, Walsh's antipathy for Luther is, as one would expect, quite strong. Yet, this antipathy leads Walsh to make a remarkable series of connections and claims. Luther is seen as an apologist for temporal rulers and the state. From the sixteenth century Walsh leaps to the twentieth century and connects Luther to Hitler. 'The effects of the Lutheran doctrine of supreme secular power,' he claims 'were not only to persist and colour the entire fabric of government in northern Germany, but may be detected in the special psychology of Germanic dictatorships from the princely butchers of the Peasants' War through Frederick the Great to Hitler and Himmler. The line runs straight' (Walsh 1948: 196). In other words, Luther was a figure who undermined the spiritual and strengthened the secular. He made a pact with the secular state which in Germany proved to be a pact with the devil, the anti-Christ.

The second identifiable 'scene' is the eighteenth-century Enlightenment and the nineteenth-century materialism and naturalism it spawned. 'The secularization of Western culture and the de-spiritualizing of its societal forms,' he suggests, 'can be traced with precision through the fabric of speculative and political thought that prevailed so widely since Machiavelli and Descartes' (Walsh 1948: 176). Machiavelli was viewed as the precursor of the growth of power politics in the nineteenth century while Descartes, though remaining a Christian, unleashed a dangerous scepticism and positivism that called inherited certitudes and beliefs into question. As Christianity began to be called into question, Western civilization began to go awry. Walsh viewed the French Revolution as an example of unbalanced enlightenment. Its historical unfolding illustrated the corruptibility of humanity, the hubris of enlightenment rationality and the dangers of social disorder and mob rule (Walsh 1948: 175–6).

The third 'scene' in the modern fall of Man is the industrial revolution. In considering American history, Walsh acknowledges the great advantages and contributions to material progress and productivity made possible by machines and modern physics, mechanics and chemistry. Yet, 'the body of mankind has benefited more than his spirit' (Walsh 1948: 289). The industrial revolution has lead directly to the horror of total warfare in the twentieth century. Technology and material progress became false gods.

> False values were created in the universal worship of mechanical achievements, and a softening of the moral fibre accompanied the modernizing of the roadbed over which humanity was proceeding. The Industrial Revolution ushered in mastery of production but it ushered out the production of masterpieces. It developed the proportion of all our senses but killed the sense of proportion.
>
> (Walsh 1948: 289)

Serial production displaced individual craftsmanship. Lost is an appreciation for the individual genius of medieval masters like Dante, Milton, Michelangelo and Shakespeare who composed not with the aid of an electric light but by the light of tallow candles. The Industrial Revolution has 'cultivated the spirit of things and discounted unduly the things of the spirit'. Its unsteady and directionless materialist imagination produced 'the absolutism of Hegel, the intellectual brutality of Nietzsche and the venom of Karl Marx' at the expense of 'the inspired humanity of Saint Francis of Assisi and the divine economy of the Sermon on the Mount' (Walsh 1948: 290). It produced Goering's aeroplanes which set the pattern for the total warfare that destroyed the craftsmanship of the Gothic Cathedrals of Europe.

The final culmination of the modern fall of Man is the era of total warfare, total power and total empire (Walsh surprisingly draws little on the concept of totalitarianism as developed by Kennan and others in the late 1940s though his ideas are quite similar; Pietz 1988; Stephanson 1989: 57, 63–4). According to Walsh's history, the era of total warfare was first conceptualized by General Ludendorff in his work *Total War* and was adopted by Hitler just as he also supposedly adopted Haushofer's global geographic strategy (Walsh 1947: 22). Warfare now embraced the whole population of states; it was no longer possible to make distinctions between combatants and non-combatants. All of the resources and technology of the state were mobilized to armaments and war. With air-power, the battle front

> has moved into every city, town and village. . . . That is one of the most calamitous consequences of the degeneration in the sense of values which began with the Industrial Revolution and culminated in the crass materialism of Communism and the cynical secularism of the Nazi philosophy of the State.
>
> (Walsh 1951: 246)

195

Total power is the form of the state realized first by the Soviet Union and subsequently by Hitler and the Nazis. It comprises a totalitarian state, modern technology and a secular ideology which is akin to a religion. The accumulation of total power by a state inevitably lead to geopolitical expansionism and world revolution.

Total power and world revolution

The culmination of Walsh's modern fall of Man meta-narrative is the Bolshevik revolution of 1917. For Walsh, this was an event of world historical significance. It provoked Walsh to write his first book in which he described it as

> not merely a revolution in the accepted sense as historically understood,
> – that is, a re-allocation of sovereignty, – but revolution in the domain of
> economics, religion, art, literature, science, education, and all other
> human activities. It sought to create a new type of humanity. . . . It was
> philosophic materialism in arms, the most radical school of thought that
> had ever come upon the stage of human affairs.
>
> (Walsh 1929: 6)

Walsh pushed its significance to even greater heights by claiming it was the most significant event in over a thousand years, an event of greater significance than even the Reformation. It was, he declared again and again, 'the most important event since the fall of the Roman Empire', a prelude to a new era of secular state religions.

> The international ramifications of the new ideology, the organised challenge, and the social upheavals consequent on the November coup d'etat all unite to equate the rise of the Communist state with the fall of the Roman Empire in the catalogue of significant world events.
>
> (Walsh 1948: 258)

The Bolshevik Revolution was a consequence of certain unfortunate and accidental historical circumstances in Russia but it was also a symptom of the much deeper crisis in Western civilization and culture. Bolshevism was a procedural development in 'a deeper cultural crisis which has been tormenting Western society since the industrial revolution. Bolshevism is not the original sin in the modern fall of man' (Walsh 1948: 256). It was the consequence of the unbalanced development of Man and the emergence of a 'despiritualized humanism' that was the legacy of Enlightenment scepticism and materialism (Walsh 1948: 259). Communism was radical in that it de-legitimated Christianity and challenged its conception of Man. It was thus always much more than a geopolitical threat for Walsh. It was an ontological threat.

196

Ironically, the way in which Walsh and the Jesuits made sense of atheistic Communism was to conceptualize it as a new religion. It was a new creed bidding for men's hearts and minds. In *The Fall of the Russian Empire* Walsh made the analogy clear:

> Lenin made Communism a religion. Karl Marx was its divinity; *Das Kapital* and *The Communist Manifesto* its inspired writings – its Bible; and he, Nicholas Lenin, was its master missionary. . . . Out of this human trilogy a faith was founded and propagated which, in its psychological reactions, supplied an earth-born substitute for that natural instinct and need which humanity feels for a divine revelation. Beginning with one pivotal dogma, – false, as most men believe, – the apostles of Communism have elaborated a set of doctrines which furnish them with weapons of daily propaganda.
>
> (Walsh 1929: 221)

For the Society of Jesus, the analogy between the struggle against Reformation Protestantism and the struggle against Communism was obvious. In a letter 'On Combating Communism' the Superior General of the Jesuits Wlodimir Ledochowski connects the two. The Society of Jesus came into existence in a time crucial for the Church. Its providential mission was to stem the tide of revolt against the Church. 'Does it not look', he asks rhetorically

> as if the present emergency entailed a fresh call to our zeal and generosity as soldiers of Christ and His Church, a call to take up arms against the great heresy of our time, more dangerous perhaps than any heresy of the past? For Communism is not merely a system of philosophy, an abstract theory fostered by scattered groups of men; it is a world force powerfully organised, and even now actively at work in various countries with incalculable harm to souls and to religion.
>
> (Ledochowski in Schmidt 1945: 907–8)

Ledochowski's letter directs each Jesuit province to appoint a director and a committee to organize anti-communist activities in that district. Information and documentation on Communism and Communists may be obtained, Ledochowski goes on to note, 'from Father Walsh of Georgetown University, who is in possession of a valuable collection of documents, books, and other material concerning Communism in theory and practice'. The anti-communist directors were also to consult Walsh on the practical steps to be taken to secure uniformity of strategy. Ledochowski's missive was written in 1934, one year after the Nazis (not the Communist) had taken over Germany.

The Catholic Church's relationship to Nazism is the subject of considerable historical controversy. Pope Pius XII was outspoken in his praise for Franco (who welcomed and gave privileges to the Jesuits after his civil war triumph) and silent on Nazism and the Holocaust. It has been suggested that with Hitler, Himmler and Goebbels all brought up in the faith, the National Socialist government was the most Catholic that Germany ever had (Mitchell 1982: 265). Himmler studied the organization of the Jesuits at length and considered the SS as a religious elite, the Nazi equivalent of the Society of Jesus. Hitler apparently made fun of Himmler's religious mania, describing him as 'our Ignatius Loyola'. It has also been alleged that Ledochowski was ready to organize some collaboration between the SS and the Jesuits against communism (Mitchell 1982: 264).

Some Jesuits, on the other hand, died at the hands of the Nazis. What can be said is that Walsh never acknowledged the ambivalences in the Catholic Church's relationship to fascism and the Nazis. For him, Nazism was the victory of a godless naturalism, an expression of a Teutonic mentality which worshipped the state. German mysticism and tribalism ('which was never wholly Christianized') triumphed over the German Christian tradition (Walsh 1948: 73). Hitler's drive for total power

> became a logical corollary in shining armour of the claims for total power advocated by a long line of pompous German philosophers in academic costume and by a flock of romanticists seeking to recapture the heroic fictions of Valhalla. At bottom, the issue was Wotan versus Christ.
>
> (Walsh 1948: 73)

Walsh makes the leaders of the German Catholic Church heroes. 'No group in Europe has been more fearless in denouncing the grave menace of racism than the German Catholic Hierarchy and', he adds in a revealingly awkward way, 'the more courageous leaders of Protestant belief in Germany' (Walsh 1947: 33). While individual Catholic leaders did speak out against the Nazis such a judgement, which completely ignores the role played by communists in challenging the Nazis, is more heroic than historical.

What is significant about Walsh's interpretation of Nazism is the equivalences he draws between it and Communism. Both were philosophies of total power and world revolution. 'These two concepts, Communism and Nazism', Walsh (1947: 36) wrote, 'included an identical objective – World Revolution. . . . Both, in their own way, accepted the belief of Hitler formulated in *Mein Kampf* (Chapter Five, page 440): "political parties are inclined to compromise; world-concepts never. Political parties count on adversaries; world-concepts proclaim their infallibility".' Interestingly, Walsh never uses the concept of 'Red

Fascism', the analytical device used in early Cold War America to map the evils of fascism onto the Soviet Union. Stalin, according to this notion, was another Hitler. One logical reason for its absence in Walsh's writings is that the Soviet Union was always the primary and overriding threat for Walsh. Rather than the Soviet Union being a version of Nazi Germany, Nazi Germany was a version of the Soviet Union for Walsh. Hitler was another Stalin, Germany a 'Brown Communism' (i.e. fascism was really a form of communism; my phrase not Walsh's). This reasoning is supported by passages in Walsh's work where he suggests that Nazism was a passing threat to democracy but that the real long-term threat remains from the Soviet Union. In a 1947 address to graduates of the FBI academy in Washington D.C. he notes that

> Hitler snatched the sceptre of World Revolution from the Kremlin and robed himself in the filched trappings of a totalitarian satrap. He strutted his little hour and passed; he was, as it were, a parenthesis in the text of History. His empire crumbled and the sceptre has now returned to Moscow.
>
> (Walsh 1947 in Watkins 1990: 64).

By the late forties and early fifties Walsh was emphasizing just how much the Soviet empire was expanding. In one his last public addresses in 1952 he figuratively pointed to the map and, in the manner common to uncritical geopolitics, concealed his interpretative politics by evoking the supposedly transparent and manifest quality of 'the facts on the map' (Ó Tuathail 1996):

> Let us look, then, to the facts. Seven years of study and planned conquest by the Soviets since 1945 have resulted in a new Communist Empire, the largest in recorded history. Some 800 million human beings are now, directly or indirectly, subjected to the control of the Kremlin; that means approximately one-third of the human race, and the end is not yet in sight. I saw some of this panorama unfolding under my own eyes in 1945 in Germany and later on in the Far East; it has evolved with foresight, forethought and geopolitical wisdom.
>
> (Walsh 1952 in Watkins 1990: 134)

Walsh knew a total empire when he saw it. The reason, he remarked, why he often insisted upon this last feature, that the empire was a conscious and planned creation, was 'because I have worked in the field of geopolitics for a good many years' (ibid.: 134). Geopolitics was Walsh's chosen field of eminence. Deciphering the geopolitical strategy of the enemy was his passion and countering their world revolutionary plans his vocation.

Propaganda, education and discernment

A fundamental founding mission of the Society of Jesus was 'the propagation of the faith by the ministry of the Word, by spiritual exercises, and by works of charity'. Special emphasis was placed on 'teaching Christianity to children and the uneducated' (Elton 1963: 200–1). From the outset, then, the Jesuits combined both a propaganda and an education mission. Both were intimately related, each involving the conquest of blank territory – pagan lands and uneducated minds – for the one true faith. In creating an extensive network of schools and universities across the globe, the Jesuits created a system of educational institutions facilitating social advancement and mobility that was unrivalled until the nineteenth century. Young minds were captivating by learning and captured for the Lord. Science and spirituality, literature and liturgy were seamlessly elements in a unified and 'balanced' curriculum. Education trained both the mind and the soul, the intellect and the spirit. A Jesuit education, it was claimed, produced men of both intellectual capability and moral character.

Father Edmund Walsh's career was part of the unfolding of this double Jesuit mission in the first half of the twentieth century. As a consequence of America's involvement in the First World War, there was a widely perceived need in the United States for education on international relations and questions of world affairs. The idea of establishing a School of Foreign Service at Georgetown University seems to have originated with Father John Creedan S.J., then President of Georgetown University. Creedan delegated responsibility for the project to Walsh who organized the curriculum of studies and opened the School in 1919 to sixty-two new students. In his opening address, attended by the Assistant Secretary of State at the time, Walsh stressed that the educational experience would provide a technical training that would rest upon a broad and liberal education, combining the best elements of age-long cultural traditions with the bracing atmosphere of individuality, characteristic of our educational institutions in the United States'. Traditional Jesuit notions of mission, responsibility and service were given a nominally secular expression. The 'high mission' of the School is to make men realize the responsibilities which they assume in a life of foreign service (Walsh, 1919 in Gallagher 1962: 202). Yet the ostensibly secular purpose of the School, the training of future diplomats and international trade officials, was enframed by certain moral requirements from students. Compliance 'with the principles of moral law is expected and required of every student', and 'failure in this regard is ground for refusal of a certificate or degree, or suspension, or even expulsion. Efficiency in studies without moral character and conduct will not entitle the student to a certificate or a degree' (Walsh 1919 in Gallagher 1962: 202).

Like Mackinder who wanted to encourage English students to think imperially and Haushofer who wanted to encourage German students to 'think in continents', Walsh had his own educational agenda which was intimately related to his political and, distinctly in his case, religious agenda (Ó Tuathail, Dalby and Routledge 1998). The Jesuit practice of discernment was fundamental to his pedagogy. One learned by confronting the teaching of one's enemies and discerning their philosophy and methods. Then one developed a counter strategy to undo the effects of their creed. Just as the early Jesuits had confronted Martin Luther and Protestantism by first studying his arguments and then constructing rigorous theological refutations of these arguments, so foreign service students needed to confront the threatening ideologies and creeds of their time, study their expression, discern their operation, and then proselytize against them. This method led Walsh to search for what he considered to be foundational documents and practical philosophical expressions of threatening creeds. He sought to identify the 'theological fathers' of these creeds and their 'biblical' ur-texts. He was particularly drawn to documents of prophetic value concerning grand strategy. Walsh then sought to discern the underlying purpose of the creed and the methods it used to advance its cause. Finally, Walsh attempted to inoculate audiences against the operation of this creed and organize campaigns of counter-propaganda.

As one might expect, Walsh's overriding preoccupation was Bolshevism and the philosophy of Marxist-Leninism. In the 'foreword' to *The Fall of the Russian Empire* Walsh describes his goal as 'supplying the perspective and understanding which becomes indispensable if one hopes to avoid the common errors fostered by propagandists, paid or unpaid, and correct the fallacies of loose thinking and still looser talking indulged in by the pamphleteers' (Walsh 1929: vii). Lenin and his lieutenants were subjected to Jesuit-style psychological discernment in this book. Lenin was a man with a 'central reservoir of hate' whose mind became a sealed book, except for three thoughts: Russia, Revolution, the World on Fire (Walsh 1929: 219–20). Walsh's second work analysed the Soviet Five Year Plan of the late 1920s somewhat hopefully (from his perspective) as a 'last stand' (Walsh 1931). In his Georgetown lectures, public speeches and later published works he discussed the philosophy of dialectical materialism and Marxism. *Total Empire* (1951) contained an appendix of quotations and Communist teachings which Walsh referred to as the 'Communist scriptures' (Walsh 1951: 268). The operation of Machiavellian and materialist philosophies in the practices of the Soviet Union and Communist parties was discussed at length. Walsh went to considerable lengths to warn his readers about the infiltration and propaganda techniques used by Communists and the varied foreign policy tactics pursued by the Soviet Union to reach their ends (Walsh 1948: 247–79; 1951: 85–165).

In the late 1930s Walsh also began keeping track of the writings of Karl Haushofer and the German School of *Geopolitik*, influenced by a Portuguese geographer Dr. Coutino whom Walsh had hired to teach at the School of Foreign Service. Coutino was a friend of Haushofer and it was he and not Walsh that first taught classes in 'political geography' and 'geopolitics' at Georgetown. Walsh, however, became interested in the new subject of 'geopolitics' and developed a number of lectures on German *Geopolitik* in 1941 (Giles interview 1999). In *Total Power* (1948: 9) Walsh describes himself as having devoted twenty years of 'attentive study' to Haushofer's activities. Like many others before and during the Second World War, Walsh believed that Haushofer's ideas represented those of Hitler and Nazi foreign policy. Haushofer was taken to be the 'brain trust' of Hitler (Ó Tuathail 1996: 111–40; Walsh 1947: 21–6 and Takeuchi in this volume). Even in 1948 after he had interviewed Haushofer and the US army had determined that he was not particularly important to the Nazi state or close to Hitler, Walsh was wont to exaggerate Haushofer's influence and his own prescience about him. The connection between Haushofer's 'apparently academic pronouncements with the concrete Nazi program for achieving total power in Europe was not apparent to the world at large' until after the outbreak of the Second World War but Walsh had discerned this connection a decade earlier. After 1939, 'the interrelation of cause and effect could no longer be disguised, as one invasion after another followed the broad pattern so long and so openly expounded in the writings and teachings of the master geopolitician' (Walsh 1948: 10).

The problem, however, is that there was no strong connection and no broad pattern. Haushofer's influence in what has been described as the 'weak dictatorship' of Hitler was marginal before the war and virtually nil during the war. Walsh demurs from the judgement not to prosecute Haushofer because it 'did not take sufficient account of the direct and influential role that Haushofer had personally played for many years in the inner counsels of the Party', nor did it 'visualize his powerful stimulus and specific activities in justifying Hitler's political and military aggressions' (Walsh 1948: 12). Yet these descriptions themselves are pale versions, as Walsh himself acknowledges, of the arguments originally made by himself and a colleague before going to Germany to interrogate Haushofer and investigate his influence. And, furthermore, even these charges exaggerate the significance and role of Haushofer.

What is interesting about this is what it reveals about the nature and accuracy of Walsh's method of discernment. Walsh was predisposed to finding prophet-leaders and conspiracies in history, visionary prophets with a plan for world revolution and radical movements with an elaborate strategy to realize the plan. Nazism and Communism were of a kind, each with their own prophets, creed and revolutionary blueprints. All were opposed by Christianity. Walsh had difficulty acknowledging the contingency and indeterminacy of history. For him, history was always deeply meaningful, the unfolding of struggles between abstract opposites. Throughout history Christianity has done battle with godlessness, spirituality with soulless materialism, democracy with totalitarianism, and freedom with empire (Walsh 1948: 102).

These were certainly the terms to be used to describe the United States's geopolitical conflict with the Soviet Union. Like many American Cold War geopoliticians, Walsh saw the Soviet Union as the inheritor of a transcendent Russian impulse towards expansionism (Walsh 1948: 268). A document purported to be the last will and testament of Peter the Great from 1757 was identified as an ur-text of Russian and Soviet geopolitics by Walsh. In *Total Power* Walsh claims that 'whatever doubts may attach to the authenticity of this remarkable document . . . no doubt can exist as to its prophetic quality. The Russian Government (*sic*) since 1939, as fact of record, has followed the Petrine pattern with obstinate fidelity' (Walsh 1948: 270). Truth for Walsh is forced to be clearer than truth. A few years later in *Total Empire* Walsh renews his obsession with this document, reproducing it as an appendix to his book while conceding that it may be inauthentic but that nevertheless it 'possesses great intrinsic interest, as embodying principles of action which have been notoriously followed out by Russia during the last hundred years, with such modifications as time and circumstances, and the variations of the European equilibrium, have rendered necessary' (Walsh 1951: 261).

Walsh's arguments echoed the Truman Doctrine and George Kennan's articulation of the policy of containment. 'The lines are drawn for a deeper conflict between absolutism and freedom of the spirit, between two antagonistic philosophies of life that can no longer be disguised. . . . As evangelist of world Communism [Russia] cannot remain static' (Walsh 1948: 318). Her efforts to ferment revolution must be resisted "by appropriate counter-policy at every point where the conspiracy becomes evident" (Walsh 1948, 329). Separating the geopolitical from the religious reasoning is impossible in Walsh for the Cold War, as he pointed out in his final speech at the Georgetown Club on the occasion of the fiftieth anniversary of his induction into the Society of Jesus, 'is a struggle between two great moral opposites' (Walsh, 1952 in Gallagher 1962: 142). The US-Soviet struggle is not only a West versus East struggle but a struggle between Christ and Karl Marx, the New Testament and the Communist Manifesto. Geopolitics is not geographical power politics; it is a omnipresent spiritual struggle.

Loyola's soldier

Characterizing the gender regime organized and represented by the Jesuits is a task beyond this chapter. Yet in any characterization of this problematic, the figure of the heroic soldier of God needs to be examined and deconstructed. As has already been noted, the Society of Jesus was founded by an ex-soldier who trans-posed his militaristic training into religious life. The Jesuits were organized as a religious elite who pledged absolute obedience to their Superior General and the Pope. For young Catholic boys down the ages dealing with their inchoate sexuality amidst Jesuits and a culture that repressed adolescent sexuality, the Order no doubt offered a clear and heroically celibate path through life. Like soldiers, the Jesuits real-ized their self-image far from the world of women, fighting in the classroom, in the public arena and on faraway frontiers for the patriarchal order of the Church. The Jesuits codified a particular masculinity that managed to be heroic at the same time as it was safe, to have a strong masculinist self-image while simultaneously managing to avoiding the embarrassments of having to negotiate the world of women. Women were potentially dangerous, so much so that Ledochowski wrote two letters of instruction for the society on the issue 'On the avoidance of long conversations with women' (1918) and 'On reserve in dealing with women' (1920) (Schmidt 1945).

While the militaristic culture of the Society of Jesus can be exaggerated, there has been a long-standing historical affinity between the Jesuits and state military organizations. The Jesuits were religious militarists and this often lead them into active support of state militarists and militarism as a means of advancing their own purposes. This fusion of spiritual and state militarism can be seen in the career of Father Walsh. What is also evident, though this question needs further research, is the operation of a particular Jesuit-style militarist masculinity that produces an aversion to diplomatic compromize in Walsh and also a dangerous attraction to the more apocalyptic forms of Cold War militarism.

Throughout his life Father Walsh had numerous social and institutional ties to the US military. Soon after obtaining an M.A. at Woodstock College, Maryland, Walsh became a member of a special commission of the War Department to administer the Student Army Training Corps (SATC). This began a long association for Walsh with the US army and its training programmes. As an 'expert' on geopolitics, he regularly lectured various branches of the US military in Washington DC and Fort Levinworth Kansas on security dangers, spiritual and geopolitical. As a prominent figure of the Washington social scene, he had good relationships with the Pentagon leader-ship, especially, it seems, with General Douglas MacArthur. In 1945 Walsh swapped his priestly attire for a US army officer's uniform to serve in Nuremberg, a position he apparently lobbied hard to obtain (Giles interview 1999). Addressing the Industrial Armed College of the Armed Forces in August 1952, he recalled lecturing to the same body decades before, especially to a young captain by the name of Eisenhower (Watkins 1990: 130). Indeed after Walsh's death, Eisenhower recalled 'the rare privilege' he once had listening to 'a magnificent lecture' by Walsh 'on the growing menace of Communism' (Eisenhower cited in Gallagher 1962: 247).

At his golden jubilee dinner in 1952, attended by amongst others General J. Lawton Collins, chief of staff of the US army, Walsh offers some rare insight into how his Jesuit training and background conditioned his approach to life:

> I thank the disciplined but patient formation of my Order, founded as it was by a soldier over four hundred years ago, which taught me to put first things first, particularly to regard no man as fit for command who has not first learned how to obey. She [sic] enjoins on all her members the obligation to weigh every challenge of life and every risk of death on the scales of eternity, make prudent election between alternatives and then fight the issue out in rank and file under unified leadership.
>
> (Walsh 1952 in Gallagher 1962: 243)

A Jesuit life for Walsh is a life of spiritual warfare. Such a determined militaristic approach to life would seem quite at odds with Walsh's position as the Regent of a School Of Foreign Service, an institution training a society of diplomats not soldiers. Diplomacy certainly required discipline and obedience but it also required a capacity for dialogue and an openness to otherness. In Walsh's world, however, the weight of eternal judgement crushed diplomacy as dialogue. The Soviet Union was an implacable moral enemy of the United States and Walsh was consequently in a permanent state of war against it. Diplomacy for Walsh was the conduct of war by other means, an attitude that probably accounts for his own lack of success as a papal diplomat in the Soviet Union in 1922–23. The Bolsheviks apparently found him 'most objectionable, proud and inclined to make a terrible scandal out of every little issue' (Fischer 1930: 522). Walsh's undisguised enmity for the Bolsheviks seems to have brought him into conflict not only with the Soviet state but also with the Vatican which was striving not to antagonise the Bolsheviks at the time (Fischer 1930: 522–3). In the end, Walsh was removed from his post and the American relief effort shut down. (Walsh rebutted Fischer's descriptions, terming him a pro-Russian 'propagandist'; see Walsh 1931.) That the Regent of a School of Foreign Service might have been a poor diplomat and a philosophical opponent of diplomacy is somewhat ironic. But Walsh's refusal to grant the Soviet Union diplomatic status and recognition was not unusual at the time. Only in 1933 did FDR finally recognize the Soviet Union, a move bitterly opposed by Walsh that triggered, as already noted, considerable protests by American Catholics and others.

Walsh never abandoned his religious war against Communism and the Soviet Union. In his later life, his crusading zeal lead him into association with some controversial figures. The charge that Walsh directly encouraged Joseph McCarthy's crusade against Communism in the United States is contentious and the product of a newspaper article by Drew Pearson, a political gossip columnist. They certainly dined together with others on 7 January 1950 in Washington before McCarthy began making his charges, a meeting at the centre of Pearson's accusation. Crosby (1978: 51) in his exoneration of Walsh from any direct connection with McCarthyism claims that while Walsh 'shows a surprising concern (one might almost say an "obsession") with the menace of external communism', 'he almost never talked about the question of subversion'. Yet Crosby, himself a Jesuit, never notes Ledochowski's 1934 directive 'On Combating Communism' which conceived the idea of a co-ordinated campaign against the subversive influence of 'the great heresy' of Communism. More strikingly, he ignores large sections of Walsh's works where he asks conspiratorially driven inquisitional questions such as: 'what makes a Communist or a sympathiser out of an American comfortably placed, sometimes a millionaire, often a successful author, a playwright, a Hollywood figure, a schoolteacher, a government employee, a well-paid labor leader, a lawyer, or a city councillor?' (Walsh 1951: 101). Also ignored are statements such as: 'Today, if Mr J. Edgar Hoover is correct, there are 500,000 John Reeds [after John Reed, "the brilliant but erratic Harvard graduate who cast his lot with Lenin's cause"] walking the streets of America' (Walsh 1951: 103). Clearly, Walsh was inclined with powerful others to see Communist conspirators and conspiracies throughout the land. Yet the idea that Walsh was the 'intellectual father' of McCarthyism is overblown, a product no doubt in part of a long-standing inclination to see the malign influence of the Jesuits in politics. Walsh was not happy at being associated with McCarthy for it threatened to damage the image he cultivated for the School of Foreign Service. He refused to respond publicly to Pearson's charge but let it be known that he considered Pearson a 'liar' (Tillman 1994: 32). His secretary Walter Giles recalled that he subsequently refused telephone contact with Senator McCarthy and that later, hospitalized after his stroke, Walsh expressed concern about McCarthyism after hearing of its effect on State Department morale from former students (Giles interview 1999).

Finally, Walsh's religious antipathy for the Soviet Union lead him to take certain positions which were surprising coming from a Catholic priest. Walsh was an enthusiastic supporter of Cold War militarism, praising in particular the building of the US navy's first nuclear powered submarines and giant aircraft carrier (Walsh, 1952 in Watkins 1990: 139). As the strongest citadel of Christian civilization, America needed to be vigilant and well armed. Addressing graduates of the FBI academy in 1947 he declared that never before was there 'greater need for clear heads, steady hands and great hearts at the controls of human destiny, for men who walk humbly in the sight of God but keep their powder dry' (Watkins 1990: 68–9).

Walsh's most stridently militarist position was his justification of a pre-emptory nuclear first strike by the United States against the Soviet Union. Writing immediately after the outbreak of war on the Korean peninsula which he interpreted as the 'final confrontation' between 'two great centers of world power whose basic and irreconcilable character' was known to Soviets decades ago, Walsh argued that all states were obligated to protect their populations from attack (Walsh 1950 in Watkins 1990: 145). Pre-emptive attacks were morally just. The United States, for example, would have been justified in intercepting and destroying the Japanese aircraft attacking Pearl Harbor. With a 'Soviet feint in some remote area of Asia or the Middle East', the US defence system had better keep its eyes fixed on the Northwest and Arctic sector for a sneak surprise attack (Watkins 1990: 149). If the US government had 'sound reason to believe (that is, had moral certitude)' that a surprise attack was being planned then President Truman was justified in 'taking measures proportionate to the danger' including use of atomic bombs (1990: 149–50). While the results would be tragic and horrific, there was no immorality in the United States government choosing the lesser of two evils. Walsh justified military force in the abstract by pointing out that '[e]ven Christ himself did not disdain to seize the lash and drive the hypocrites out of the Temple' (Watkins 1990: 150). Writing on this same issue of the atom bomb and the Christian conscience in *Total Empire* he ends the book with the rather ominous sentence: 'The debate is not whether we can afford to do the necessary things for the defense of Christian civilization – but can we afford not to do them?' (Walsh 1951: 259).

Conclusion: religion and geopolitics

This chapter has been no more than an introduction to Father Edmund Walsh and Jesuit anticommunism. I have concentrated on providing a portrait of Walsh's geopolitical thinking and the larger context within which he worked. Further research on all the relationships and issues explored in the chapter is needed. What is hopefully apparent, though, is that the relationship of religion to geopolitics is a significant problematic in critical geopolitical studies. In any consideration of traditions of geopolitics, the complex interrelationships between geopolitics and religion need to be acknowledged and investigated. Geopolitics is far from being the secular science and practice it is sometimes represented as being.

Today, geopolitics and religion remain interpenetrated in ways that are sometimes recognized and acknowledged but at other times not. Many contemporary global conflicts are represented in predominantly religious categories such as those in Bosnia, Northern Ireland and Israel-Palestine. Western security discourse regularly identifies threats in religious terms, most especially in terms of Islamic fundamentalism (Esposito 1992; Huntington 1996). The papacy of John Paul II has been credited with helping precipitate the downfall of Communism in Eastern Europe and many hope it will have a similar effect in Cuba (Bernstein and Politi 1996). Sometimes, the religious faith of particular Western leaders like Jimmy Carter or Ronald Reagan is acknowledged as an important element of their geopolitical philosophy and approach to global crises and problems. Yet, though a hegemonic 'Judeo-Christian tradition' has conditioned Western foreign policy practices for some time, the influence of religion on how the United States and its Western allies approach and conceptualize the world is rarely acknowledged and problematized. Indeed, amongst some there is a perception that the West's approach to the world has ignored religion altogether and become much too secular. The West needs to begin to re-acknowledge religion as the 'mission dimension of statecraft'.

In a book with this very title, *Religion: The Missing Dimension of Statecraft*, edited by figures with associations to the Georgetown University's Center for Strategic and International Studies, Edward Luttwak (1994) argues that the Enlightenment-era prejudice against religion has contributed to a 'secularizing reductivism' in US foreign policy and a failure to appreciate the limits of materialistic determinism. US foreign policy professionals and the media tend to reduce conflicts and crises to secular terms, like 'Right' and 'Left', which ignore the complexity and importance of religious motivations in political conflicts. Also, these same actors tend to slight spiritual motivations in analysing crises and situations. The results are serious foreign policy intelligence failures and geopolitical setbacks, such as the fall of the Shah in Iran. Ostensibly motivated by the need for better diplomatic training and foreign policy analysis – Luttwak (1994: 16) proposes, for example, that 'religious attachés' could be assigned to diplomatic missions – Luttwak's arguments have an implicit message; the Western foreign policy community needs to become more sceptical of its own secular reasoning – especially neoliberalism – and to begin to recover a sense of the power of religious consciousness.

Much more explicit is the message of the coalition seeking to enact the Freedom From Religious Persecution Act in Congress which if passed would trigger US sanctions against states judged by a White House Office for Religious Persecution Monitoring to be tormenting their citizens on the basis of their religion. The coalition, made up of Protestant evangelicals, Jewish activists and others with broad support from the National Conference of Catholic Bishops, also considers religion the 'missing dimension of statecraft' and through the legislation seeks to mandate that the foreign policy bureaucracy of the United States rediscover their own religious vision and power. Their efforts have already yielded some results. Former US Secretary of State Warren Christopher established an Advisory Committee on Religious Freedom Abroad to study the question. Current US Secretary of State Madeleine Albright, a former faculty member at Georgetown University, has mandated U.S. diplomats to provide frequent and thorough reports on the status of religious freedom in the countries to which they are accredited. She has also spotlighted the issue in the State Department's annual report on human rights though she opposes the proposed Congressional legislation because of the foreign policy complications it undoubtedly would create.

The United States is perhaps the most enduringly religious state in the post-modern world. In order to understand its political culture and geopolitical practices one needs to appreciate how religion provides certain narratological resources and discursive strategies for its leaders to represent and interpret the world. Heroic religious stories of transcendent struggles against evil, heresy and godlessness have long been mapped onto the world political map by religious leaders, politicians and strategists deeply socialized by these narratives. In many instances, these scripts are exceedingly dangerous for they refuse the complexity of international affairs and falsely reduce it to predetermined moral categories. Part of the task of developing critical geopolitics, therefore, involves a struggle to deconstruct the orders of power/knowledge found in (con)fusions of geopolitical and religious traditions. This task has only begun.

Acknowledgements

Many thanks to the staff of the Georgetown University Special Collections library, Robert Gallucci, Dean of the Edmund A. Walsh School of Foreign Service and especially to Walter Giles for an interview on 6 February 1999.

Bibliography

Agnew, J. and Corbridge, S. (1995) *Mastering Space*, London: Routledge.

Berstein, C. and Politi, M. (1996) *His Holiness : John Paul II and the Hidden History of Our Time*, New York: Doubleday.

Campbell, D. (1992) *Writing Security: United States Foreign Policy and the Politics of Identity*, Minneapolis: University of Minnesota Press.

Cosgrove, D. (1999) 'Baroque Geography and Enlightenment', 33–66 in D. N. Livingstone and C. W. J. Withers (eds) *Geography and Enlightenment*, Chicago: University of Chicago Press.

Crosby, A. (1997) *The Measure of Reality*, New York: Cambridge University Press.

Crosby, D. (1978) *God, Church and Flag: Senator Joseph R. McCarthy and the Catholic Church 1950–1957*, Chapel Hill: The University of North Carolina Press.

Elton, G. R. (1963) *Reformation Europe, 1517–1559*, Glasgow: Fontana/Collins.

Esposito, J. (1992) *The Islamic Threat: Myth or Reality?*, New York: Oxford University Press.

Fischer, L. (1930) *The Soviets in World Affairs: A History of Relations Between the Soviet Union and the Rest of the World*, vol. II, London: Jonathan Cape.

Gallagher, L. (1962) *Edmund Walsh, S.J.: A Biography*, New York: Benziger Brothers.

Halberstam, D. (1972) *The Best and the Brightest*, New York: Penguin.

Huxley, A. (1941) *The Grey Eminence: A Study in Religion and Politics*, New York: Harper.

Kovel, J. (1997) *Red Hunting in the Promised Land: Anticommunism and the Making of America*, London: Cassell.

Lacouture, J. (1995) *Jesuits: A Multibiography*, Washington D.C.: Counterpoint.

Luttwak, E. (1994) 'The Missing Dimension', 8–19 in D. Johnston and C. Sampson (eds) *Religion, the Missing Dimension of Statecraft*, New York: Oxford University Press.

McDonough, P. (1992) *Men Astutely Trained: A History of the Jesuits in the American Century*, New York: Free Press.

Mitchell, D. (1980) *The Jesuits: A History*, London: Macdonald Futura.

Ó Tuathail, G. (1996) *Critical Geopolitics*, Minneapolis: University of Minnesota Press.

Ó Tuathail, G., Dalby, S. and Routledge, S. (1998) *The Geopolitics Reader*, London: Routledge.

Pietz, W. (1988) 'The "post-colonialism" of Cold War discourse', *Social Text* 19/20: 55–75.

Schmidt, A. G. (ed.) (1945) *Selected Writings of Father Ledochowski*, Chicago: Loyola University Press.

Shapiro, M. (1992) *Reading the Postmodern Polity*, Minneapolis: University of Minnesota Press.

Stephanson, A. (1989) *Kennan and the Art of Foreign Policy*, Cambridge, Mass.: Harvard University Press.

—— (1995) *Manifest Destiny: American Expansion and the Empire of Right*, New York: Hill and Wang.

Tillman, S. P. (1994) *Georgetown's School of Foreign Service: The First 75 Years*, Washington: Georgetown University Press.

Walsh, E. (1929) *The Fall of the Russian Empire*, Boston: Little, Brown and Company.

—— (1931) *The Last Stand: An Interpretation of the Soviet Five Year Plan*, Boston: Atlantic Monthly.

—— (1947) 'Geopolitics and international morals', 12–39 in H. Weigert and V. Stefansson (eds) *Compass of the World: A Symposium on Political Geography*, New York: Macmillan.

—— (1948) Total Power: *A Footnote to History*, New York: Doubleday.

—— (1951) *Total Empire: The Roots and Progress of World Communism*, Milwaukee: Bruce.

Walsh, J. (1934) *American Jesuits*, Freeport, N.Y.: Books for Libraries Press.

Watkins, A. (ed.) (1990) *Footnotes to History: Selected Speeches and Writings of Edmund A. Walsh S.J. Founder of the School of Foreign Service*, Washington, D.C.: Georgetown University Press.

9

REPRESENTING POST-COLONIAL INDIA

Inclusive/exclusive geopolitical imaginations

Sanjay Chaturvedi

Introduction

Once the British started constructing 'their India' during the later nineteenth century, as an integral part of the larger Enlightenment project, which through observation, study, counting and classification attempted to understand and control the world outside Europe, they also set out to 'order' in their geopolitical imaginations the peoples who inhabited their new Indian dominion. Territorial annexation of over 60 per cent of the territory of the Indian sub-continent from 1757 to 1857 by the English East India Company (Fisher 1993) was followed by the annexation of 'Indian' in imperial knowledge systems and expropriation of Indian civilization by the British Crown (Cohn 1997). Accordingly, categories such as caste and tribe were placed at the heart of the Indian social system, along with the idea of two opposed and self-contained religious communities of the 'Hindus' and the 'Muslims' (Pandey 1990: 23–65). It was the centrality of religious community, along with that of caste, which for the British marked out India's distinctive status as a fundamentally different land and peoples. Despite its inconsistencies and subordination to the needs of colonial rule, the British ethnographic enterprise had far-reaching consequences. For, these very categories informed the ways in which the British, and in time the Indians themselves, conceived of the basic structure of their society (Metcalf 1995: 114).

Mahatma Gandhi paid the British imperial rule a handsome tribute in the early 1930s by conceding that 'the Indian nation was a creation of the empire-builders' (Sen Gupta 1997: 299). With independent India inheriting the colonial nation, much in the British ideology of 'difference' also survived and flourished, leaving its mark above all in the conception of a society informed by a passionate commitment to community, defined essentially in religious terms, and of the public arena as a site where those communities engaged in relentless pursuit for power. The

211

amorphous structure of Indian civilization with a remarkable capacity to accommodate a multiplicity of social and linguistic identities, sometimes in a cluster of regional polities, and on other occasions, in a somewhat fragile pan-Indian polity was also replaced by a uniquely colonial construct of the centralized state with an administrative bureaucracy and a standing army in particular, and an attendant ideological trappings of 'ordered unity', 'indivisible sovereignty' and the like (Kumar 1997). The historical circumstances of partition, dissidence, insurgency and war, ensured further that the apparatuses left behind by the colonial state would not be dismantled, but actually reinforced (Kaviraj 1997a: 233).

When the political elite of post-Colonial India began constructing the 'nation-state', it was compelled to negotiate a geopolitical disjuncture between an acknowledgement of diversity and an insistence upon the 'consciousness' of India as a single geopolitical entity, characterized by an organic unity (Parker 1988). It was modern, anti-colonial nationalism which was responsible for producing 'national unity' in a way that was largely alien to India's historical experience, ably assisted by the so-called 'nationalist interpretations' of Indian history which focused on unity, rather than differences and discord within India (Sen 1998a). To some extent this was a reaction to the colonial thesis that India was diversity and it had no coherent communality except that given by British rule under the integrating system of the imperial crown. The nationalist counter-argument was that despite the diversity, there was an essential unity; and that this unity was not accidental, but reflective of the unifying tendency in Indian culture and civilization as the ultimate foundation of nationalism.

It is against such a backdrop that this chapter purports to investigate how post-Colonial as well as post-Partition India has been represented by groups and individuals, political elites, and the dominant institutions and intellectuals of statecraft. In other words, it intends to show how a variety of geopolitical imaginations have come to compete with one another, as different actors seek to impose maps of meaning, relevance and order onto the highly complex and dynamic political universe they inhabit, observe, try to understand, and sometimes even desire to dominate. It begins with a comparative assessment of the manner in which geopolitical idioms, myths and representational practices have been used by the so-called 'Secular Nationalists' and the 'Hindu Nationalists' to inscribe something called India and endow that entity with a content, history, meaning and trajectory. This is followed by a critical examination of how, more recently, a homogenous and monolithic Hindu identity is being carved out of a remarkably diverse and eclectic cultural tradition on the sub-continent by the protagonists of 'Hindutva' or 'Hindu nationalism' in India. Finally, the chapter critically examines the geopolitical fallout of the five nuclear tests conducted by India in May 1998; especially the manner in which 'legitimacy' is being sought by the government for its decision through formal, practical and popular forms of geopolitical reasoning.

India's sacred geography, national unity and cartographic anxieties

In India, as shown by Ashutosh Varshney (1993), one could possibly talk of two principal geopolitical imaginations about national unity and national identity: the 'secular nationalist' – combining territory and culture – and the 'Hindu nationalist' – combining religion and territory. The defining principle of national identity for both, however, is *territory*. In the secular imagination, the territorial notion of India, emphasized for twenty-five hundred years since the times of the *Mahabharata*, is of a land stretching from the Himalayas in the north to *Kanya Kumari* (Cape Comrin) in the south, from the Arabian Sea in the west to the Bay of Bengal in the east. India is not only the birthplace of several religions (Hinduism, Buddhism, Jainism, and Sikhism), but during the course of its history it has also received, accommodated and absorbed 'outsiders' (Parsees, Jews, and 'Syrian' Christians, followers of St. Thomas, arriving as early as the second century, thus reaching India much before they reached Europe). What make Indian civilization unique, therefore, are the virtues of syncretism, pluralism and tolerance reflected in the cultural expression: *Sarva Dharma Sambhava* (equal respect for all religions).

One good example of the secular nationalist construction of India's national identity is Jawaharlal Nehru's *Discovery of India* (1981). For Nehru, 'some kind of a dream of unity has occupied the mind of India since the dawn of civilisation'. He 'discovers' India's unity as lying in culture and not in religion: hence no 'holy land' in his mental mapping of the country. For him the heroes of India's history – Ashoka, Kabir, Guru Nanak, Amir Khusro, Akbar and Gandhi – subscribe to a variety of Indian faiths and it is Aurangzeb, the intolerant Moghul, who 'puts the clock back'. India's geography was sacred to Nehru not literally but metaphorically (Varshney 1993: 236).

Nehru's secular nationalist construction of India stands in sharp contrast to the religious notion of India as originally the land of Hindus, and it is the only land which the Hindus can call their own (Pattanaik 1998: 43–50). According to V. D. Savarkar, the ideological father of Hindu nationalism,

> a Hindu is he who feels attached to the land that extends from Sindhu to Sindhu as the land of his forefathers – as his Fatherland; who inherits the blood of the great race whose first and discernible source could be traced from the Himalayan altitudes of the Vedic *Saptasindhus* [the land of seven rivers] and which enabling all that was assimilated has grown into and come to be known as the Hindu people.
>
> (Savarkar 1969: 100)

What unites the Indian landscape is the sacred geography of Hindu holy places

(Benaras, Triputi, Rameswaram, Puri, Haridwar, Badrinath, Kedarnath, and now Ayodhya) and the holy rivers (Cauveri, Ganga, Yamuna, and the confluence of the last two in Prayag).

It is important to note that the boundaries of India as suggested by the secular-nationalist are coterminous with the 'sacred geography' of the Hindu nationalist whose hallowed pilgrimage sites mark off essentially the same boundaries of the country, although the Hindu nationalist would go much further into mythic history of more than two and a half millennia to date the origin of these sites. As Varshney remarks:

> Since the territorial principle is drawn from a belief in ancient heritage, encapsulated in the notion of 'sacred geography', and it also figures in both imaginations [secularist and nationalists] it has acquired political hege-mony over time. It is the only thing common between the two competing nationalist imaginations. Therefore, just as America's most passionate political moment concerns freedom and equality, India's most explosive moments concern its 'sacred geography', the 1947 partition being the most obvious example. Whenever the threat of another break-up, another 'partition' looms large, the moment unleashes remarkable passions in politics. Politics based on this imagination is quite different from what was seen when Malaysia and Singapore split from each other, or when the Czech or Slovak republics separated. Territory not being such an inalien-able part of their national identity, these territorial divorces were not desecrations. In India, they become desecrations of the sacred geography.
>
> (Varshney 1993: 238)

As a result, both the secular-nationalists and the Hindu-nationalists share in common what Sankaran Krishna (1994: 508) has termed 'cartographic anxiety'; the anxiety surrounding questions of national identity and survival. According to Krishna, the term 'cartography' implies more than the technical and scientific mapping of the country; it refers to 'representational practices that in various ways have attempted to inscribe something called India and endow that entity with a content, history, a meaning and a trajectory'. Krishna argues that cartographic anxiety is a facet of larger post-colonial anxiety syndrome manifested by a society that perceives itself as suspended forever in the space between the 'former colony' and 'not-yet-nation'. 'This suspended state can be seen in the discursive production of India as a bounded sovereign entity and the deployment of this in everyday poli-tics and in the country's violent borders'. The critical question then becomes: is there anything post about the post-colonial at all? According to Krishna,

> if we examine the degree of anxiety revealed by the state over matters of

cartographic representation, the inordinate attention devoted to notions of security and purity, the disciplinary practices that define *Indian* and *non-Indian*, *patriot* and *traitor*, *insider* and *outsider*, *mainstream* and *marginal*, and the *physical* and *epistemic* violence that produces the border – the answer to this question is negative.

(Krishna 1994: 508)

Jawaharlal Nehru too had his moments of cartographic anxiety. Nehru, despite his firm belief in the timeless existence of a spiritual and civilizational entity called India, begins his modern reconstruction of India's past in *Discovery of India* (1981) with a graphic description of the country's 'natural' frontiers. 'Nehru's imaginative geography depicted impassable mountain ranges, vast deserts, and deep oceans that produced a natural "cradle" for what became India' (Krishna 1994: 509). And in Nehru's *Autobiography* (1936) also, one finds occasional evidence of his anxiety over safeguarding the physical boundaries of the nation; tracing the country's downfall to porous frontiers and, more importantly, to an unfortunate timing by which a disunited and fragmenting India encountered and succumbed to the cresting and united civilization of the British. Nehru's recollection of the partition of India (1956: 247) also reflects his anxieties over breakdowns of all kinds; 'but above all, what was broken up which was also of the highest importance was something very vital and that was the body of India' (ibid). As Dijkink (1996: 129) puts succinctly, 'what Nehru sought to overcome in his historical perspective spanning 2,500 years or more, was 200 years of British rule. His idea of unity was none the less conditioned by the era he wanted to wipe out.'

Early on, the actual foreign policy of an independent India was shaped by Nehru himself and his chief foreign policy adviser, Krishna Menon (Brecher 1968). Their distinctive world-view had emerged out of the long struggle with the British and out of the wider process of decolonization then taking place, and India was seen by them 'as offering an alternative world vision which was co-operative rather than confrontational in stance' (Parker 1998: 129). A fairly illustrative account of Jawaharlal Nehru's geopolitical vision can be found in his book *The Discovery of India* (1981) especially in the chapter entitled, 'Realism and geopolitics, world conquest or world association: the USA and the USSR'. What appears as most striking in Nehru's geopolitical thinking is the curious mix of 'idealism' and 'realism', 'internal' and 'external' on which he relies to construct his vision of the post-war world and India's role in it. Nehru is critically dismissive of the geopolitical theories of Mackinder and Spykman, which, according to him, are the pseudo-scientific justifications for the 'power quest', 'power politics' and 'world domination'. To quote Nehru:

Geopolitics has now become the anchor of the realist and its jargon of 'heartland' and 'rimland' is supposed to throw light on the mystery of national growth and decay. Originating in England (or was it Scotland?), it became the guiding light of the nazis, fed their dreams and ambitions of world domination, and led them to disaster . . . And now even the United States of America are told by Professor Spykman, in his last testament, that they are in danger of encirclement, that they should ally themselves with a 'rimland' nation, that in any event they should not prevent the 'heartland'(which means now the USSR) from uniting with the rimland.

(Nehru 1981: 539)

Behind Nehru's world-view and underlying his reflections on the desirability and feasibility of a just and peaceful post-war world order, one finds a much larger and deeper geopolitical concern with India's internal-domestic situation, and with the need for a foreign policy that would safeguard a 'dream of unity' that according to Nehru had occupied the mind of India since the dawn of civilization. He is thoroughly convinced about the unity of India: within its British Imperial borders. To quote Nehru:

thus, we arrive at the inevitable and ineluctable conclusion that, whether Pakistan comes or not, a number of important and basic functions of the state must be exercised on an all India basis if India has to survive as a free state and progress. The alternative is stagnation, decay and disintegration, leading to the loss of political and economic freedom, both for India as whole and its various separated parts.

(Nehru 1981: 533)

Views about a future division of India are subjected to the same criticism that Nehru levels at traditional realist power politics: the small-nation state is a phenomenon of the past and territorial division of India would soon reveal how dependent both new units are on each other and would immediately raise the need for a federal association. Besides, any acceptable territorial division would leave the Muslims with a territory that was both smaller and economically less viable. The prospect for the Indian sub-continent was, according to Nehru (1981: 533) either 'union plus independence or disunion plus dependence'. The tragic events of the separation of India and Pakistan, as Dijkink (1996: 121) points out, 'suggest that Nehru was actually engaged in "constructing" rather than "discovering" India.'

To Krishna Menon, 'external affairs' were 'only a projection of internal or national policy in the field of International Relations' (Brecher 1968: 4). And to a large extent, this was reflected in India's policy of non-alignment. The Indian foreign policy elite, according to Jean Houbert (1989: 202), opted for non-alignment India also because it enhanced the ability of the Indian state in 'containing' communism in its internal

216

politics. Given that the Communist movement in India was neither homogenous nor united in one well-organized party, any challenge on its left could be tackled by the Congress without the Soviet Union intervening. On the contrary, there was every probability that Moscow would use its considerable influence with the pro-Soviet parties to get them to support the regime in non-alignment. Moreover, non-alignment provided India with the means to save on military spending and put the priority on social and economic development, 'thus winning the allegiances of the electorate and cutting the grass from under the feet of the communists' (ibid.). External as well as internal security considerations too were likely to be better served by non-alignment than by joining the Western military alliances. But there was more than security and ideology to India's support for non-alignment; it

> would also enhance the power of India . . . by harnessing the moral dimension of international politics non-alignment provides India with more power in its weakness than alignment would do . . . thus the ideology of peace was bound together with India's security interests and power considerations in the non-aligned policy.
>
> (ibid.)

Yet another example of how the sense of commitment to 'national unity' and 'national identity' of post-colonial India dominated the priorities and policies of its political elite is Sardar Vallabhbhai Patel, independent India's first Home Minister, hailed by many as the Iron Man of India, who is said to have even outdone Bismarck. In the words of K.P.S. Menon, the former Foreign Secretary of India,

> when the British left India the unity of even divided India was in danger. Some 560 Princely States had been left in the air. It was open to them to adhere to India, to accede to Pakistan, or to remain independent . . . It almost looked as if India was going to be balkanised. But this danger was averted by the firm handling of the Princes by a man of iron, Sardar Vallabhbhai Patel.
>
> (cited in Krishna 1995: 433)

To Sardar Patel, as Krishna points out,

> Hyderabad seemed to have mattered much more than Kashmir. Situated as it was in India's belly, he naturally asked: 'How can the belly breathe if it is cut off from the main body?' That would have sounded the death-knell of Patel's dream of One India; and the cancer of disunity and divisiveness would have spread to totally balkanise the country.
>
> (Krishna 1995: 398)

SANJAY CHATURVEDI

Even a casual observer of contemporary India cannot fail to notice an increasing obsession with the threats — real and/or imaginary — emanating from across the border to the 'unity and integrity' of India; the 'alien' infiltration with shadowy 'foreign hand' is out to destabilize and destroy the body as well as soul of the nation. The Indian state has inherited its discourse as well as practices on borders or borderlands from the imperial powers. We may remind ourselves in passing that the geo-histories of the borders of what would come to be known later as 'South Asia' were being written by those who were 'creating' or 'constructing' them in the first place for their own power-political purposes. As a result, the maps that were drawn by the imperial power were both too static and too simple to capture the diversity and the dynamism of the borderlands, and as Paula Banerjee (1998: 11) points out, 'based on these maps and with an admixture of territorialism, carto-graphic absoluteness and frontierism the South Asian nation-states came into being, or to use a more current phraseology, were *constructed*'. It is largely to Curzon perhaps that the credit goes for coupling innate strategic and geographic senses and setting the stage for the appropriation of borders monopoly by government agents, military personnel and secret service agents. The remarkable influence of Curzon's legacy on the Indian state can be well gauged from the fact that it is still unable to get out of the mind-set that it has inherited, so much so that even today it continues to deny its own citizens access to maps of the border region, even outdated ones.

The persistent concern, occasionally bordering on obsession, of the Indian state with the 'territorial integrity' of its geopolitical realm, and the corresponding acute cartographic anxiety this generates, are also reflected in the manner in which the Indian state relates itself to its immediate neighbours on the sub-continent, especially Pakistan. According to Ayesha Jalal (1995: 5), in the dominant geopolitical discourse of India the country's geographical size and an ideal of its unity, albeit largely mythical and symbolic, are often cited as key differences between India and Pakistan, a fabrication of political necessity split into two parts separated by a thousand miles. Going by the stereotypes of the dangerous 'Other' which the vested interests promote through the press and televisual media in both the countries, Pakistan is represented by the dominant political discourse in India as inherently hostile, monolithic, identity-crisis ridden society populated and run by fanatics, who would like to 'crush' India, and who would not mind risking yet another war (even nuclear war) over Jammu and Kashmir to complete the 'unfinished task of partition'. Today, just about every other act of subversion in India is blamed on the elusive but omnipresent Pakistani ISI (Gill 1998). Whereas India ('Bharat') is portrayed by the dominant geopolitical discourse in Pakistan as (mis)governed by a Brahmin-dominated elite in New Delhi, permanently hostile not only to the existence of the polity and people of Pakistan but also to its Muslim 'minority'.

218

The cartographic anxiety of the Indian state is at its best – or perhaps at its worst – in Jammu and Kashmir, the northernmost state of the Indian Union. This major bone of contention between India and Pakistan is a good example of how peoples and places with distinctive histories, cultures and ethno-linguistic identities can be reduced to the status of mere 'issues' in the threat perceptions of the intellectuals and institutions of statecraft. At the heart of the dominant Indian discourse on Kashmir lies the polemical two-nation theory. While India is said to have somehow reconciled to the theory as an 'inevitability' (Dixit 1995: 199), Pakistan's rigid position is believed to be that 'the partition of the sub-continent will remain incomplete till all Muslim-majority areas of India either become part of it or are independent Muslim political entities' (ibid.). India's commitment to principles of plurality, synthesis and co-existence – transcending the factors of ethnicity, language, religion and sub-regional identities – is contrasted sharply to Pakistan's devotion to religious homogeneity of Islam as the sole basis for national and territorial identity. So any prospect of Kashmir breaking away from the Indian union is dismissed outright as a challenge to its territorial and ideological integrity. Attempts have also been made by certain ruling regimes in both Pakistan and India to re-write their respective countries pasts to suit their present political ideologies (Behera 1998). Pakistan's argument, that Kashmir was a part of Muslim kingdoms over the last 1,200 years and more, is strongly contested by the Indian historians on the ground that history neither commences nor ends abruptly. If Kashmir was a part of Muslim kingdoms and empires, then, it was also a Hindu and Buddhist polity at some stage. India's position is that if the political and territorial affiliations were to be based on religious and historical arguments, the political map of not only the Indian sub-continent but of the whole world would have to undergo changes.

According to Sumantra Bose, the Kashmir uprising:

> supplied the Hindutva movement with an unrivalled propaganda weapon. For here was 'evidence' of the diabolical designs of a group of Muslims living in India to destroy India's unity in conjunction with the historical enemy Pakistan. The 'enemy within, enemy without' (where the Indian Muslim is the fifth-columnist for Pakistan) conspiracy theory, a long-standing staple of the 'Hindu nationalist' world-view, was ostensibly finding some vindication.
>
> (Bose 1997: 144)

Once physical preservation of the national borders is held as synonymous with the very existence/survival of the state of the Indian union by one and all across the ideological spectrum in the country, the perceived indispensability of 'secure' or 'inviolable' borders for national unity, national development, and a coherent and cogent National Identity itself, diverts attention from the violence that produces the border (Krishna 1994: 511). A classic example of how a 'production of border' can lead to a senseless costly 'war' over the frozen wastes is that of the conflict between India and Pakistan over the Siachen Glacier, which begins at 12,000 feet above sea level, in the Slatoro Range of the Himalayas and ascends to 22,000 feet.

The centralized state's hegemonic project to infuse local and regional political economies and cultures with a singular and monolithic idiom of national unity has somehow failed to displace the widespread sense of attachment to 'place' in India, which is as potent as attachment to group, and the two are closely intertwined (Weiner 1997: 248–9). This is yet another major source of cartographic anxiety for the Indian state. Groups often regard the territory in which they live as the site of their exclusive history, a place in which great events occurred and sacred shrines are located. Tribal and linguistic groups often regard a homeland as exclusively their own and would, if they could, exclude others or deny others the right to enjoy the fruits of the land or employment provided within the territory. Hence, India's linguistic minorities define themselves as 'sons of soil' with group rights to employment, land and political power denied to those who come from 'outside'. According to Rajani Kothari (1997: 51), 'the more exploited segments of the Dalits [backward castes] and Adivasis [tribals], on the one hand, and of several territorial units and sub-units on the other, have given voice to a rising discourse of alienation'.

India's 'Northeast' is one good example of how and why it is not sufficient that the groups would be content with simply the occupation of the territory that they consider as their homeland; they also desire and seek to exercise some political control over it. To quote B.G. Verghese:

> It [Northeast] was long secluded, then politically 'excluded', only to find itself overnight reduced to a distant appendage at Partition, landlocked, its economy disrupted , overwhelmed by migrant waves and utterly bewildered by the suddenness and rapidity of change. Uncertainty, frustration and misperceptions in the minds of a series of hitherto sequestered communities awakening to a new identity consciousness created some distance first from Assam and then from 'India' through a process of differentiation. Sullenness turned to agitation, struggles and a series of insurgencies, most with external connections, six of which are still extant in some measure or other in Nagaland, Manipur, Assam, 'Bodoland', Tripura and Meghalaya, the number of armed groups involved being much larger.
>
> (Verghese 1996: 393)

Other movements having a bearing on India's national unity are the 'separatist'/ 'secessionist' movements in Panjab (Sikhs) and Kashmir (Muslims) and the regional Dravidian movement in Tamilnadu. As Gertjan Dijkink (1996: 131) points out succinctly,

> although the separatist movements in the north concern small minori-
> ties, their impact on national unity is huge, first, because they threaten
> India's 'sacred geography' and, because foreign actors may be involved
> (like Pakistan in Kashmir and Panjab), introducing a dangerous element
> of external insecurity. The Dravidian movement also obtained an inter-
> national dimension with its connection with Tamil separatism in Sri Lanka
> (the Tamil Tigers).
>
> (Dijkink 1996: 131)

Anxieties are also mounting in India over 'illegal' Bangladeshi immigrants or 'infil-
trators', alleged to be more than ten million in the country, half of whom alone
are in West Bengal. It is feared that

> it may effect a demographic change in Indian areas around Bangladesh so
> much so that they may one day cease to be Indian territory. The idea of
> such demographic aggression against India has been there since 1958, but
> it has been put into practice with intensity after the emergence of
> Bangladesh.
>
> (Maheshwari 1998: 1)

Last but not the least, one finds in the predominant Indian geopolitical imaginations a strange idealization of 'our' land, which has given birth to the cult of
the Mother in a social set-up which otherwise remains, by and large, highly
hierarchical and patriarchal (Mahanta 1977; Sarkar 1996: 162). In the nationalist discourse, the secularized concept of 'Bharatmata' or Mother India,
representing the toiling masses of rural India, and bound in chains, served as
the powerful tool for arousing the nationalist sentiment. 'Bharatmata Ki Jai' –
Victory to Mother India – became the rousing battle cry which appealed as
much to men as to women. As Sugata Bose (1998: 54) points out, 'thoroughly
"Westernised" nationalists like Jawaharlal Nehru relied heavily on the metaphor
of sexual aggression and rape in their critiques of violence perpetuated by the
colonial masters'. Even now the concept of the country as mother continues to
serve as a potent rallying cry for nationalist or sub-nationalist sentiment, as in
the Assam movement (1979–1985) when the battle cry was 'Joi Ai Asom',
Victory to Mother Assam (Mahanta 1997: 93). However, the equation of nation
with Mother Goddess, as attempted by many Hindu nationalists, leaves many
Muslims cold; leaving no space whatsoever for the accommodation and expression
of religious diversity of India (Kaviraj 1997b).

The crisis of the 'nation-state' and the rise of a 'Hindutva' geopolitics

The rise of 'Hindutva' or 'Hindu nationalism' in India has been attributed by some, and perhaps rightly so, to the overarching 'organic' crisis of the Indian state, having reached its zenith in the 1990s. The various overlapping causes as well as consequences of this crisis have been ably discussed elsewhere (Kothari 1998; Sumantra Bose 1997; Hasan 1996; Kaviraj 1994), and need not detain us here. The issues that are of more immediate concern to this paper relate to the 'geopolitical raw material' used in the construction of a new Hindu identity as a part of a larger ongoing project of making India more explicitly 'Hindu': the 'Hindu Raj'. According to Sumantra Bose (1997: 160–1), 'the entire project of 'Hindutva' is reducible to two complementary core ideas: repudiation, denial, suppression or neutralization of the manifold forms of diversity, conflict, cleavage and oppression in Indian society on the one hand; and a glorification of the monolithic, organic unity of the 'nation' (preferably in its 'natural' hierarchy, though this is sometime oblique and qualified), and a concomitant deification of indivisible and unitary state power, on the other.'

Today a new Hindu identity is under construction in many parts of India, especially the northern and central states (Ludden 1996; Nandy, Trivedi and Yagnik 1995). This process is undoubtedly assisted by the fact that this identity is also the basis of political mobilization by the party in power in New Delhi, the Bharatiya Janata Party (BJP) (Malik and Singh 1994). The BJP is the only cadre-based party in India in the real sense of the term. Unlike the Communist parties and the Congress, having their front organizations with distinctive identities, BJP is a political arm of the *Rashtriya Swayamsewak Sangh* (RSS, National Voluntary Corp), meant to implement its programme. The RSS, also known as *Sangh Parivar* (family), has emerged since its inception in 1925 as *the* organization articulating Hindu revivalism, especially among the youth, devoted to the establishment of a 'vibrant Hindu nation' with the ethos of the alleged Golden Vedic Age at its core (Graham 1993).

Once a local, nativist party in Bombay (Mumbai), the *Shiv Sena* (taking its name from a seventeenth-century Maratha warlord who successfully fought the Mughals), now finds itself the dominant political force in the state of Maharashtra – in alliance with the BJP – with a ready capacity to play the 'Hindu card' and court the loyalties of Marathi-speaking male youth for a holy war (*dharm yudha*), champion a patriotism that demonizes 'anti-national Muslims ('those who have their heart in Pakistan' and those who set off firecrackers of victory when Pakistan defeats India in stadiums of Bombay); exert immense mobilization power (both electoral and on the streets), incite widespread violence, extract rents, and shape public policy and legislative initiatives (Katzenstein, Mehta and

Thakkar 1998; Gupta 1995). Repeatedly, Shiv Sena conjures up images of Muslim treachery and betrayal. To quote Shiv Sena chief, Bal Thackeray, who at times even 'speaks admiringly of Hitler's love of nation' (ibid.),

> Muslims revolt in their own areas. They beat Hindus, demolish temples and attack the police. The government is appeasing these traitors. It is learnt that Pakistan has manufactured seven bombs. But the bomb that has been made in India with the blessings of Pakistan is more dangerous. Now Pakistan need not cross the borders for launching an attack on India. Twenty five-crore (ten million) Muslims loyal to Pakistan will stage an insurrection. One of these seven bombs made by Pakistan lies hidden in Hindustan.
>
> (Katzenstein, Mehta and Thakkar 1998; 224)

According to C.P. Bhambhri,

> a particular kind of Hindu ethos is being promoted by the BJP Government where temples, rituals, priests, tilak on the forehead of ministers and MPs are on public display . . . the Hindu saints in politics in the 1980s and 1990s are legitimising a politics of militant Hindutva directed against every minority community in the country . . . *sant sammelans* [gathering of the saints] are being used today for targeting Muslims by raising the issue of 'liberation of temples' from 'Government control' or for using violence to halt the so-called invasion of foreign culture, Muslim, Christian or a vaguely defined Western.
>
> (Bhambhri 1994: 4)

An insightful analysis of how Hindu – and for that matter Muslim – identity is formed in popular geopolitical reasoning by rumour, religion and bigotry, and how they are fuelled by nostalgic histories, periods of violence and peace between the two communities and the anxieties and uncertainties produced by the process of modernization is to be found in Sudhir Kakar's, *The Colours of Violence* (1995: 196). Kakar critically examines the constructed nature of the revival of Hindu identity, by choosing as his text a speech by *sadhavi* Rithambra, one of the most articulate speakers for the RSS. The prefix *sadhavi* is the female counterpart of *sadhu* (monk), a man who has renounced the world in search of personal salvation and universal welfare within the Hindu religion's world-view. The broader geopolitical context of the speech is the mobilization of Hindus by the BJP on the issue of constructing a temple to the god Rama at Ayodhya, his reputed birthplace, and the destruction that would follow of the Babri Masjid on 6 December 1992 by literally thousands of *kar sevaks* (Sangh volunteer workers) led by *Vishva Hindu Parishad* (VHP), established in the 1970s to launch anti-Christian Missionary Campaigns in the north-east, and the *Shiv Sena* and spurred on by major leaders of RSS and the BJP, followed by large-scale killings of the Muslims in many parts of the country.

Kakar shows how, in Rithambra's rhetoric, a 5,000 year old 'Hindu' religion, with a traditional lack of central authority structures such as a church, with a diffused essence, and with a variety of sects with diverse beliefs, is 'personalized' around randomly chosen gods and saints from Indian history, ancient and modern, and 'communalized' in terms of 'ego ideals', to be shared by members of the community in order to bring about and maintain group cohesion. To Rithambra, identity implies definition rather than blurring, solidity rather than flux or fluidity, and this makes boundaries of a group extremely paramount. To quote Kakar,

> it is Rithambra's purpose to include all the Hinduism spawned by Hinduism. The presiding deity of the Shaivite sects, Shiva, is hailed, as is Krishna, the most popular god of the Vaishnavas. The overarching Hindu community is then sought to be further enlarged by including the followers of other religions whose birthplace is India. They are the Jains, the Buddhists, and the Sikhs, and Rithambra devoutly hails Mahavira, Buddha, and the militant last guru of the Sikhs, Guru Gobind Singh who, together with Banda Bairagi, has the added distinction of a lifetime of armed struggle against the Mughals . . . The Harijans or 'scheduled castes' the former 'untouchables' of Hindu society, are expressly acknowledged as a part of the Hindu community by hailing Valmiki, the legendary author of the Ramayana who has been recently elevated to the position of the patron saint of the Harijans . . . the immortal gods and mortal heroes from past and present are all the children of Mother India, the subject of final invocation, making the boundaries of the Hindu community coterminous with that of Indian nationalism.
>
> (Kakar 1995: 200–1)

Rithambra, however, is not content with making her 'Hindu' audience aware of their collective cultural-national consciousness. To her, identity is not a 'product' but a process constantly under threat by hostile forces both from within and outside. In the militant Hindu discourse, of which Rithambra's speech is typically representative, power and violence express themselves most clearly in the politics of space. 'The rituals and riots are intricately linked to the specific rituals of violation of the "territories of the self" by the threatening "Other"; a slaughtered cow in the Hindu sacred space or a slaughtered pig in the Muslim sacred space' (van der Veer 1996: 259). For Rithambra, 'Our [Hindu] civilisation has never been one of destruction . . . wherever you find ruins, wherever you come upon broken monuments, you will find the signature of Islam. Wherever you find creation, you discover the signature of the Hindu. . . . We have always been ruled by the maxim, the world is one family' [*vasudhe kuttumbkam*]. Whereas the Hindus are being idealized by Rithambra as creative, compassionate, insightful and having religious tolerance in their bones, the Muslims are not only demonized as inherently destructive, but also warned to 'behave like sugar in the milk' or else face the fate of a lemon which is 'cut, squeezed dry and then thrown on the garbage heap'. What Rithambra is trying to put across is the Hindu nationalists' stand that non-Hindu groups can be a part of India, but only by assimilation. Hinduism to Hindu nationalists is the source of India's identity, and it alone can provide national cohesiveness. Such a view inevitably raises the question: who is a Hindu?

According to Savarkar, as mentioned earlier in the paper, 'A Hindu means a person who regards his land . . . from the Indus to the Seas as his Fatherland as well as his Holyland' (cited in Varshney 1993: 231). In order to qualify as a 'Hindu' a person or a group must meet three criteria: territorial (land between the Indus and the Seas), *genealogical* ('fatherland') and *religious* ('holy land'). Hindus, Sikhs, Jains and Buddhists can be part of this definition, for they were born in India and meet all three criteria, whereas Christians, Jews, Parsees (already assimilated) and Muslims meet only two. India is not their 'holy land'. If the Muslims wish to become a part of the Indian nation, they must stop insisting on their distinctiveness and agree to the following essential requirements for complete assimilation: first, accept unconditionally the centrality of Hinduism to Indian civilization; second, acknowledge key Hindu figures such as Ram as civilizational heroes, and not simply religious personalities of Hinduism; third, admit that Muslim rulers ('invaders') in various parts of India (between roughly 1000 to 1857) destroyed the pillars of Indian civilization, especially Hindu temples; and, fourth, withdraw all claims to special privileges, such as the maintenance of religious personal laws and special state grants for their educational institutions.

More recently, a comprehensive programme of Hindu mobilization has been worked out by the *Sangh Parivar* by setting Bajrang Dal Units in all 750 districts and 7,531 blocks in the country. 'These units of the Bajrang Dal *bal upasana kendras* (centres to pay obeisance to strength) are where the Hindu youth is trained in judo, karate and other martial arts to give a fitting reply to the Pakistan Inter-Services Intelligence agents, Christians and carriers of cultural invasion' (Bhambhri 1998: 4). Both the VHP and the Bajrang Dal, not everyone's idea of the traditional and tolerant face of Hinduism, are reported to be contributing to 'restlessness in border states' (*The Tribune* 1998: 8); as in several towns and villages in Gujarat, and to some extent in neighbouring Rajasthan, both Christians and Muslims have complained of both physical attacks and social-cultural harassment. The various facets of the 'cleansing operation' include accusing Muslim youths of kidnapping girls from other communities and converting them to Islam, launching a state-wide campaign against inter-community marriages and religious conversions, asking the majority community to boycott the services of Muslims, and distributing bright 'OM' stickers to Hindu rickshaw pullers to set them apart from the 'others'. In parts of Rajasthan, younger members of the minority community are reported to have 'reacted by setting up a fundamentalist outfit and a war of words rages below the surface' (ibid.). Such a 'restlessness' in borderlands generates tremendous anxiety in media and elsewhere because 'both Gujarat and Rajasthan have common borders with Pakistan and the ISI is active. Restlessness born out of a real or imagined sense of harassment suits the ISI, rather than the long-term interests of Secular India' (ibid.).

Geopolitics of 'nuclear nationalism': exploding geopolitical imaginations

In the aftermath of the five nuclear tests conducted by India on the 'auspicious' day of *Buddha Purnima*, on 11 May 1998, the moral arguments or morality-based politics are not considered relevant any more; *Realpolitik* seems to have replaced *Moralpolitik*. Unlike the first test in 1974, there is no apologetic suffix of 'peaceful' to Pokharan II. To the contrary, there is a deliberate attempt to flaunt the yet-to-be acquired weaponry and glorify the importance of 'peace-keeping bombs' for imparting to Indians a sense of security and self-confidence. According to India's Prime Minister Mr. Atal Behari Vajpayee, India has exercised its nuclear option as a deterrent against any design that external powers might have on the country. In his view, 'the entire country's power was manifested in the success of the tests which took years to achieve' (Vajpayee 1998). In response to the observation that the bomb was described by some as the 'Hindu bomb', Mr. Vajpayee has said, 'such rumours are being spread to divide the country. People from different religions were involved in the test-firing. Dr A.P.K. Kalam is a Muslim. The bomb is for the country's protection' (*The Hindu*, New Delhi, 31 May 1998: 1). The touchstone that has guided India in conducting nuclear tests, according to the Prime Minister, was 'national security'. As a result,

> India is now a nuclear weapon state. This is a reality that cannot be denied. It is not a conferment that we seek, nor is it a status for the others to grant. It is an endowment to the nation by the scientists and engineers. It is India's due, the right of one-sixth of humankind. Our strengthened capability adds to our sense of responsibility. We do not intend to use these weapons for aggression or for mounting threats against any country; these are weapons of self-defence to ensure that India is not subjected to nuclear threats or coercion. We do not intend to engage in arms race.

> (Vajpayee 1998: 3)

The nuclear debate in India, at least for the time being, is clearly tilted in favour of the 'hawkish' loyalists; predominantly male doomsayers, expounding on macho-sounding, but rather obscure, terms of hard-core realism, and proudly flagging their anti-west patriotism. The basic rationale or 'compulsion' behind India's nuclear-testing initiative is described in the practical geopolitical reasoning of India's foreign policy elite as being alert to its security environment, being responsive to its threat perceptions and being conscious that there is no substitute for self-reliance for ensuring the country's territorial integrity and security. According to J.N. Dixit, a former foreign secretary of India, 'an answer' to all the critics of India's nuclear weaponization and the tests of 11 and 13 May

> lies in the security environment around India stretching from Diego Garcia in the west in an encircling arc right up to Pakistan, the Gulf and the Straits of Hormuz and then on to the South China Sea. A number of countries have a nuclear presence in this entire region, one of whom, Pakistan, has threatened the use of its nuclear and missile capacities against India more than once.
>
> (Dixit 1998: 16)

Whereas those with a dissenting voice, tirelessly reminding the Indian political elite of formidable internal sources of insecurity, such as communalism, criminalization of politics, corruption, poverty, hunger and unemployment, growing socio-economic polarization, institutional decline and decay, dismal record of human resource development, political killings and group massacres, violence and counter-violence etc. (D'Monte 1998), fear that they might be condemned by the 'pro-nuke' lobby as *deshdrohis* (traitors), west-inspired anti-nationals or simply CIA agents. Amidst national euphoria, ably assisted by a battery of tele-hawks, a 'manufactured consensus' has emerged in India; a government-inspired sameness of view in favour of the nuke establishment which then, due to a very weak anti-nuclear movement in the country and a widespread ignorance about the scale of devastation that nuclear arms can unleash, becomes public discourse. *The Sunday Observer Special* (New Delhi, Mumbai, 24–23 May 1998) brought out a four-page 'felicitation' supplement entitled 'Mr. Prime Minister We Are With You', carrying congratulatory messages from various private companies (automobiles, jewellers, cargo agencies, printers, taxi services, industrial security services, etc.) and individuals.

> *Today, we feel even more proud to be Indian.* By successfully exploding five indigenously developed nuclear devices, our scientists have shown that *we are as good as the best in the world.* By giving the go-ahead for the explosions, Mr. Prime Minister, you have shown the world that we are capable of defending our nation, *no matter what the consequences.*

Soon the world will realise the rightness of our stand. we love *ahimsa* [non-violence], but we know how to defend ourselves when attacked; we believe in peaceful co-existence, but we are prepared for war; we hate nuclear weapons, but we will give up ours only when the rest of the world does so too (emphasis given).

(Sunday Observer Special 1998)

Apparently, the sponsors and advocates of 'nuclear nationalism' in India have little concern to express over the ecological and health 'consequences' for the tribal population of the uranium mining belt of Jaduguda in south-eastern Bihar, who are exposed to grave dangers but still keep the country's nuclear programme going (Sarin 1998). Also silenced in pro-nuke assertions are the radical concerns and reservations, such as those expressed by human rights activist Saumen Guha:

The military–industrial–academic complex had long been at work. Homi J Bhabha, India's pioneering nuclear scientist who headed the Atomic Energy Department in the 1960s, was pro-bomb, as are several retired generals and many hawkish defence analysts associated with policy think-tanks. In 1974, when India first tested a nuclear device in Pokhran, Prime Minister Indira Gandhi was facing increasing trouble at home . . . the test was used to consolidate her power base by appealing to shallow nation-alism. Subsequently, over the years, the Tatas and later, Larsen and Tourbo, both of which won large defence contracts, emerged as the industrial component of the military-industrial-academic complex pushing India into the nuclear orbit. The current BJP government is shaky and the nuclear tests were just what was required to gain a measure of stability. And the BJP has always proclaimed that it wants India to have the bomb.

(Banerjee 1998: 9)

The media-intellectuals and media-'experts' too have come forward with a most imaginative concoction of geopolitical arguments/reasoning in support of the nuclear blasts by India (Karlekar 1998; Ghatate 1998; Prakash 1998; Singh 1998; Gupta 1998; Bhargava 1998; Mehta 1998). To cite just a few examples, 'those who criticise India', are reminded that a Communist China has

surrounded India with its nuclear might in alliance with Pakistan [whose very survival is said to rest on its 'hate India' ideology] and Mynamar . . . threatening India's trade routes in the Indian Ocean. It has also deployed nuclear missiles in Tibet against India. Chinese technology has enabled Pakistan to test-fire the *Ghauri* missile, a weapon aimed solely at India.

(Prakash 1998)

Whereas those who equate Pakistan with India in terms of nuclear weapon capability are being told that this 'artificial parity' does not simply hold because, 'India is no territorial, ethnic or religious creation; it is heir to a mighty, ancient and living civilisation', while 'Pakistan is a recent off-shoot of India, about one-seventh its size, with a poor tradition of democracy and pretensions to equality' (Jain, 1998).

An editorial by Ranjan Gupta, a freelance 'foreign' correspondent and author, for *The Pioneer*, one of the widely read national dailies (Gupta 1998: 8), shows how the sophisticated geopolitical reasoning deployed in support of Pokahran II by the foreign policy spokesmen and experts of the BJP-led coalition government at the centre is being circulated among the masses of a 'peace-loving nation' through media. According to Gupta,

> by choosing nuclear power as its expression of *greatness*, India has chosen the hard *road* to success. India's defiance has to be seen in historical terms, and only a party believing in *national destiny* like the BJP could have taken this hard decision . . . as a *hard state*, it [India] will get more respect but less love. . . . Acquisition of nuclear capability has placed India in the same category of major Asian powers as China. It would have taken a century of liberalised economics to reach this status (emphasis given).
>
> (Gupta 1998: 8)

To Gupta, India's true friends are those who not only support the nuclear blasts, important as they may be internationally, but also 'show a commitment to Indian greatness'. Accordingly, concludes Gupta,

> *France has turned out to be India's best friend in its hour of need.* . . . A Pakistani-Islamic bomb is a security threat to Israel as much as it is to India. . . . India and Israel have common geostrategic needs. Militant Islam is as much a threat to India as it is to Israel. *The Hindu has been squashed as much as the Jew. Both Hindus and Jews have suffered mental suffocations, persecution and lack of appreciation.* . . . India must reward its friend and punish its foes. . . . This may be a good time to cut loose from all ties with defunct countries like Britain who have cashed in on colonialism to hang on to India.
>
> (Gupta 1998: 8)

The geopolitical significance of Pokharan II is being conveyed by Gupta to his readers in geo-historical terms. They are being asked to rejoice over the 'fact' that

> *At last India has found its destiny.* . . . The tests are more than nuclear tests. Their importance lies in fighting global inequality, in challenging Anglo-American domination. . . . *A new sinister plan to squeeze India dry on the tests*

> *which in time will lead to the break up of the country is on the anvil. Some Western*
> *nations think this may be a good time to balkanise India so that it never again chal-*
> *lenges Western hegemony.*

(Gupta 1998: 8)

And the 'unpatriotic' Indians who have failed to grasp the historic significance of the event as well as the political parties that have criticized the tests, both in and outside Parliament, on 'narrow partisan considerations' are said to be the victim of the same 'narrow considerations that lead countless Mir Kasims to sell India to Englishmen'. This sense that India was defying the Western nations is also demonstrated in the cartoon 'Not one of us' from the *Sunday Pioneer*, 26 July 1998 (Figure 9.1). A figure representing India is threatened by intimidating figures that represent Tony Blair, Jacques Chirac and Bill Clinton, the leaders of the Western Nuclear powers who want to maintain their hegemony over nuclear power.

The nuclear tests have also set off an explosion of geopolitical imaginations among the Hindu nationalists as well. In the wake of India becoming a 'nuclear weapon state' the Vishva Hindu Parishad (VHP) has proposed to carry Pokhran dust – sacred soil – in sanctified vessels across the country in yet another set of yatras, this time in the name of *guarav* or pride. It has been proposed by the same organization that a temple to the bomb should be built at the site of five nuclear explosions – a *shaktipeeth*.

Conclusions

This chapter has shown that representing post-Colonial India and endowing this 'Community Ship' of 'communities within communities within communities' (Larson 1997: 285) with a monolithic content, linear history, and uncontested meaning and trajectory continue to pose a formidable intellectual as well as political challenge to the two principal geopolitical imaginations about India's national identity, namely the secular-nationalists and the Hindu-nationalists. While drawing upon – though not necessarily agreeing upon – a whole range of other discourses, assumptions and beliefs that are specific to religious-cultural moorings and manifestations of Indian civilization, both resist disintegrative-centrifugal tendencies in the country while converging around a firm commitment to maintaining 'territorial integrity' of India's geopolitical realm – though they seek to do it in different ways. Apparently such imaginations or ideas do not exist 'out there' or simply in religious or secular texts and governmentalized policy documents, but emanate from and feed into practices or social action (Agnew 1998: 125).

The secular-nationalist and the Hindu-nationalist, however, neither represent nor exhaust the entire range and variety of geopolitical imaginations to be found across the 'sacred' but profoundly diverse human-cultural geographies of India.

230

Figure 9.1 'Not one of us'
Source: 'Cole' in *The Sunday Pioneer*, 26 July 1998

The sheer number and latitude of regional and local imaginations, from Kashmir to Northeast for example, their inter-generational character, the myths they generate, the manner in which they are passed on from one generation to another as form of common sense or guide to action, the way they adapt to challenges and changed historical contexts, and the intensity with which they often compete with one another, further compound the task of carving out neatly a 'secular' or for that matter a 'Hindu' national identity.

Of late, however, the Hindu nationalist discourse has been on the rise and the proximate reasons seem to have been supplied by a variety of factors, especially the mounting anxiety over 'separatist' agitations of the 1980s in Panjab and Jammu and Kashmir, deepening the crisis of political governance and ideological vacuum caused in Indian politics by the 'crisis of secularism' (Bhargava 1998). Hindu nationalists present themselves as an institutional and ideological alternative to the alleged 'pseudo-secularism' practised thus far in the country since the partition of the country. The discourse and the representational practices of 'Hindutva' reveal how the concepts of power, religion and territory are bound

231

together in an intricate and intimate manner. It is confirmed that a geopolitical discourse signifies much more than the identification of specific geographical influences upon a particular domestic or foreign policy situation and 'to identify and name a place is to trigger a series of narratives, subjects and understandings' (Ó Tuathail and Agnew 1992: 195). To designate a community or the area it inhabits as 'Hindu' for example amounts to not only a naming ritual, but also to enframing it terms of its 'sacred geography', 'authentic politics' and the type of foreign policy that its 'nature' demands. Also, geopolitical imaginations are rendered meaningful and 'legitimate' largely through popular geopolitical reasoning, which seems to rely more on religious sentiments and narratives than on formal geopolitical reasoning.

Our analysis of the various forms of geopolitical reasoning offered in support of the nuclear blasts by the BJP-led coalition government in New Delhi supports Neumann's contention (1997: 148) that 'geopolitical shifts [as currently observed in the case of post-Pokharan II South Asia] are not only, or perhaps not even first and foremost, about changes in the balance of power: they are about changes in the balance of threats and challenges as these are constructed in dominant political discourse'. Whereas the discussion on India's policy of non-alignment has shown that the practices related to the production of knowledge to aid the practice of statecraft are also a part of the larger geopolitical project to further the power of the state.

Bibliography

Agnew, J. (1998) *Geopolitics: Re-visioning World Politics*, London, Routledge.

Banerjee, Paula (1998) 'To re-instate historians in the history of border', paper presented at a seminar on 'Asian geopolitics: borders and transborder flows' at New Delhi, 23 and 24 March, 1998, organized by Maulana Abul Kalam Azad Institute of Asian Studies, Calcutta.

Banerjee, Partha (1998) 'The bomb is not everything', *The Statesman*, New Delhi (editorial), 29 May: 9.

Behera, N. C. (1998) 'Perpetuating the divide: political abuses of history in South Asia', *Indian Journal of Secularism*, 1(4): 53–71.

Bhambhri, C. P. (1998) 'BJP agenda: hidden or real?', *The Pioneer,* New Delhi, 15 April.

Bhargava, G. S. (1998) 'Journalists' euphoria over Pokharan', *Mainstream*, 36(23): 8–9.

Bhargava, R. (ed.) (1998) *Secularism and Its Critics*, Delhi, Oxford University Press.

Bose, Sugata (1997) 'Nation as mother: representations and contestations of "India" in Bengali literature and culture', 50–75 in S. Bose and A. Jalal (eds) *Nationalism, Democracy and Development: State and Politics in India*, Calcutta: Oxford University Press.

Bose, Sumantra (1997) '"Hindu Nationalism" and the crisis of the Indian state: a theoretical perspective', 104–64 in S. Bose and A. Jalal (eds) *Nationalism, Democracy and Development: State and Politics in India*, Calcutta: Oxford University Press.

Brecher, M. (1968) *India and World Politics: Krishna Menon's View of the World*, London: Oxford University Press.

Cohn, B. S. (1997) *Colonialism and Its Forms of Knowledge: The British in India*, Delhi: Oxford University Press.

D'Monte, D. (1998) 'After the hangover: the costs of nuclear club membership', *The Indian Express*, Chandigarh, 29 May: 8.

Dijkink, G. (1996) *National Identity and Geopolitical Visions: Maps of Pride and Pain*, London:Routledge.

Dixit, J. N. (1995) *Anatomy of a Flawed Inheritance: India-Pakistan Relations 1971–94*, Delhi: Ajanta Publications.

—— (1998) 'Blasting a straitjacket', *Outlook*, New Delhi, 1 June.

Fisher, M. H. (1993) *The Politics of the British Annexation of India 1757–1857*, Delhi: Oxford University Press.

Ghatate, N. M. (1998), 'No first use', *The Pioneer*, New Delhi, 19 May: 9.

Gill, K. P. S. (1998) 'Give internal security the top priority', *The Pioneer*, New Delhi, 13 June: 8.

Graham, B. (1993) *Hindu Nationalism and Indian Politics: The Origins and Development of the Bharatiya Jana Sangh*, Cambridge: Cambridge University Press.

Gupta, D. (1995) 'The political jungle: Shiv Sena tiger roars to success', *The Times of India*, New Delhi, 12 August.

Gupta, R (1998) 'India's metamorphosis into a hard state', *The Pioneer*, New Delhi, 12 June.

Hasan, M. (1996) 'The myth of unity: colonial and national narratives', 185–208 in D. Ludden (ed.) *Making India Hindu: Religion, Community, and the Politics of Democracy in India*, Delhi: Oxford University Press.

Houbert, J. (1989) 'India between land and sea', *Current Research on Peace and Violence* XII, 4: 201–11.

Jain, S. (1998) 'Pakistan's "cricket match" mindset', *The Pioneer*, New Delhi, 18 June: 8.

Jalal, A. (1995) *Democracy and Authoritarianism in South Asia: A Comparative and Historical Perspective*, Cambridge: Cambridge University Press.

—— (1997) 'Exploding communalism: the politics of Muslim identity in South Asia', 76–103 in S. Bose and A. Jalal (eds) *Nationalism, Democracy and Development: State and Politics in India*, Calcutta: Oxford University Press.

Kakar, S. (1995) *The Colours of Violence*, New Delhi: Penguin Books.

Karlekar, H. (1998) 'Are they suffering from a death wish?', *The Pioneer*, New Dehi, 19 June: 8.

Katzenstein, M. F., Mehta, U. S. and Thakkar, U. (1998) 'The rebirth of Shiv Sena in Maharashtra: the symbiosis of discursive and institutional power', 215–38 in A. Basu and A. Kohli (eds) *Community Conflicts and the State in India*, Delhi: Oxford University Press.

Kaviraj, S. (1994) 'Crisis of the Nation-State in India', *Political Studies*, XLII: 115–29.

—— (1997a) 'The modern state in India', 225–50 in M. Doornbos and S. Kaviraj (eds) *Dynamics of State Formation: India and Europe Compared*, New Delhi: Sage.

—— (1997b) 'On the structure of nationalist discourse', 298–335 in T. V. Sathyamurthi (ed.) *State and Nation in the Context of Social Change*, Delhi: Oxford University Press.

Kothari, R. (1997) 'Fragments of a discourse: towards conceptualization', 38–54 in T. V. Sathyamurthi (ed.) *State and Nation in the Context of Social Change*, Delhi: Oxford University Press.

—— (1998) *Communalism in Indian Politics*, Ahmedabad: Rainbow.

Krishna, B. (1995) *Sardar Vallabhbhai Patel: India's Iron Man*, New Delhi: Indus.

Krishna, S. (1994) 'Cartographic Anxiety: Mapping the Body Politic in India', *Alternatives* 19: 507–21.

Kumar, R. (1997) 'State formation in India: retrospect and prospect', 395–410 in M. Doornbos and S. Kaviraj (eds) *Dynamics of State Formation: India and Europe Compared*, New Delhi: Sage.

Larson, G. J. (1997) *India's Agony Over Religion*, Delhi: Oxford University Press.

Ludden, D. (1996) (ed.) *Making India Hindu: Religion, Community, and the Politics of Democracy in India*, Delhi: Oxford University Press.

Mahanta, A. (1977) 'The Indian state and patriarchy', 87–131 in T.V. Satyamurthi (ed.) *State and Nation in the Context of Social Change*, Delhi: Oxford University Press.

Maheshwari, A. (1998) 'The Face Behind the Mask', *Hindustan Times Sunday Magazine*, New Delhi, 8 August: 1.

Malik, Y. K. and Singh, Y. B. (1994) *Hindu Nationalists in India: The Rise of the Bharatiya Janata Party*, New Delhi: Vistaar Publications.

Mehta, R. (1998) 'Wholly positive development from nationalist point of view', *Mainstream* 36(23): 10.

Metcalf, T. R. (1995) *Ideologies of the Raj*, Cambridge: Cambridge University Press (The New Cambridge History of India).

Nandy, A., Trivedi, S. Mayaram and Yagnik, A. (eds) (1995) *Creating a Nationality: The Ramjanmbhumi Movement and the Fear of the Self*, Delhi: Oxford University Press.

Nehru, J. (1936) *An Autobiography*, London: John Lane.

—— (1956) *Independence and After: A Collection of Speeches 1946–9*, Delhi: Government of India.

—— (1981 [1946]) *The Discovery of India*, New Delhi: Jawaharlal Nehru Memorial Fund, Oxford University Press.

Neumann, I. B. (1997) 'The Geopolitics of Delineating "Russia" and "Europe": The Creation of the "Other" in European and Russian Tradition', 147–73 in O. Tunander, P. Baev and V. I. Einagel (eds) *Geopolitics in Post-Wall Europe: Security, Territory and Identity*, London: Sage.

Ó Tuathail, G. and Agnew, J. (1992) 'Geopolitics and foreign policy: practical geopolitical reasoning in American foreign policy', *Political Geography* 11: 190–204.

Pal, R. M. (1998) 'Human rights and nuclear explosions at Pokharan', *Mainstream*, 26(23): 6–7, 12.

Pandey, G. (1990) *The Construction of Communalism in Colonial North India*, Delhi: Oxford University Press.

Parker, G. (1988) 'Geopolitical perspectives on India and Indian foreign policy', in *The Ford Foundation Lectures in International Relations Studies*, Department of Political Science, The Maharaja Sayajirao University of Baroda.

—— (1998) *Geopolitics: Past, Present and Future*, London: Pinter.

Pattanaik, D. D. (1998) *Hindu Nationalism in India: Conceptual Foundation*, New Delhi: Deep & Deep Publications.

Prakash, P. (1998) 'South Asian arms race began in 1964', *The Pioneer*, New Delhi, 20 June: 8.

Sarin, R. (1998) 'Inside radiation zone', *The Indian Express*, Chandigarh, 21 June: *Express Magazine*.

Sarkar, T. (1996) 'Imagining Hindurastra: the Hindu and the Muslim in Bankim Chandra's writings', 162–84 in D. Ludden (ed.) *Making India Hindu: Religion, Community, and the Politics of Democracy in India*, Delhi: Oxford University Press.

Savarkar, V. D. (1969) *Hindutva*, Bombay: Savarkar Prakashan, fifth edn.

Sen. A. (1998a) 'On interpreting India's Past', 10–35 in S. Bose and A. Jalal (eds) *Nationalism, Democracy and Development: State and Politics in India*, Calcutta: Oxford University Press (Oxford India Paperbacks).

—— (1998b) 'Secularism and its discontents', 454–85 in R. Bhargava (ed.) *Secularism and its Critics*, Delhi: Oxford University Press.

Sen Gupta, B. (1997) 'India in the Twenty-First Century', *International Affairs* 73(2): 297–314.

Singh, S. (1998) 'India must speak in one voice now', *The Pioneer*, New Delhi (editorial), 29 May.

The Tribune (1998), Chandigarh, 4 August, 1998: vol. 118, no. 214, city edition.

Vajpayee, A. B. (1998) 'Rationale for the government's decision on nuclear tests', *Mainstream* 36 (23): 3, 33.

van der Veer, P. (1996) 'Writing violence', 250–69 in D. Ludden (ed.) *Making India Hindu: Religion, Community, and the Politics of Democracy in India*, Delhi: Oxford University Press.

Varshney, A. (1993) 'Contesting meanings: India's national identity, Hindu nationalism, and the politics of anxiety', *Daedalus* 122, 3: 227–61.

Verghese, B. G. (1996) *India's Northeast Resurgent: Ethnicity, Insurgency, Governance, Development*, New Delhi: Konark Publishers Private Limited.

Weiner, M. (1997) 'Minority identities', 241–53 in S. Kaviraj (ed.) *Politics in India*, Delhi: Oxford University Press.

Part 3

RECLAIMING AND REFOCUSING GEOPOLITICS

10

HÉRODOTE AND THE FRENCH LEFT

Paul Claval

Introduction

In the after aftermath of the Second World War, French geographers showed little formal interest in geopolitics (*géopolitique*) and political geography (*géographie politique*) because of the damaging associations with German geopolitics and Nazi Germany. The 'geopolitical' had become a pejorative adjective and interest in political geography, which had been strong in the inter-war period, when French geographers tried to build a 'geopolitics of peace' and prompt the development of a united Europe, had completely disappeared. In the inter-war period, French geographers had contributed to international debates on geopolitics through their attempts to promote international equilibrium and global peace rather than colonial expansion and national power (Parker 1985; Claval 1994; Muet 1997). French authors such as Albert Demangeon, for instance, were alive to the dangers posed by German *Geopolitik* and its obsession with space and the earth at the expense of human environments. After the Second World War, geopolitical research disappeared in France even though some scholars maintained an interest in the field. Few if any geographers, however, sought to publish on geopolitics because they wanted to avoid the harsh criticisms proffered by Marxists against all forms of geopolitics, regardless of inspiration and intent. The only geographer who escaped this ban was Jean Gottmann who, since he was Jewish and had to flee to the United States during the Second World War, could not be accused of sympathies with Nazism. In 1952, he published his fundamental book on *La Politique des Etats et leur géographie* (Gottmann 1952). However, for the next twenty-five years, political geography and geopolitics disappeared from French geographical publications, with the exception of Claude Delmas, who worked for NATO and specialized in geostrategy (Delmas 1971).

As a consequence, even though a few geographers tried to modernize their discipline through the use of spatial economics and modern conceptions of ecology, most French scholars remained faithful to the Vidalian tradition of 'Geography'. Therefore, post-war French geography was predominantly concerned with the investigation of rural landscapes in France and with overseas tropical geography. From the start of the 1960s, this Vidalian tradition was increasingly displaced as scholars developed new research interests in European industrial problems, regional economics and urban networks. However, it was apparent that major methodological innovation was sparse and perhaps the most neglected element of French geography was political geography and geopolitics in particular.

The events of 1968 had more influence on French collective behaviour and social structures than on its political regime, which withstood the crisis. The period gave a major impetus to a process of intellectual revival which had started a few years before. French geography was altered by a range of intellectual and political transformations which included the introduction of spatial economics, urban theory and quantitative methods. In this spirit of radical change, a number of geographers such as Armand Frémont gave a fresh impulse to local and regional studies through his emphasis on the lived experience of space (*espace vécu*) whilst Georges Bertrand and Gabriel Rougerie tried to build closer links between natural and human geography through the analysis of landscape. Elsewhere, Roger Brunet developed a new systematic approach to regional studies which appeared timely given the political and economic changes.

It has to be acknowledged, however, that geography as taught in the majority of departments was not at the vanguard of radical French intellectual culture and the events of 1968. The new, emergent forms of geography outlined above contributed little to the cultural and political sea-change of the era. In spite of their radical intent, neither the young geographers under the supervision of Professor Pierre George, nor left-wing geographers in general were on the frontier of French geographical research in the early 1970s. In contrast, Yves Lacoste, a Professor of Geography at the University of Vincennes, rejected the accusation that geography was cretinous and reactionary and proposed a new approach which remained sympathetic to the principles he had learned from Pierre George and Jean Dresch, and the tradition of situation analysis which George had inherited from the Vidalians. For many colleagues who were reluctant to invest heavily in the new quantitative and theoretical developments of French Geography, Lacoste appeared to be a radical and heroic iconoclast. The emergence of the geopolitical journal *Hérodote* in January 1976 was the culmination of a long intellectual journey which sought to answer the simple but fundamental question: what is geography for?

The subsequent sections of this paper seek to evaluate the intellectual and political trajectories of Lacoste and his colleagues on the Left in the development of this particular version of geopolitics. The initial section considers the events leading up to the creation of *Hérodote* and the various conceptions of geopolitics employed by Lacoste. Thereafter, our attention turns in some detail to the various studies of geopolitics within *Hérodote*, ranging from the analysis of war, religious fundamentalisms, nationalism and changing world orders. The final part of the chapter investigates the public and academic reactions to *Hérodote* in France and in the wider world.

Yves Lacoste, geopolitics and *Hérodote*

In the foreword of his *Dictionnaire de géopolitique* (Lacoste 1993a), Yves Lacoste outlined his current conception of the nature and remit of geopolitics. His aim was to study both geopolitical situations and geopolitical ideas, with analysis of each being grounded in particular geographical situations:

> Whatever its territorial extension . . . and the complexity of geograph-ical data . . . a geopolitical situation is defined, at a given time of an historical evolution, by rivalries of powers of more or less large scale [generally at the international scale], and by relations of forces which are located on the different parts of the concerned territory.
>
> The rivalries of powers are first those of States, large or small, which quarrel over the possession or control of certain territories. . . . Rivalries of powers, official or not, can also exist *within* numerous states, whose people, often minorities, claim either their autonomy or inde-pendence. . . . Lastly, within a nation, geopolitical rivalries exist between the main political parties which try to extend their influence in a given region or agglomeration, and attempt to conquer or keep the control of electoral constituencies.
>
> In order to show the distribution of these diverse forces, included in relatively limited spaces, clear and suggestive maps are necessary, in particular, historical maps, which allow an understanding of the evolu-tion of a specific situation and an appreciation of the 'historical rights' which are claimed in contradiction by several states over the same territory.
>
> (Lacoste 1993a: 3)

Yves Lacoste went a step further and proposed that:

> In order to understand a geopolitical conflict or rivalry, it is insufficient merely to determine and map what is at stake, instead it is necessary to understand the *reasons* and the *ideas* of the main actors – states' rulers, leaders of regionalist, autonomist, independent movements etc., each

translating and influencing the same public opinion he [*sic*] represents. The role of ideas – even wrong ones – is central in geopolitics, since they explain the projects and determine, as much as material data, the choice of strategies.

(Lacoste 1993a: 4)

In his foreword, Lacoste presented a history of the evolution of geopolitical ideas, from Kjellén to the present. He highlighted the ban imposed by the communist regimes after the Second World War and the role of the Cold War, its reappearance at the end of the 1970s at the time of Cambodia's war, and its triumph in the 1980s and 1990s. According to Lacoste, the successful re-emergence of geopolitics could be explained by focusing on the emergence of democratic regimes, the triumph of the idea of self-determination for peoples, and the impact of the modern media:

The analysis of the many geopolitical conflicts . . . which recently appeared in Europe, and the attention paid to the equally geopolitical debates they induced both between nations, and in each of them, confirm . . . that territorial rivalries are specifically geopolitical when they form the subject of contradictory representations widely diffused today by medias, and which provokes political debates among the citizens, if there is some freedom of expression.

(Lacoste 1993a: 17)

In less than fifteen years then, the old conception of Geography as the analysis of situations was completely renewed. The Vidalians' naturalistic approach to human geography, that Lacoste and his colleagues had inherited from the French tradition, was increasingly complemented by modes of investigation concerned with social actors and the politics of representation. And throughout this period, in which Lacoste recast his perspective upon geopolitics, he was constantly influenced and informed by the wider cultures of French academia and his immediate intellectual circle associated with *Hérodote*.

Yves Lacoste and the origins of *Hérodote*

The journal *Hérodote* was strongly shaped by the ideas of Yves Lacoste and the earlier academic evolution of the discipline of geography, although the creation of the journal coincided with a more general reorientation of French geography and its analysis of political problems in the 1970s and 1980s. As a young academic, along with Jean Dresch and Pierre George who were considered by many to be the leading figures in French geography, Yves Lacoste belonged to an active group of geographers all of whom were members of the French Communist Party. It was not so much an academic circle with an interest in epistemological problems, as a group of militant communists armed with orthodox Marxism which prevented them from developing new theoretical approaches in geography (Varii Auctores 1991). The members of this group based their work upon the conception of geography developed by Pierre George during the 1950s. Most of his inspiration was derived from the French geographical tradition, but he was critical of some Vidalian concepts which he considered obsolete for industrial societies such as the *genres de vie* [ways of life] notion. *Genres de vie* were substituted by *modes of production* in order to address the complexities and the problems of the modern world. As for Pierre George and his followers, there was no need for an epistemological break and a scientific revolution in order to modernize geography; it was sufficient to displace the descriptive Vidalian *genres de vie* by a Marxist concept. Thus, George's geography took its main features from Vidal de la Blache, but refused to incorporate the kind of descriptive and naturalistic rural studies which had been prominent for two generations in French geography. There was no intellectual contact with the new quantitative geography which was becoming popular in Sweden, the USA, Britain and among some colleagues in France. It was 'new' only in so far as it was both a Marxist and a neo-Vidalian interpretation of the discipline.

By comparison to their Anglophone counterparts, French geographers had not yet developed theoretical research programmes. In contrast, French geographers considered their role to be based on explaining real problems observed either in a particular country or in the international relations of countries. The analysis of such situations was central to Pierre George's geography just as it was to the Vidalian geography of the first half of the present century (Claval 1998). It started with the idea that each case had to be examined at different scales; the area had then to be studied as a set of *milieus*, stressing ecological relations and constraints; the last phase was centred on the description and role of flows, that is, *circulation*. The subsequent interplay of the *milieu* and *circulation* helped to create particular *genres de vie* which in turn were reflected in the evolution of states in the case of political geography. Yves Lacoste was faithful to Pierre George's mode of analysis and during the 1960s he worked mainly on the problems of decolonization and development in the Third World, relying on the analysis of geographical situations as a research procedure. He pursed his field studies in Latin America and West Africa, but focused mainly on French-speaking North Africa (most of these studies have been reprinted in: Lacoste 1980) because he had been raised in Morocco, where his father, a geologist, was responsible for oil prospecting. Therefore, his childhood experiences in North Africa were later to exert an important influence on his analyses of colonialism and domination because, when reading geographical studies dealing with colonization, he found that the economic conditions were over-emphasized. In his memory, the French protectorate of Morocco relied first and foremost on the presence and strength of the French army (Lacoste, personal communication)!

It was only at the end of the 1960s that Yves Lacoste began to suspect that there were problems with this kind of Geography. Since he considered that the methodological practices of the discipline were sound, the only weaknesses stemmed from the subjects covered by geographers and the way in which the discipline was taught in secondary schools and universities. In Lacoste's view, the crisis of geography resulted from the refusal by geographers to confront the real problems of the world. To help address these problems, Lacoste undertook the responsibility of producing a series of new textbooks for secondary schools for the publishing company, Nathan (Guglielmo, Lacoste and Ozouf 1965). In these publications, Lacoste stressed both the visual dimension of geography (lessons were based on a high quality commentary of synthetic landscapes, which summarized all the features of the problem or region under scrutiny) and the processes and events which were shaping contemporary problems: industrialization, urbanization, Third World development, and areas of conflicts. In this way, Lacoste introduced the politics of geography into the school classroom.

Lacoste's endeavour to explore the connections between the state and geography was strengthened by his mission as an invited expert in Vietnam in the early 1970s. In 1966, and again in 1972, the Vietnamese had accused the American air force of deliberately bombing the dikes which protected the paddy-fields of the Tonkin River delta. The Vietnamese Communist Party asked Jean Dresch to study the evidence of this allegation. Since the earliest research experiences of Yves Lacoste had been on the alluvial levees of the Gharb plain in Morocco, Dresch asked him to analyse this geographical issue. The material transmitted by the Vietnamese was initially unconvincing. However, after a season of field work (in the summer of 1972), Lacoste concluded that the Vietnamese were right, and that the Americans had tried to systematically destroy the farming basis of the Vietnamese economy through their strategic bombing. His report had much impact on international opinion and a paper he had written for Le Monde (6 June 1972), before moving to Vietnam, attracted the attention of left-wing French groups.

As a consequence of his Vietnamese experiences and in the light of his growing conviction about the importance of interrogating the roles of geography in statecraft, Lacoste decided to stress the significance of political and military factors in Geography. The publisher, François Maspéro, accepted his proposal to launch a new journal on the relationship between Geography and power. The title, Hérodote, had been chosen in 1972, following market research prior to the launch of the journal. The title of the journal was derived from the ancient Greek geographer and historian, Herodotus. The first issue eventually came out in 1976 and included a long report on the Red River delta bombing (Lacoste 1976b).

Hérodote: a new orientation in left-wing French geography

Hérodote was launched twenty-three years ago and the first issue was published in January 1976. It marked a radical departure from the long-standing Vidalian legacy of Francophone political geography. Ironically, the first issues of *Hérodote* in the mid-1970s were not devoted to geopolitics. The term was not incorporated into the journal's sub-title until 1983 (*Revue de geographie et de géopolitique*). In his first editorial, Yves Lacoste clearly explained that the aim of the new review was to cover the highly political problems of military power and its role in the evolution of the political map of the world. During the first years, many articles dealt with Revolutionary Wars (in the Cuban Sierra Maestre (Lacoste 1977) and other theatres of civil or revolutionary wars), or analysed the adventures and defeat of Che Guevara in Bolivia (Varlin 1977) and considered the strategic legacies of Clausewitz (Lacoste 1976c). They also included a reflection on the nature of nations, nationalities and national movements. In doing so, Lacoste and his colleagues were seeking to overturn the traditional agendas of Geography which had tended to concentrate on the state and the national territory at the expense of focusing on ecological, anti-colonial and revolutionary subject matter. The journal was also a radical departure from the less political aspects of French geography, as indicated by *Hérodote's* sub-title: *Stratégies, géographies, idéologies*. In 1996, Lacoste explained the significance of these terms:

> The association of these three terms expressed adequately our preoccupations: the plural indicating that, if there are multiple strategies and diverse ideologies, there are also different ways of being a geographer and hence, different geographies, according to their strategic functions and ideological roles. In order to make people conscious of what was really at stake in geography, we wanted to underline in particular the contrast between what we called the geography of teachers and the geography of military staff.
>
> (Lacoste 1996: 7)

A few months after the first issue of *Hérodote*, Yves Lacoste published a short and very polemical book which clarified his intellectual and political objectives: *La géographie, ça sert, d'abord, à faire la guerre* (Lacoste 1976a), the purpose of geography is, above all, the making of war! The publisher, François Maspéro (the name of his firm became 'La Découverte' in 1983), was the same as for *Hérodote*. The short book was a kind of pamphlet which emphasized the role of geography in supporting the power of the state, and the reasons for Lacoste's criticisms of the traditional role of geography and the geographical establishment (see also Parker 1998: 51–2). Yves

Lacoste wished to expose geography to new curiosities and research areas because the military aspects of conflicts, and in particular the nature of revolutionary wars, was considered too restricted. For the first few years, he explored a range of different possibilities.

Yves Lacoste worked with a small editorial board and one of its members left it rapidly: the talented scholar, Maurice Ronai. Yet from the onset, other colleagues played a dominant role in the team. Béatrice Giblin prepared a doctoral dissertation on Elisée Reclus (Giblin-Delvallet 1976) and this research was considered a model for *Hérodote*. The methodology employed by many of the earliest contributors was still derived from Vidalian approaches, although they preferred to refer to the great anarchist Reclus as a source of inspiration because of his active political involvement in ecological affairs. Similar political themes were pursued by Michel Foucher, a cautious scholar who was originally from Lyon (he used the pseudonym of Thomas Varlin, a well known anarchist of the Paris Commune, in 1871, in two of his early contributions to *Hérodote*). He set very high standards for fieldwork when preparing reports on case studies (for instance, Foucher 1979–80) and his articles were judged on their empirical strength rather than the ideological position of the author: a position that was thoroughly endorsed by Yves Lacoste. At the end of the 1970s and the beginning of the 1980s he developed an interest in boundaries as a geographical problem and became a respected expert on these problems with special reference to South America (Foucher 1983 and 1986).

Yves Lacoste did not rely only on geographers for the definition of the editorial line of *Hérodote*. Michel Korinman taught German at the University of Vincennes and prepared a doctoral dissertation on Haushofer's geopolitics (Korinman 1990, 1991). This research on *Geopolitik* enabled French geographers to discover that German political geography and geopolitics were more complex than was generally thought, and that some of their insights were still valuable (Korinman 1984). In his subsequent research, he demonstrated the fundamental impact of Ratzel's writings on politics and the environment in the German-speaking world (Korinman 1990). Lacoste was also a friend of François Châtelet, a well known left-wing philosopher who worked at the University of Vincennes where he had taught an introductory course on the philosophy and methodology of geography jointly with Lacoste in 1969–1970. François Châtelet wrote a paper for *Hérodote* (Châtelet 1976) and he later arranged a memorable encounter between Michel Foucault and Yves Lacoste. Lacoste interviewed Foucault about his conceptions of space (Lacoste 1976d) which then resulted in Foucault preparing several questions to be asked of geographers (Foucault 1976). Both the interview and the questionnaire were published by *Hérodote*, but the collaboration was short-lived. Lacoste was left with the impression that Foucault had no real interest in geography, and that spatial problems were not central at that time to his research (Lacoste, personal communication). Camille Lacoste-Dujardin, his wife,

a well known ethnologist, researched the Kabyl culture in Algeria and her influence was critical in ensuring that *Hérodote* was conscious of the similarities and
differences in the way geographers and anthropologists practised fieldwork.
Camille Lacoste-Dujardin had close connections with Pierre Bourdieu, who had
prepared his doctoral dissertation on Kabyl culture. She presented Bourdieu's
research and ideas to the readers of *Hérodote* (Lacoste-Dujardin 1976) and demonstrated the inter-disciplinary connections that *Hérodote* would pursue in its
examination of space, knowledge and power.

During the first years of publication, *Hérodote's* empirical material was mainly
devoted to the post-colonial wars and revolutionary movements in the Third
World. There were, however, other interests which were critical to the emerging
editorial policy. The interest in political problems more generally was motivated,
for Yves Lacoste, by a desire to change the content, intent and applied nature of
geography. *Hérodote* was not conceived, therefore, to only study the competition
of super-powers, the consequences of decolonization and the problems of development. It also had to contribute to the modernization and reformulation of
geography, and not only political geography, but the entire discipline.

From 1978 onwards, each issue was devoted to a special topic and for about five
years Lacoste tried to cover all the various aspects that composed the modern geography he wished to promote. The journal covered issues of landscape studies ('A quoi
sert le paysage', *Hérodote*, no. 7, 1977; later: 'Paysages en action', *Hérodote*, no. 44,
1987); fieldwork ('L'enquête de terrain, I', *Hérodote*, no. 8, 1977; 'L'enquête de
terrain, II', *Hérodote*, no. 9, 1978.); the conception of time developed by geographers ('Le temps des géographes', *Hérodote*, no. 20, 1981); cartography and its use
as a tool of power ('Dominer: cartes et quadrillages', *Hérodote*, no. 13, 1979.); physical geography and ecology ('La géographie et sa physique', *Hérodote* no. 12, 1978;
'Ecologie/géographie', *Hérodote*, no. 26, 1982); and the geography of natural
hazards ('Terres à hauts risque's, *Hérodote*, no. 24, 1982). In Lacoste's opinion, there
was no need to develop general and/or universal statements on the nature of human
or political geography because according to the neo-Vidalian line he had inherited
from Pierre George, the aim of the discipline was to analyse geographical *contexts*,
and not to look for scientific regularities. Rather, the historical nature of human societies meant that there was no sense in looking for abstract laws in specific locations
and contexts. The main function of geography was to give focus for action: to be
'active', according to Lacoste's own words. Within eight years of the journal's
formation, the main phase of reflection on the nature of geography concluded with
the International Geographical Union Congress in Paris. A special issue prepared for
that event, 'Les géographes, l'action et la politique' (*Hérodote*, no. 33–4, 1984),
summarized the positions of the *Hérodote* group.

In 1976, Lacoste was clearly willing to break with the predominant apolitical
stance of academic geography; however, he was still not sure of the precise

PAUL CLAVAL

orientation to give to *Hérodote*. Seven years later, the editorial board he chaired
had developed clearer views. The military aspects of international life were only
one facet of geography and half the papers published on this field by *Hérodote*
between 1976 and 1996 were issued during these first five years. Later, intellectual attitudes changed as the label 'geopolitics' become more popular in the
French and English-speaking worlds. The term was frequently used in journals
and the media including the left-wing newspaper *Le Monde*. Marie-France
Garraud, a former political adviser of Georges Pompidou and Jacques Chirac,
had also launched a journal called *Géopolitique*. As a result of this development
and the rehabilitation of the term geopolitics, the sub-title *Revue de géographie et
de géopolitique* was substituted by *Stratégies, géographies, idéologies*. The change
occurred for issue no. 27 in 1982, 'La Méditerranée américaine'. From that
time, the majority of issues were devoted to topics like 'German geopolitics'
(*Hérodote*, no. 28, 1983), 'Geopolitics in the Near East' (*Hérodote*, no. 29–30,
1983), and 'Geopolitics of the sea' (*Hérodote*, no. 32, 1984). Thereafter, Lacoste
revealed an ongoing concern to consider and question the roles of geography
and geopolitical thought within society.

Lacoste wrote in 1984:

> In order for geography to be recognised by the scientific community as a
> knowledge . . . as necessary as medicine or agronomy, geographers . . . have
> to become conscious of the fact that the reason for them to exist in the
> society is *to know how to think spatially in order to allow for more efficient action*
> (added emphasis).

(Lacoste 1984: 19)

Relying on the analysis of situations, Lacoste defined the art of 'thinking space'
in this way:

> In order to be efficient, the geographer has to start with the principle
> that each phenomenon which is isolated by thought has its peculiar spatial
> pattern, which corresponds on the map to a peculiar *spatial set*. The
> number of spatial sets which intermix on the surface of the globe is
> therefore immense. Their classification is operated by distinguishing
> different *magnitudes*. . . .
>
> Geographical observation is actually conducted at very different
> levels of analysis, from a world level to a level suitable for the inventory
> of the characteristics of a small place. There are, roughly, as many levels
> of analyses as there are magnitudes in the dimensional range of spatial sets
> as distinguished by geographers.

(Lacoste 1984: 19–20)

Lacoste went on: to 'reason geopolitically is to show the complexity of the relations between politics and geographical patterns' (Lacoste 1984: 30). For him at
that time, the role of geographers was to critically analyse the geopolitics imagined
and practised by rulers or politicians or other figures in authority:

> In order to be clearer and more explicit, to expose hidden strategies, we
> must go back to the map, not only examine and show one map, but maps
> which, prepared at different scales, allow us to understand the inter-
> twined nature of problems and forces of power according to the size of
> territory. In this domain, the way geographers 'think' demonstrates its
> complete usefulness.
>
> (Lacoste 1984: 31–2)

Yves Lacoste stressed the value of geopolitics, but did not conceive it as
fundamentally different from geography. Geopolitics had to rely on cartographic
analysis in order to show how the dialectics of scales played out in particular places.
It had to measure the natural constraints of a situation and their role in the conflicting
interests at stake. It also had to describe settlement patterns, urban hierarchies, and
the flows that prevailed before conflicts developed, and which so often underpinned
and motivated them. Thus, the geopolitical was grounded in the geographical. What
was new, however, was the attention devoted to actors and the ways in which the
analysis of social and political situations was carried out because traditionally this
relied upon exhaustive inventories of the natural or social forces at work. By the mid-
1980s, the situation was increasingly conceived as an arena in which rulers, political
parties and interest groups competed for influence and power. The analysis of
situations ceased to be based on a paradigm borrowed from the natural sciences. The
emphasis now centred upon social architecture and social interaction rather than on
universal laws and regularities. Amidst this intellectual seachange, *Hérodote's* focus was
increasingly placed upon the study of geopolitics. Eight years after the first issue, the
agenda of the journal had been clearly defined. From 1984, the style of the review
did not change much. Each issue covered, and still covers a theme, generally the
geopolitical problems of a particular area. Even if the editorial line was not altered,
the nature of the 'geopolitical' in the journal continued to be adapted.

New attitudes towards development and international order

From the beginning of the 1980s, *Hérodote* was increasingly devoted to the
analysis of the geopolitics of the conflict areas of the world. Many of the journal's
issues dealt with the political problems or crises of the time, either civil wars

(Foucher and Pichol 1978), or international conflicts. There were papers on the conflicts of Nicaragua (Foucher 1979), Chile (Santibañez 1977), Angola (Anonymous 1976) and Cyprus (Péchoux 1976) in the 1970s, Afghanistan (Gentelle 1980) or Israel and Palestine in the 1980s ('Géopolitiques au Moyen-Orient', with papers by Nadia Benjelloun-Ollivier, Michel Foucher, Michel Korinman, Peter Demant, Maxime Rodinson, *Hérodote*, no. 29–30, 1983). Others covered wider areas and provided a broader view of the evolution of regional problems. This type of analysis started in 1983 with a volume on the geopolitics of German partition ('Géopolitiques allemandes', *Hérodote*, no. 28, 1983); the geopolitics of Africa ('Géopolitiques en Afrique', *Hérodote*, no. 46, 1987); the Soviet Union ('Géopolitique de l'URSS', *Hérodote*, no. 47, 1987); Central Europe ('Europe médiane', *Hérodote*, no. 48, 1988); Monsoon Asia ('Géopolitique en Asie des Moussons', *Hérodote*, no. 49, 1988); or Australasia ('Australasie', *Hérodote*, no. 52, 1989). At the time when the law of the sea was under discussion, *Hérodote* presented an overview of the consequences of the new rules of national appropriation of maritime areas ('Géopolitiques de la mer', *Hérodote*, no. 32, 1984). Problems of insularity were also treated ('Ces îles où l'on parle français', *Hérodote*, no. 37–8, 1985).

A surprising omission in the journal's content for the 1980s was that there was no study on the armaments race, Reagan's 'Star Wars' initiative and the Second Cold War. The transition towards a market economy in China was ignored as was the building of the European Economic Community. Instead, Africa, South America, the Mediterranean, south-west, southern and south-eastern Asia were the main areas covered by the review and the focus of *Hérodote* remained resolutely upon the developing world. Having said that, *Hérodote's* attitudes towards development changed progressively. Because the evolution of *Hérodote* was a gradual one, the changing position of the journal did not trigger direct reaction in the press. However, when Lacoste summarized his new understanding concerning the process of development and the attitudes of developed countries relative to the Third World in a typically polemical and short book published in 1985 *Contre les anti-tiers-mondistes et contre certains tiers-mondistes* (Lacoste 1985), the publication triggered strong reactions, especially in the left-wing humanities and social sciences journals.

Yves Lacoste was deeply concerned with the problems of development. He was critical of the right-wing journalists and politicians who pleaded for a more self-interested attitude towards the developed nations and he disagreed with the policies of 'disengagement' concerning the Third World. Left-wing commentators were not surprised by this part of his book, yet the other argument sounded heretical because Lacoste was as critical of many *Tiers-mondistes* (the supporters of the Third World and of its attacks against new forms of imperialism) as against *anti-Tiers-mondistes*. Lacoste argued that people in poor and under-developed countries had to adopt responsible attitudes towards development. The discourses of many

Tiers-mondistes contributed to passive attitudes among the rulers and elites of the newly independent countries, he maintained. He considered it too simplistic to persuade peoples that colonizers and imperialists were the only culprits of all the ills and weaknesses of their countries. By contrast, Lacoste argued for a more geographically and historically informed perspective.

The past of the developing countries was essential to an understanding of their contemporary situations he claimed, as the peoples living in Third World countries were not all exploited in the same way at the time of European expansion. Yves Lacoste wrote:

> Amongst the horrors of the colonial system, the most abominable was undoubtedly the slave trade in Africa. In fact, it was not the Arab or European slave traders who captured these men and women tearing them from their villages, but local potentates who then sold the captives along the coast to foreign traders. . . . These manhunts organised by African people against African people explain to a large extent the tensions that exist between ethnic groups within many states.
>
> The responsibility of privileged native minorities are generally kept silent in the tragedies which the people of the Third World have experi-enced since the European expansion, and in the identification of factors which today determine their situation of underdevelopment.
>
> (Lacoste 1985: 36)

He argued that some Third World leaders who complained that their countries had been ruined by European colonialism and imperialism were members of groups or families who were complicit with the slave traders and, later, colonizers.

> It appears that the contemporary problems of the Third World have much older origins than it is generally thought. The causes of its strong demographic expansion, the weakness of the stock of productive equipment at its disposition to confront it, and the contradictions which result from this unbalance, the prime causes for these ills are ultimately to be found in the past well beyond the colonial era, because it is in fact these pre-colonial structures which made possible the European domination, after having in effect slowed down the development of the productive forces.
>
> (Lacoste 1985: 134)

In Lacoste's opinion, Third World problems were severe ones, but they resulted from a wide variety of situations, even if demographic expansion was a symptom shared by almost all the developing countries. Since the causes of belated development lay in the distant past and varied according to different countries, each nation had to resolve specific problems and devise its own path to development. The solutions proposed by left-wing *tiers-mondistes* – the abolition of private property and the generalization of international assistance – appeared ineffective solutions by the mid-1980s. Lacoste's conclusion was that development was a long-term process, in which simple and uniform solutions did not work; instead progressive peoples had to take charge of their destiny.

Many left-leaning journalists considered that Yves Lacoste had, thus, broken his links with the French left. Others were more receptive to his analysis and his sensitive appreciation of the policies which had to be used in order to insure sustainable development in the Third World. In that way, Lacoste and *Hérodote* were instrumental in accelerating the rejection of the modernist and universal attitudes applied at the time of decolonization. More attention was given to local conditions and the evaluation of the national policies of the new independent states became more genuinely critical. Right-wing politicians or journalists ceased to be the only ones to question international subsidies when they were misused.

Geopolitics at the scale of cities, regions and nations

With the passing of time, other aspects of the geopolitics used by Yves Lacoste and the editorial board of *Hérodote* changed. In the late 1970s and early 1980s, the journal dealt mainly with international problems. *Hérodote* had, however, always displayed an interest in the domestic political problems of nations: in 1976, Lacoste had prepared for discussion a paper on 'Selling off geography. . . selling off the idea of nation' (Lacoste 1976e). Béatrice Giblin had equated nation and landscape (Giblin-Delvallet 1977) and Marie-Claude Maurel had presented the way governments organized territorial administration as a geopolitics of territory (Maurel 1984). Yet internal French political problems remained beyond the early remit of *Hérodote*.

The situation changed in 1983–4. When preparing the issue published for the Paris International Congress of the IGU, there was a discussion among the editorial board about the political orientation of some regions in France. Béatrice Giblin, who came from the *Nord* region, mentioned that the eastern part of the coal basin, in the department of *Nord*, was traditionally communist, whereas the western part, in the department of *Pas-de-Calais*, was socialist. There was no objective reasons for such differences in the political involvement of colliers and in attempting to address this question the members of *Hérodote's* staff developed a field of *géopolitique interne* or domestic geopolitics. Although such concerns had first been explored by a geographer, André Siegfried (Siegfried 1913), they had later been abandoned by the discipline and from the 1950s practically all electoral studies were undertaken by specialists in political science. François Goguel, who was the Director of the *Institut d'Etudes Politiques* in Paris for a long time, was the most distinguished specialist in the field (Goguel 1970).

Béatrice Giblin and Yves Lacoste developed a different conception of electoral studies to mainstream Anglophone political geography. Instead of focusing their studies on the votes and the social composition of constituencies, they were fascinated by the role of local political leaders, the local strategies developed by political parties, and the way they listened to local complaints, proposed different solutions for the different groups, and were represented by deeply rooted local political cultures. Electoral geography was really then a component of geopolitics and Lacoste decided to exploit this vein. He knew that there was a keen interest in electoral problems in all French regions as it was the period when the *Front National* was gaining a national audience for the first time. It emerged that this party was thriving in the parts of France which experienced the highest rates of growth from the late 1940s to the early 1970s. In the suburban areas of these regions, young communities had been struck in the 1980s by the rapid rise of unemployment. The presence of numerous immigrants combined with problems of urban violence, created a dramatic atmosphere in cities such as Marseilles that the Front exploited effectively.

The publishing house 'La Découverte' undertook to publish a huge study on the domestic geopolitics of France. Many geographers and some sociologists or political scientists contributed to the book, which was organized on a regional basis. The collective work was prepared in a very short time. It was published by 'Fayard', after 'La Découverte' withdrew from the project. The manuscript had to be shortened and the editing of the texts would have been improved had Yves Lacoste and Béatrice Giblin enjoyed more time to finalize the different chapters (Lacoste 1986). However, the result was a huge three-volume collection which was well received by French public opinion. As in many countries, the electoral geographies of France displayed a high degree of stability: from the beginning of the Third Republic in the 1870s, for example, Western France was a stronghold of right-wing parties. Yet the emergence of the *Front National* was something new. Western France did not vote for the *Front National*, except at La Trinité, in Morbihan, the home town of Jean Marie Le Pen. In fact the highest ballots for the *Front National* came from areas where the left had traditionally obtained high votes and it was in constituencies dominated by the Communist Party that Le Pen's party had often made its major inroads. Lacoste's book tackled these apparent differences and it was especially interesting for left-wing political leaders, since they were directly threatened by the new ultra-right party.

From 1984 onwards, questions of French national geopolitics became increasingly important in *Hérodote*. Issues were devoted to France in 1986 ('Géopolitique de la France', *Hérodote*, no. 40, 1986) and 1988 ('La France, une nation, des citoyens', *Hérodote*, no. 50–1, 1988). At the beginning of the 1980s, the urban issue was treated by *Hérodote* only in the Third World ('L'habitat sous-intégré', *Hérodote*, no. 17, 1980). However, in 1983, *Hérodote* delved into land speculation and the political problems of cities. In 1986, the journal touched directly on the geopolitics of cities, in an issue on the evolution of the former suburban belts of communist majorities in France ('Après les banlieues rouges', *Hérodote*, no. 43, 1986). Once more, Lacoste and *Hérodote* had shown a willingness to respond to the contingencies of modern France.

Geopolitics at a time of fundamentalisms, independence movements and nationalisms

Finally, international problems were also changing in nature during the 1980s and this evolution was also reflected by *Hérodote*. The initial volumes devoted to geopolitical problems were concerned mainly with the areas defined as shatterbelts by Saul B. Cohen (Cohen 1973). By the mid-1980s onwards, therefore, new factors had intruded into the international scene: economic interests and the struggle between the capitalist and the socialist versions of Western civilization had ceased to be the only global forces at work. A growing number of people reacted against processes associated with cultural globalization and fought for the preservation of their culture and traditions. This transformation started with the emergence of religious movements as a new factor on the international political scene.

In fact, this factor had started at least sixty years earlier, with the foundation of the Muslim Brothers in Egypt during the 1920s. For a long time, such movements had only limited impact in a few countries. Religious movements were often active in the Independence wars, but they were less organized, efficient and conspicuous than the left-wing revolutionary groupings. The situation changed with the 1979–1980 Shi'ah revolution in Iran. Within a few years, Islamic fundamentalism became a major factor in political calculations from East Africa to Central Asia, and from Indonesia to Morocco and the Sahel region of West Africa. *Hérodote* covered this evolution thanks to two issues published in 1984 ('Géopolitique des islams. I: les islams "périphériques"', *Hérodote*, no. 35, 1984) and 1985 ('Les centres de l'Islam. II: Géopolitiques des islams', *Hérodote*, no. 36, 1985): the first concerned the peripheries of Islam, the second dealt with its centre. Religious influences were visited again using a more general approach in 1990 ('Eglises et géopolitique', *Hérodote*, no. 56, 1990). By the mid 1990s, France was directly affected by the rise of fundamentalism in its Muslim communities and *Hérodote* returned again and again to this problem, with two special issues ('Maîtriser ou accepter les islamistes', *Hérodote*, no. 77, 1995; 'Périls géopolitiques en France', *Hérodote*, no. 80, 1996).

In addition to religious movements, in recent years the journal has also responded to the collapse of the Soviet-influenced regimes and the spectacular renewal of nationalisms that occurred on the world political scene. In many cases, the arguments developed by the new national movements had been imagined among exiles and had been preserved and sharpened by socialist rule. Using another perspective, these national movements could be analysed as forms of protest against the cultural globalization process. Thus, regionalisms and independence movements in Western democracies, nationalisms in former Eastern bloc countries and fundamentalisms in the developing world were all diverse expressions of a common rejection of some aspects of modernization. As a result, the political map of the World began to change more quickly than ever before during the present century as the Soviet Union, Czechoslovakia and Yugoslavia were dismantled. The disintegration process was a violent one in Bosnia, Armenia and Azerbaijan, Chechnya, and Tajikistan. The *apartheid* regime in South Africa did not survive the end of the cold war environment, while elsewhere in Africa, many states lost their infrastructural power and credibility now that the political and economic support of the superpowers had evaporated. Instability threatened governments and the general probability of dramatic changes of boundaries in the near future increased rapidly.

Hérodote covered all these transformations. It published fascinating case studies on the shatterbelts of the contemporary world, and these dramatic changes continue to occupy the pages of the journal. However, because it also excels in the analysis of situations at different scales, it is able to address the rise of new nationalisms and of fundamentalist affiliations among the Muslim minorities of Western Europe, that render long-established nation-states suddenly more fragile. The long process of fragmentation, which was so important in the Balkans at the end of the nineteenth and beginning of the twentieth centuries, has resumed its momentum ('Balkans et balkanisation', *Hérodote*, no. 63, 1991). Hence the development of *Hérodote* 's reflections upon the idea of nation.

The idea of the nation

On several occasions from 1990 onwards, *Hérodote* began to problematize the nation-state and examined the problems linked with nationalisms, the creation of new national States and the international crises which resulted from this process. In 1991, the special issue on 'The territories of the nation' was centred upon the evolution of France.

The question of the nation and that of its territory are important in

France, but in a very particular way, more, we might say, as an internal, central problem than as one of its periphery or border zones.

(Lacoste 1991a: 12)

There are regionalist and independence movements in France, but except in Corsica, they lack the strength they display in Britain, Ireland or Spain. The decentralization policy initiated in 1982 had created new forms of relations and tensions between the local state and Paris, but the conflicts were limited. Lacoste commented: 'It is mainly, however, the so-called problem of immigration which leads many Frenchmen, either right or left wing, to worry about the destiny of the nation' (Lacoste 1991a: 13).

The high unemployment rate is partly responsible for the tensions between citizens and France's immigrant communities, but Lacoste suggested that the problem is actually more complex. Economic forces were responsible for the concentration of lower income groups in Council Housing (HLM):

The problem of integration is not only an economic or socio-cultural problem, it is also a geopolitical problem, when the fundamental question of ghettos is raised. . . . Today it is possible to wonder whether some political forces, the Muslims notably, do not deliberately seek the formation of ghettos, that, if not ethnically homogeneous, at least have a Muslim majority.

(Lacoste 1991a, p. 18)

It meant, according to Lacoste, that the problem of integration is not fundamentally an economic one. It had a political dimension:

integration is mainly and above all a political venture, particularly given the roles in the nation . . . of all those who are (or soon will be) legally French citizens, but who still consider themselves as immigrants, and think that France is not really their country.

(Lacoste 1991: 19)

There are psychological obstacles to integration:

Such an undertaking must overcome many obstacles, preconceptions and racism, which exist as much amongst the 'immigrants' as amongst the French people.

But, for the 'immigrants' to be really integrated as citizens of the French nation, this effort has to bring them something important, and not only the material advantages they already possess.

256

. . . For that reason, Frenchmen [*sic*] have to be proud again to be French. This remark may appear to many as a naïve patriotic statement, but, in these problems, geopolitical representations are of great importance and, in the past, a large number of immigrants, and notably prestigious immigrants . . . have become French because they were proud of it.

(Lacoste 1991a: 19)

As a result, Lacoste considered that the integration of immigrants into the French nation depended in many ways upon a certain degree of national pride, and the attitudes of Frenchmen [*sic*] concerning the French nation.

This argument, central to the 1991 special issue on 'The territories of the nation', was developed and elaborated again and again in the 1990s. In 1993, Lacoste wrote:

The political situation in France is thus dangerous: unemployment is certainly the fundamental and enduring factor, but it can be worsened by geopolitical factors within or outside the nation, and by the reactions of public pinion. That is why the analysis of geopolitical representations is necessary more than ever, and above all, the different conceptions of the nation require study. It is important that this key idea ceases to appear as the privilege of nationalist groups of the far right, and becomes again a democratic project of all the citizens.

(Lacoste 1993b: 7)

Yves Lacoste stressed also the need to study the way nations are conceived in the regions where states disintegrated after the fall of the Berlin wall. National groups simultaneously maintained different conceptions of that national space. This includes the territory in which the group dominates, but also regions where minorities of this group are mixed with other populations, and areas which the members of the group had to leave in the past, but where they left manifest traces of their former cultural role:

These countries, even if they are not officially claimed, are still perceived as a part of the national patrimony. This is the case, for example, for German peoples, of the former East Prussia, Pomerania or Silesia: regions that some nationalist groups consider as the real Eastern Germany.

(Lacoste 1991b: 7)

Hence the complexity and potential danger of nationalist representations:

If we accept that the same nation can 'have' these different types of territory, these form an array of overlapping and intersecting territories as soon as we envisage several rival neighbouring nations. Their state territories, limited by official boundaries, are juxtaposed on a map in a relatively simple way, [but it is] . . . a far more complicated map which represents the intermingling of the other types of territories of varied nations, but it is this map which allows a better understanding of the tensions between the peoples and to which extent geopolitical situations can be muddled.

(Lacoste 1991b: 13)

Just as with the problems of development in the 1980s, Yves Lacoste recently summed up *Hérodote*'s reflection on national problems by the publication of a book, *Vive la Nation! Destin d'une idée géopolitique* (Lacoste 1996). The lesson is apparently simple:

Finally, my purpose is not to evoke an ideal of the nation, but a new way to define it, taking into account common characteristics of the main nations of Western Europe, and that, in the perspective of the European Union. . . .

For a large number of French people, the idea of the nation is not only a question of its territory, its language, its history but also of the state. The idea of the nation, for each of us . . . includes a kind of intimate relationship with all of this. If the nation is such a key idea, it is because it mobilises deep personal feelings. . . . Because it refers to a territory, a language, a history and a State, the nation is a geopolitical representation, and moreover, it is for many much more sentimentally internalised and individualised.

(Lacoste 1997: 329)

As a result, it is important to retain the idea of nation (hence the title of the book: *Long Live the Nation!*). It offered a space so that the sense of collective responsibility, which is central to the survival of democracies, could thrive. For Lacoste, it would be impossible to have European citizens in the European Union if they were not, first and foremost, citizens of different European nations. The danger is to prevent a monopolization of the idea of the nation by far-right movements, when it is in fact the basis of all modern democratic experiences. Once again, Yves Lacoste offers a critical analysis of prevailing conceptions, and suggests interpretations which incorporate elements which are often exclusively claimed by right-wing parties. He shows that they are fundamental for building a modern democratic society, either from a conservative or a liberal perspective.

The audience of *Hérodote* in France

Hérodote differs from many geographical journals due to the public it tries to attract. *Hérodote* is very much Lacoste's own journal, he always includes an editorial and from the start he imposed a few rules on the contributors: papers have to be well written in the sense that they most be clearly expressed and avoid technical terms. They have, therefore, to be accessible for the non-specialist reader. The review is written for geographers, but it is not conceived as a publication reserved for academics. The main ambition of Yves Lacoste is that his journal is read by all those who teach geography in secondary schools, and some of their pupils, especially those in the sixteen to eighteen years old bracket. Underlying this aspiration lies the belief that the main function of geographers in the universities is to provide a valuable training for the secondary school teachers of tomorrow. In doing so, secondary school geographers should deliver good lessons, which means well informed ones and at the same time confront the real problems of the contemporary world which are relevant to the modern citizen. *Hérodote* has a civic mission: hence the attention devoted to the problems of the nation in general and to its political life more particularly, and in Lacoste's reformulation of geopolitics, a concern for the politics and future of France.

However, Lacoste is not pursuing a solo agenda because one of the characteristics of French geography since the 1970s has been the desire of many university teachers to influence secondary school level teaching. Roger Brunet has also sought to influence schools geography through his tenure as editor, since 1972, of *L'Espace géographique*, an academic geography journal. Later, he was instrumental in creating in 1986 a quarterly cartographic journal called *Mappemonde*. It has two major features: the charting of modern developments in this discipline (automatic cartography, remote sensing), and the varied applications of cartography (its use in teaching more particularly). The reading public of *Mappemonde* is up to a point the same as that of *Hérodote*, but the methodologies differ. Since the 1970s, Brunet has developed a cartographic interpretation of human geography. For him, it was easy to represent each geographic situation by a simplified cartographic sketch. Reality was reduced to its significant geographical components, points, lines, areas and their combinations which were called *chorems*. Brunet thought it was thus possible to anchor the analysis of geographical situations into the developments of the 'new' geography and systems analysis. His interpretations encouraged many secondary school teachers, who discovered that the simplified sketches based on *chorems* were valuable didactic tools.

From the early 1980s, there were thus two rival interpretations of modern geography which tried to modify the way geography was taught in the French secondary school system, and which vied for the same public. Lacoste had always been critical of Brunet's conception of cartography. His hostility grew in the 1990s when his interest in the idea of the nation developed. According to Lacoste,

Brunet's method of mapping reality is incapable of representing the diversity and complexity of the individual's conception of territories. Lacoste wrote in 1993:

> In fact, on these questions (with regard to the idea of a nation), two streams of thought are in conflict among geographers; to simplify, it is possible to say that a first group, who wish to analyse concrete geopolitical situations, publish in *Hérodote*, while another group forbids all references to geopolitics and, in order to reveal and justify a certain 'order' or 'laws of space' which would be established by a would-be new science, the *chorematics*. And all of this has more importance than it appears at first glance and does not concern only geographers, but all citizens, since it is in fact a question of organization of space and territorial planning.
>
> (Lacoste 1993b: 7–8)

This problem appeared important enough for Lacoste to prepare a whole issue of *Hérodote* in which *chorematics* is critically, and mercilessly, evaluated ("Geographers, science and illusion", *Hérodote*, no. 76 1995).

On another occasion, *Hérodote* has been attacked from sections of the French Left. When the journal first appeared, *L'Humanité*, the newspaper of the French Communist Party, published a sharp critique of *Hérodote* written by a young geographer, Jacques Lévy. Yves Lacoste recently recalled:

> But *Hérodote* did not leave the political groups indifferent and, paradoxically, it was in *L'Humanité* that the most virulent critiques appeared in two long articles, in succession, shortly after. They denounced this new way, more or less 'Third-World' to deal, geographically, with political questions without referring to Marxism. In the eyes of the 'orthodoxies', the fact that we could not be challenged with any reference to geography in the Marxist holy writ made our writings suspect as they risked signalling this omission [of the geographical] by the founding fathers. This was for us a chance for freedom.
>
> (Lacoste 1996a: 7)

Jean Dresch, an influential figure in French communism, reacted rapidly and explained to the editors of *L'Humanité* that *Hérodote* was by no means a right-wing journal. The attacks against the new review ceased immediately. They were, however, symptomatic of the reactions of many left-wing geographers who did not understand the reasons why geographers had to move back to political geography and geopolitics after more than thirty years of either harsh criticism against this branch of the discipline, or simply neglect. In fact, the renewed interest in political geography was not limited to left-wing geographers, for in the period between 1975 and 1980, for instance, new contributions to the field of geopolitics were proposed by André-Louis Sanguin (Sanguin 1977), Paul Claval (Claval 1978) and Claude Raffestin (Raffestin 1980). However, this distance demonstrates not only the extent to which *Hérodote* is embedded in French geopolitical discourse, but also its impact outside academia, in the public realm.

When trying to measure the impact of *Hérodote* on French public opinion, it is important to remember that its main target is not the enlightened French left wing intelligentsia, but all those who teach geography, and pupils and students interested in the problems of today. This orientation explained why *Hérodote* has had a limited impact on the daily or weekly newspapers (except perhaps *Libération*), but its influence remains important. Its circulation (about 6,000 copies) is high for a learned journal in the French-speaking world. Part of the influence of *Hérodote* came from the former members of its editorial board who left to develop their own field of activity. Michel Foucher, for instance, created a consultancy dealing with geopolitical problems. He works for banks and business corporations, and played an important role, at certain times, at the *Quai d'Orsay* (the French Ministry of Foreign Affairs).

The reception of *Hérodote* abroad

The influence of *Hérodote* was also strong abroad, especially in the Latin-speaking countries. French geographers had been active in the development of the discipline in countries such as Brazil and Argentina. They served as models in Spain, Portugal, and up to a point in Italy (where German geography was also well known). Many of these countries had tragic fascist experiences, either before the Second World War, during the War or after. The majority of their geographers looked to France for anti-fascist and anti-imperialist directions and this optimism appeared well grounded given that the French geographers associated with Pierre George provided such a focus. The ideas of Yves Lacoste and *Hérodote* grew on ground prepared by such intellectual seeds. In fact, the impact came mainly from the initial pamphlet – *La Géographie, ça sert, d'abord, à faire la guerre* – and from the articles of the first five years (1976–1981). It means that the more progressive changes in the conceptions of geopolitics, imperialism and nations since then remained ignored, especially in Spain, Brazil and Spanish-speaking Latin American countries.

The influence of *Hérodote* in Italy took another form. Lacoste's *La géographie, ça sert, d'abord, à faire la guerre* had been translated as: *Crisi della geografia, geografia della crisi* (Lacoste 1978). The review *Hérodote/Italia, Strategie, geografie, ideologie* was launched in 1978 by the publisher Bertani, in Verona. The political orientation was however different from the French one: it was a Marxist review. In 1983, the title became *Erodoto. Problemi di geografia* but the success was limited and the review disappeared in 1984, after the publication of six issues. However,

geopolitics had been present in Italy during the Fascist period, with the publication of the review *Geopolitica* in Trieste (Atkinson 1995; Antonsich 1996). After
the war, the term and the field were shunned in the country. The Italian edition
of *Hérodote* appeared too early.

Ten years later, the situation was different because discussions concerning
geopolitics really started amongst Italian geographers in 1983–1985. Officers of
the Italian Army and Navy such as General Carlo Jean began to publish on problems of geostrategy and geopolitics. In November 1991, *Micromega*, a cultural review
of the Italian left, published an issue on geopolitics, with papers from some members
of *Hérodote*. It raised great curiosity in Italy. Lucio Caracciolo, who had had
contributed to *Micromega*'s review, contacted Michel Korinman who had connections
in Milan. They gathered a group of influential military officers (Carlo Jean), political
scientists, economists, journalists, and some geographers (Gaetano Ferro and Maria
Paola Pagnini), and in March 1993, launched *Limes, Rivista Italiana di Geopolitica*. The
first issue dealt with the geopolitics and then dissolving boundaries of Yugoslavia. It
reached a large audience, partly because it departed from the moral condemnation of
all forms of power which was central to *Hérodote/Italia*. It was considered a left-wing
journal, but without any Marxist involvement. Its authors came in fact from various
political backgrounds. One of the major achievements of *Limes* was to reintroduce the
theme of the nation as a major element in political discussions among left-wing intellectuals in Italy. An annual French version of *Limes* was launched in 1996, but it
appeared more superficial than *Hérodote*, and did not meet as much success in terms
of impact and circulation as its Italian model.

An English translation of articles of *Hérodote* by P. Girot and E. Kofman was
published by Croom Helm in 1985 (Girot and Kofman 1985) . Øyvind Østerud
presented a short evaluation of *Hérodote* in the paper he wrote on 'The uses and
abuses of geopolitics' (Østerud 1987). More recently, Geoffrey Parker's new
survey of the western geopolitical tradition provides a thorough analysis of French
political geography and the role of *Hérodote* (Parker 1998). As a result, *Hérodote*'s
intellectual trajectory is quite well known in the English speaking world and
recent contributions to Anglophone critical geopolitics have acknowledged the
contribution of Lacoste's examination of the relationship between geography and
state power (Ó Tuathail 1996). However, it would be fair to note that active
engagement with the literature generated by *Hérodote* has been relatively limited
in Anglophone political geography in spite of the obvious intellectual influence of
French thinkers such as Foucault, Derrida and Baudrillard.

Conclusion

How does one summarize the main orientations of *Hérodote* and evaluate its
influence? Lacoste succeeded in creating a permanent and widely circulated

journal. Its audience is large among French students in geography, secondary school teachers and a wider public of left-wing sensibility. *Hérodote* contributed substantially to the transformation of left-wing attitudes: thanks to its analyses, representations of the developing world have ceased to be over-simplified and imperialism is not read as the only source of evils in the poor countries. Twenty years ago, the majority of left wing intellectuals were very critical of the idea of the nation. As a consequence of Lacoste's influence, judgements are now more diverse: many people think that the nation had been, and may remain, a privileged signifier of identity and citizenship.

Yet there are also shortcomings in *Hérodote*. In the 1970s, Lacoste was critical of the over-emphasis placed upon economic factors in many publications of the time. A large part of the contemporary political evolution of the world results, however, from economic factors. In its twenty-three years of publication, *Hérodote* has offered no analysis of economic globalization and its political correlates. Braudel had friendly relations with Yves Lacoste who wrote an interesting study of him (Lacoste 1990). The great historian heartily supported *Hérodote* . The kind of geopolitics based on the idea of a world-economy that he promoted had however no success in France. In a similar vein, the major powers of the world were analysed by *Hérodote* only in the 1990s. The collapse of the Soviet Union accounts for the number of studies on the new borders and states which were generated from that time within its former boundaries. There were special issues on Japan in 1995 and the United States in 1997. In the last case, the journal only covered the problems of minorities, racism and the inner equilibrium of the country and ignored NAFTA and the prospects for Pan-American economic union. For a publication specializing in geopolitics, an assessment of the role of the United States in the post-Cold War era is still missing. Likewise, *Hérodote* had only given over a few pages to China's domestic problems and its foreign policy.

The paucity of information on these important regions and problems is partly explained by the composition of the editorial committee: the majority of its members were mainly familiar with Europe, the Mediterranean, the Middle East, Africa and South America. *Hérodote* had its main intellectual connections with the Latin-speaking countries in Southern Europe and Latin America, and with intellectuals and geographers fluent in French in Eastern Europe, the Arabic World and the Middle East. Its links with Germany relied essentially on Michel Korinman. The relations with English-speaking geographers remained scarce until the recruitment in the editorial committee of young scholars partly trained in the US, Frédérick Douzet for instance. Yves Lacoste knew perfectly well that he had to develop studies of North America, but it was only recently that he found competent collaborators in this area. With regard to China, he is still searching for appropriate experts. Yet the reasons for the imbalance went deeper than a simple lack of expertise, for in the main Lacoste had a tendency to overemphasize the role of political factors and to neglect wider cultural, economic and social processes in the non-Francophone world.

Nevertheless, *Hérodote* played a decisive role in the rise of geopolitics in contemporary France where the discipline has ceased to be associated with imperialism, power politics and ideological biases. At the University of Paris-VIII, Lacoste has created a doctoral programme in geopolitics, which is the most active in France and attracts many well qualified geographers and political scientists. In this way too, he contributes to the quality of contemporary geopolitical discussions in France. Many French geographers are now working on geopolitics although there is a considerable amount of intellectual diversity. They have developed new connections with the English-speaking world (André-Louis Sanguin, Paul Claval) or Germany (Richard Kleinschmager). French geopolitical writers are also interested in the problems of minorities and multicultural situations (Sanguin 1993; Goetschy and Sanguin 1995), the growing role of great metropolises on the world political scene (Claval and Sanguin 1997); and the political role of diasporas (Prévelakis 1996). They also tend to spend time exploring the roots of modern geopolitical thought (Claval 1994; Muet 1997; Raffestin, Lopréno and Pasteur 1995), while Georges Prévelakis has revived Jean Gottmann's idea of iconography in order to build a coherent system of geopolitical interpretation.

Elsewhere, at the *Fondation Nationale des Sciences politiques*, a group of geographers has developed a range of comparative approaches to political geography and geopolitics (Lévy 1996; Durand, Retaillé and Lévy 1992; Retaillé 1997). They share with *Hérodote* the will to strengthen the sense of democracy and the responsibility of citizens through geopolitics. Geopolitics is also practised by an increasing number of non-geographers. Some are journalists (Chaliand 1990; Chaliand and Rageau 1983), generals (Gallois 1990; Poirier 1985) political scientists (Moreau-Desfarges 1994; Lorot 1995; Thual 1996; Joyaux 1991–1993). The most original of these authors is certainly Bertrand Badie, a political scientist, who explores the consequences of the decline of the type of national state first developed in Western Europe after the seventeenth century, and later exported everywhere (Badie 1996; Badie and Smouts 1992). Undeniably, therefore, new directions in French geography are developing in contemporary France and Yves Lacoste's *Hérodote* has been the catalyst for the developments of this discipline since 1976. In this respect, Lacoste deserves a special place in any history of the term 'geopolitics'.

Acknowledgement

The editors owe a debt of thanks to Sylvie Gray for her assistance in translating and editing the French extracts of *Hérodote* into English.

Bibliography

Antonsich, M. (1996) 'Geografia politica e geopolitica in Italia dal 1945 ad oggi', *Quaderni del dottorato di ricerca in Geografia politica*, Trieste, 2: 19–53.

Anon. (1976) 'Des Cubains en Angola, mais aussi des Angolais à Cuba', *Hérodote*, 2: 23–9.

Atkinson, D. (1995) 'Geopolitics, cartography and geographical knowledge: envisioning Africa from Fascist Italy', 290–332 in M. Bell, R. A. Butlin and M. J. Heffernan (eds) *Geography and Imperialism, 1820–1940*, Manchester: Manchester University Press.

Badie, B. (1996) *La Fin des territoires*, Paris: Fayard.

Badie, B. and M.-C. Smouts (1992) *Le Retournement du monde. Sociologie de la vie internationale*, Paris: Presses de la Fondation Nationale des Sciences Politiques et Dalloz.

Châtelet, F. (1976) 'Hegel et la géographie', *Hérodote*, 2: 78–94.

Chaliand, G. (ed.) (1990) *Anthologie mondiale de la stratégie. Des origines au nucléaire*, Paris: Robert Laffont.

Chaliand, G. and J.-P. Rageau (1983) *Atlas stratégique: géopolitique des rapports de force dans le monde*, Paris: Fayard.

Claval, P. (1978) *Espace et pouvoir*, Paris: PUF.

—— (1994) *Géopolitique et géostratégie*, Paris: Nathan.

—— (1998) *La Géographie française depuis 1870*, Paris: Nathan.

Claval, P. and Sanguin, A. L. (eds) (1997) *Métropolisation et politique*, Paris: L'Harmattan.

Cohen, S. B. (1973) *Geography and Politics in a World Divided*, London: Oxford University Press.

Delmas, C. (1971) *Armements nucléaires et guerre froide*, Paris: Flammarion.

Durand, M.-F., Lévy, J. and Retaillé, D. (1992) *Le Monde. Espaces et systèmes*, Paris: Fondation Nationale des Sciences Politiques et Dalloz.

Foucault, M. (1976) 'Des questions de Michel Foucault a *Hérodote*', *Hérodote*, 3: 9–10.

Foucher, M. (1979–1980) 'Enquête au Nicaragua, I', *Hérodote*, 16: 5–35; 'Managua, ville éclatée, Enquête au Nicaraqua, II', *Hérodote*, 17: 32–51.

Foucher, M. (1983) 'Israël-Palestine: quelles frontières', *Hérodote*, 29–30: 95–134x.

Foucher, M. (1986) *L'invention des Frontières*, Paris: Fondation pour les Etudes de Défense Nationale.

Foucher, M. and Pichol, M. (1978) 'Territoire à prendre, territoire à défendre: le Larzac', *Hérodote*, 10: 91–115.

Gallois, P. M. (1990) *Géopolitique. Les voies de la puissance*, Paris: Plon.

Gentelle, P. (1980) 'Afghanistan: Russes et Asiatiques dans le piège', *Hérodote*, 18: 57–85.

Giblin-Delvallet, B. (1976) 'Elisée Reclus: géographie, anarchisme', *Hérodote*, 2: 30–49.

—— (1977) 'La nation-paysage', *Hérodote*, 7: 148–57.

Girot, P. and Kofman, E. (eds) (1985) *International Geopolitical Analysis. A Selection from Hérodote*, London: Croom Helm.

Goetschy, H. and Sanguin, A.-L. (eds) (1995) *Langues régionales et relations transfrontalières en Europe*, Paris: L'Harmattan.

Goguel, F. (1970) *Géographie des élections françaises sous les IIIe et IVe Républiques*, Paris: A. Colin.

Gottmann, J. (1952) *La Politique des Etats et leur géographie*, Paris: A. Colin.

Guglielmo, R., Lacoste, Y. and Ozouf, M. (1965) *Géographie, classe de Première*, 'Cours Ozouf', Paris: F. Nathan.

Joyaux, F. (1991–1993) *Géopolitique de l'Extrême-Orient*, Brussels: Editions Complexes, 2 vols.

Korinman, M. (1984) 'Friedrich Ratzel, Karl Haushofer, "Politische Ozeanographie"', *Hérodote*, 32: 144–57.

—— (1990) *Quand l'Allemagne pensait le monde. Grandeur et décadence d'une géopolitique*, Paris: Fayard.

—— (1991) *Continents perdus. Les précurseurs de la géopolitique allemande*, Paris: Economica.

Lacoste, Y. (1976a) *La Géographie, ça sert, d'abord, à faire la guerre*, Paris: Maspéro.

—— (1976b) 'Enquête sur le bombardement des digues du fleuve Rouge (Viêtnam, été 1972)', *Hérodote*, 1: 86–117.

—— (1976c) 'A propos de Clausewitz et d'une géographie', *Hérodote*, 3: 65–75.

—— (1976d) 'Questions à Michel Foucault sur la géographie', *Hérodote*, 1: 77–85.

—— (1976e) 'Brader la géographie . . . brader l'idée nationale ', *Hérodote*, 4: 9–55.

—— (1977) 'Fidel Castro et la Sierra Mestra', *Hérodote*, 5: 7–33.

—— (1978) *Crisi della geografia, geografia della crisi*, Milano: Feltrinelli.

—— (1979) 'A bas Vidal, viva Vidal !', *Hérodote*, 16: 68–81.

—— (1980) *Unité et diversité du Tiers Monde*, Paris: Maspéro, 3 vol.

—— (1984) 'Les géographies, l'action et la politique', *Hérodote*, 33–4: 3–32.

—— (1985) *Contre les anti-tiers-mondistes et contre certains tiers-mondistes*, Paris: La Découverte.

—— (ed.) (1986) *Géopolitiques des régions françaises*, Paris: Fayard, 3 vol.

—— (1990) 'Braudel géographe', in Y. Lacoste (1990) *Paysages Politiques*, Paris: Livre de poche, 83–149.

—— (1991a) 'Editorial: les territoires de la Nation', *Hérodote*, 62: 12.

—— (1991b) 'Editorial: Balkan et balkanisation', *Hérodote*, 63: 3–13.

—— (ed.) (1993a) *Dictionnaire de géopolitique*, Paris: Flammarion.

—— (1993b) 'Editorial: Démocratie et géopolitique en France', *Hérodote*, 69–70: 3–8.

—— (1996) '1976–1996, Hérodote à 20 ans', *Hérodote. Vingt ans de géopolitique 1976–1996*, May.

—— (1997) 'La République et la nation: quelques réflexions géopolitiques', *Géopolitique* 60: 60–5.

Lacoste-Dujardin, C. (1976) 'A propos de Pierre Bourdieu et de 'l'Esquisse d'une théorie de la pratique", *Hérodote*, 2: 103–16.

Lévy, J. (1996) *Le Monde pour Cité*, Paris: Hachette.

Lorot, P. (1995) *Histoire de la géopolitique*, Paris: Economica.

Maurel, M.-C. (1984) 'Pour une géopolitique du territoire, le maillage politico-administratif', *Hérodote*, 33–4: 131–43.

Moreau Defarges, P. (1994) *Introduction à la géopolitique*, Paris: Seuil.

Muet, Y. (1997) *Les Géographes et l'Europe. L'idée européenne dans la pensée géopolitique française de 1919 à 1939*, Geneva: Institut Européen.

Østerud, Ø. (1987) *The Uses and Abuses of Geopolitics*, Department of International Relations, Australian National University, Research School of Pacific Studies.

Ó Tuathail, G. (1996) *Critical Geopolitics,* London: Routledge.

Parker, G. (1985) *Western Geopolitical Thought in the Twentieth Century*, Beckenham: Croom Helm.

—— (1998) *Geopolitics: Past, Present and Future,* London: Mansell.

Péchoux, P.-Y. (1976) 'Les dimensions géographiques d'une guerre localisée: Chypre, 1974–1976', *Hérodote*, 3: 11–44.

Poirier, L. (1985) *Les Voix de la stratégie*, Paris: Fayard.

Prévelakis, G. (ed.) (1996) *Les réseaux de diasporas. The Networks of Diasporas*, Nicosia: Kykem, Paris: L'Harmattan.

Raffestin, C. (1980) *Pour une géographie du pouvoir*, Paris: Litec.

Raffestin, C., Lopréno, D. and Pasteur, Y. (1995) *Géopolitique et histoire*, Lausanne: Payot.

Retaillé, D. (1997) *Le Monde du géographe*, Paris: Presses de Sciences Po, Paris.

Sanguin, A.-L. (1977) *La géographie politique*, Paris: PUF.

—— (ed.) (1993) *Les Minorités ethniques en Europe*, Paris: L'Harmattan.

Santibañez, R. (1977) 'Contrôle de l'espace et contrôle social dans l'Etat militaire chilien', *Hérodote*, 5: 82–107.

Siegfried, A. (1913) *Tableau politique de la France de l'Ouest sous la IIIe République*, Paris: A. Colin.

Thual, F. (1996) *Méthode de la géopolitique. Apprendre à déchiffrer l'actualité*, Paris: Ellipses.

Varii Auctores (1991) *Autour de Raymond Guglielmo. Géographie et contestation*, Paris: CREV.

Varlin, T. (alias Michel Fouchet) (1977) 'La mort de Che Guevara. Le problème du choix d'un théâtre d'opérations en Bolivie', *Hérodote*, 5: 39–81.

11

GÉOPOLITIQUES DE GAUCHE

Yves Lacoste, *Hérodote* and French radical geopolitics

Leslie W. Hepple

Introduction

Prologue

In 1976 the world of French geography was shaken by two publications from the experimental university at Vincennes, renowned focus for post-Marxist and anarchist thinking. The two texts were the first issue of the enigmatically named journal *Hérodote*, and Yves Lacoste's *La géographie, ça sert, d'abord, à faire la guerre* (Lacoste 1976a). Lacoste's book, in the pocket-sized edition and distinctive blue cover of the Petite Collection Maspero (which specialized in leftist analyses), was seen by many as a revolutionary manifesto for geography, comparable to Mao's 'little red book'. The historian of French geography Numa Broc wrote that 'the University of Vincennes is launching a flame-thrower (literally 'fire-ship', *brûlot*) against the well-raked flower-beds of university geography' (Bro, 1976). The present writer remembers being regaled at a conference at Montpellier in 1978 with the tale of these militants at Vincennes and being shown the explosive tracts, an occasion which started a long-term following of the group's work.

The central thesis of Lacoste and the *Hérodote* group – a thesis that will be examined in more detail below – was that geography was a form of strategic and political knowledge, central to military strategy and the exercise of political power, but that this strategic discourse had become hidden behind the 'smokescreen' (*rideau de fumée*) of academic geography. Geographers needed to cast off the limitations of their 'mystified and mystifying discourse', and become militant and critical analysts of strategy, working to unmask the geographical structuring of power and assisting in the development of counter-strategies. Unlike many of the extreme claims and positions taken in the decade after Paris 1968, the Lacoste-*Hérodote* project has flourished, expanding and evolving into a major school of geography-geopolitics. The project has undoubtedly transformed itself, and Lacoste himself has moved, as one recent interview essay puts it, to 'a rebel transformed into a mandarin of geopolitics' (Duroy 1998). The journal *Hérodote* has flowered into the largest-circulation French geographical journal and, over twenty years later, the current volume is issue 92 (spring 1999). Lacoste and the *Hérodote* group have written numerous books, both academic and polemical, and Vincennes has become the base for CRAG (*Centre de Recherches et d'Analyse Géopolitiques*), with its publications and doctoral programme in geopolitics.[1] Today, the journal *Hérodote* is the largest and most substantial body of contemporary geopolitical analysis in the world, and, together with the other writings of Lacoste and the *Hérodote* group, it represents a major and highly coherent geopolitical tradition. Indeed the cohesion and identity of this tradition is significant enough to allow us to refer to a corporate 'Lacoste-*Hérodote*' with regard to much (but not all) of the writings of the group.

Interpreting Lacoste-Hérodote

This essay interprets the main strands and arguments of the Lacoste-*Hérodote* school, and particularly locates them in relation to ideas and developments in Anglophone geography and geopolitics. To date such linkages have been very limited. The Anglophone geographical and geopolitical communities tend to *know of* the existence of the *Hérodote* school, but there has been remarkably little critical engagement with its ideas or referencing of *Hérodote* sources in substantive geopolitical analyses. In many quarters *Hérodote* is best, and perhaps solely, known through the interview with Michel Foucault in the first issue (Foucault 1976a), an interview which appeared in translation in Foucault's widely-read *Power/Knowledge* (Foucault 1980). Although a selection of *Hérodote* papers was translated and published in book form (Girot and Kofman 1987) (perhaps unfortunately without any introductory, intepretative essay), and brief surveys have been included in reviews of progress in French geography (Buleon 1992; Clout 1985) and by Parker in his entries for the *Dictionary of Geopolitics* (O'Loughlin 1994; see also Parker 1998) the only critical analysis relating *Hérodote's* perspective to current Anglophone perspectives has been that of Ó Tuathail (1994, 1996).

This neglect appears remarkable when one starts to list some of the features of the Lacoste-*Hérodote* school: the early work by Lacoste in the Marxist tradition; the construction of *Hérodote's* analysis within the radical, post-Marxist culture of Vincennes (where Foucault and Deleuze, with other influential thinkers, worked for periods); and a direct engagement with Foucault in the first issues of *Hérodote* (Foucault 1976a, 1976b; Bernard *et al.* 1977). Lacoste-*Hérodote* also developed a critical and radical regeneration of geopolitical discourse several years before the Anglophone construction of 'critical geopolitics' by Dalby, Ó Tuathail, Agnew,

269

Taylor and others. There was also emphasis on complexity, difference and a scepticism towards metanarratives, again well before such arguments became fashionable (again) in Anglophone geography; and a recognition of the ideological and strategic role of cartography and geographical surveillance that foreshadowed the Anglophone work of Harley and others. Indeed, in the early years of *Antipode*, there was some recognition of *Hérodote's* relevance for contemporary Anglophone debates: a version of Lacoste's analysis of the Red River bombings in Vietnam appeared there in 1973 (Lacoste 1973b, reprinted in 1977a), three years before the main French-language analysis was published in the first issue of *Hérodote* (Lacoste 1976b); and an extract of his writing on underdevelopment appeared four years later (Lacoste 1977b). However, despite all of this, *Hérodote* remains enigmatic to most Anglophone geographers.

What might explain this lack of attention to the Lacoste-*Hérodote* work? The first reason is quite simply that few Anglophone geographers read much French geography, or indeed (to our shame) any non-English language sources: the neglect of *Hérodote* is not specific, but part of a wider disengagement. Although French social theory is *à la mode* in Anglophone geography, it is usually in translation. Undoubtedly there is a linguistic arrogance and imperialism at work here – a perspective that we will later see in more directly geopolitical terms – but there is also a straightforward lack of language skills. Although based upon anecdotal evidence, the second reason for neglect, I would argue, is that when Anglophone geographers – particularly radical and critical geographers interested in the perspectives of critical geopolitics – do explore *Hérodote*, they find its approach and emphasis puzzling and difficult to comprehend. *Hérodote* appears to be very regional in its emphasis, oriented around empirical case-studies, quite conservative geographically and with little theoretical content. The gap is further emphasized by the publication histories of Lacoste's books: his 1976 blue book *La Géographie, ça sert, d'abord, à faire la guerre* was translated into Italian, Spanish, and Portugese, and it is referenced in critical/radical geographical papers as far afield as Brazil and Argentina, but an English-language translation has never appeared. Nor have his best-selling books on underdevelopment (Lacoste 1959, 1965 and later editions, 1980, 1985b) been published in English, though again several have been translated into a number of other languages. The ways in which this divergence of perspectives has emerged, and the extent to which it reflects real disagreements, is a theme I shall address in a later section.

Yet if Anglophone geography has neglected Lacoste-*Hérodote*, the neglect is certainly reciprocated. Right from its beginnings *Hérodote* has been embedded within the Francophone world and within the debates of French geography. There has been little referencing of English-language sources (and those references mainly in *Hérodote* essays by authors from outside the editorial core-group), and certainly none to the growth of Anglophone geopolitical studies and critical geopolitics from the 1980s onwards. Few Anglophone geographers have written for *Hérodote*: Baker and Clout appeared in an issue on historical geography (Baker 1994; Clout 1994), and Agnew has also contributed, but writing upon electoral geography in Italy (Agnew 1998a) rather than on his more global, critical geopolitics (Agnew and Corbridge 1995; Agnew 1998b). The Lacoste-*Hérodote* analysis of the development and shaping of geography and geopolitics is also set almost exclusively within the French context (with some references to the earlier history of German geography). Likewise, the contemporary debates it includes – with Lévy on Marxism, Brunet on *chorèmes*, and Raffestin on geopolitics – are Francophone debates.

The lack of engagement is thus mutual: two communities address apparently similar concerns but with little reference to each other's work or perspectives. The scale and achievements of the Lacoste-*Hérodote* school are too substantial for this neglect to continue. One aim of this essay is therefore to take the 'situated knowledge' of Lacoste-*Hérodote* and interpret it and its French academic context for Anglophone geographers. But a second aim is to demonstrate that many of the arguments within French geography and the Lacoste-*Hérodote* perspective can only be evaluated from a wider, multilingual and international context. Arguments, developments and debates, such as those in the history of geography and the recent revival of geopolitics, are often presented by Lacoste-*Hérodote* as specific to French or Francophone circles. However, in several cases they are much more general and international in their scope, and looking for specific French causations may be too limiting. The present essay can only explore a few aspects of this 'school' of work, and I would want to emphasize that the richness of the Lacoste-*Hérodote* tradition deserves sustained and detailed attention from Anglophone geography.

The origins of *Hérodote*

Contexts: institution and discipline

Yves Lacoste was thirty and already an established geographer when the post-1968 reforms brought him to the new 'experimental' University of Paris-VIII at Vincennes. Lacoste was a specialist in the geography of underdevelopment, with particular expertise in North Africa. He had been born in Fez (Morocco) and his wife, the ethnologist Camille Dujardin, also specialized in North Africa. His pre-Vincennes work followed the French marxist tradition of Dresch, Tricart and George, and he had written the underdevelopment section for George's *La Géographie Active* (George, Guglielmo, Kayser and Lacoste 1964) and influential texts on underdevelopment (Lacoste 1959, 1965, and Lacoste, Nouschi and Prenant 1960). Lacoste had also written a study of the Arab historian Ibn Khaldun, and this remains the only book by Lacoste to be translated into English (Lacoste 1966). It is conventional for writers from both left and right to belittle the intellectual impact of Marxism on French geography in this period, and to argue that it

271

left conservative, Vidalian practices *in situ*. However, this is an over-simplification: Marxism had profound impacts in a number of areas, one of which was the study of development issues. Lacoste's texts on underdevelopment reflected contemporary Marxist perspectives on dependency, but, along with colleagues such as Dresch, he developed important arguments on the role of class-structure *within* the underdeveloped countries and the importance of local bourgeoisies and landowners in colluding with colonial and global capitalism. He and his French Marxist colleagues also emphasized the role of colonial capitalism in harmful environmental impacts. Again these themes and linkages were only followed by Anglophone geographers considerably later.

At Vincennes Lacoste found not only a chaotic university disrupted by aggressively militant students, but also colleagues who included Foucault (at least for a time), Deleuze, Chatelet, Serres and Poulantzas (Eribon 1991). Debates and negotiations with students challenged not only traditional academic approaches (such as Vidalian human geography) but also the tenets of Marxism. In the early 1970s, just as Anglophone human geography was becoming excited by the ideas of Althusser and French structural Marxism, Lacoste and his Vincennes colleagues were moving to a post-Marxist analysis. Therefore, although still radical they moved towards an approach that was profoundly sceptical of metanarratives. The first lengthy statement of Lacoste's new analysis came in his contribution on 'Geography' to the multivolume *Histoire de la Philosophie*, edited by Francois Chatelet (Lacoste 1973a). In it he presents the core analysis that later appeared in his 1976 book, in the 'autocritique' to the third edition of *Géographie du Sous-Développement* (Lacoste 1976c), and in the first issue of *Hérodote*. Both *La Géographie, ça sert, d'abord, à faire la guerre* and *Hérodote* were published by the left-wing publishing house of François Maspero, and Lacoste formed a core 'editorial group' of colleagues and former students at Vincennes, perhaps most notably Béatrice Giblin (now Giblin-Delvallet), who has been part of the group since its foundation and is today Director of CRAG.

The central thesis of Lacoste's analysis, and the tradition that emerged from these texts, was that a serious absence of 'epistemological reflection' had blinded French geography to the ways in which the subject had been constructed in a narrowed and emasculated fashion. He argued that there were two distinct geographies: that of the military and political staff-officers and that of the academy and the school. This 'bifurcation of geographical knowledge' (*dedoublement du savoir geographique*) took place, Lacoste-*Hérodote* argued, at the end of the nineteenth century with the establishment of Geography as a school and university discipline. Lacoste-*Hérodote* emphasized that mapping and practical geographical knowledges had long been important aspects of military, political, colonial and commercial strategies, extending from Herodotus in classical times to the present, not least through the twentieth-century geopolitics of Haushofer, Mahan and Mackinder. 'General Pinochet is also a geographer', wrote Lacoste (1976a: 10) to demonstrate this point, referring to the Chilean dictator's role as a geopolitics professor in the national military academy and author of a text entitled *Geopolitica*. Similarly, Lacoste's own analysis of the US bombings in North Vietnam, argued that the bombs were targeted strategically to damage the dyke-system and to flood the adjacent region and its populace. With such examples, Lacoste and his colleagues demonstrated compellingly that 'geographical reasoning' remained an active and central aspect of the practicalities of politics and warfare. By contrast, the structuring of school and academic geography systematically excised and censored any acknowledgment of such political knowledge. Instead, academic geography was constructed as limited, apolitical, 'useless', and even 'silly' (*bonasse*). Lacoste-*Hérodote* argued that this role was profoundly ideological, serving to provide the citizenry with basic geographical facts about their country and the world, but also serving to mystify and mask the 'higher' role of geography as political-strategic knowledge. In this process

> Geographers have been the instruments of this mystification, but in the process they have themselves been mystified. And what one calls today the crisis of geography corresponds, for the most part, to the progressive discovery by geographers themselves of the extent of the mystification of which they are the agents and themselves the object.
>
> (Lacoste 1973a, 294–5)

Academic geographers needed to engage in epistemological reflection to unmask this ideological inheritance, and to do this they needed to confront several 'intellectual stumbling-blocks or impediments' (*concepts-obstacles*).

The central '*concept-obstacle*' targeted by Lacoste-*Hérodote* was the concept of the Vidalian region. Lacoste's concern was that the focus on the particular scale of 'the region', and the methods used to study society and landscape at that scale, helped ensure the obfuscation and erasure of political-strategic understanding. The geography of the professors acted as a 'smoke-screen'. Lacoste complained further that this was paralleled by an emphasis on scientific status, and (in contrast to its sister discipline of History) Geography had a worrying absence of polemic argument. Yet Lacoste-*Hérodote* also argued that neither Marxist geography nor the 'New Geography' of quantification that was beginning to travel into France from Anglo-America in the early 1970s provided a viable way forward. Marxist theorising neglected space, imposing aspatial theories on the diversity of geographical spaces. Here Lacoste moved away from (and criticized) his own former positions. His own field-experiences, now extended from North Africa to also include Vietnam and Cuba, reinforced this change, a change amply reflected in his study *Unité et Diversité du Tiers Monde* (Lacoste 1980). Equally, Lacoste thought that the new quantitative and applied geography was both a handmaiden of state, commercial and bureaucratic power (rather than a critical analysis of such power) and also suppressed geographical diversity in its emphasis on spatial order and positivistic theory.

The programme of Lacoste-*Hérodote* was built on this critique, but it also, and fundamentally, involved developing a more encompassing form of geographical reasoning that overcame the limitations of both conventional and newer approaches. This geographical reasoning, based on studying the intersections of geographical phenomena at different scales of analysis, will be examined in detail below, but amongst its objectives was not only the critique of existing power-geographies, but also the construction of counter-strategies by militant and oppressed groups. As the final chapter title of Lacoste's little blue book puts it, geography should be 'knowledge of how to think about space in order to know how to get organized there, to know how to fight there' (*Savoir penser l'espace pour savoir s'y organiser, pour savoir y combattre*) (Lacoste 1976a: 163).

The basic division made by Lacoste-*Hérodote* between a 'headquarter's geography' and an 'academic geography' has not gone unchallenged. Broc's response to *Hérodote* issue 1 argued that the intertwining of political-military power and geographical knowledge before 1900 was much more complicated and less one-sided than Lacoste allowed. Broc claimed that geographers (or those pursuing geographical knowledge), for example, often tried unsuccessfully to persuade the powerful in society of geography's relevance (Broc 1976). This convoluted archaeology of geography as a 'social technology of power' (to adopt Ó Tuathail's expression) requires much excavation, but the Lacoste-*Hérodote* group have only returned to it in terms of the emergence and shaping of French academic geography. Lacoste and Giblin re-examined the geographical writings of the anarchist socialist Elisée Reclus, emphasizing how his work, notably his *Géographie Universelle*, embraced both geography and politics (Giblin 1981; Lacoste 1981a; see also Reclus 1982). Lacoste studied the ways in which Reclus' grand, encompassing view of geography was narrowed and constrained in the construction of academic geography. He also claimed Vidal de la Blache, the 'father of modern French geography', had written politically (on Alsace-Lorraine), but that French geography had been systematically depoliticized in later accounts (and Reclus' work and Vidal's political dimension excised from the standard accounts). Lacoste argued that historians, keen to defend their own intellectual territory, were key actors in this depoliticizing of human geography, and that Lucien Febvre was especially notable with his advocacy of 'an unpretentious geography' (*une géographie modeste*) (Lacoste 1979, 1985a). It should be noted that these interpretations of the history of French geography have again not gone unchallenged; thus, on Vidal de la Blache, see Sanguin's biography and analysis (Sanguin 1993).

Lacoste's rethinking of basic geographical assumptions also saw him coin two useful neologisms at this time, *'géographisme'* and *'géographicité'*. *Géographisme* designates a figure of speech, very widely employed, where 'one makes a proper name of a territory, an actor from a certain number of political actions or economic operations' (Lacoste 1993a: 685): for example 'Paris' voted for Mitterand, 'Lorraine' struggled against plant-closures. Subsequent writing has examined the competing geopolitical representations encapsulated in such terms (e.g. *Hérodote* 14–15, 1979, on Euro-géographismes). The second term, *géographicité*, means 'that which geographers consider geographical and worthy of their interest' (Lacoste 1993a: 676), by analogy with the term *'historicité'* (*historicity*). Lacoste's whole project is, of course, an argument for an enlarged *géographicité*, especially in terms of politics. Both terms have proved very useful, and might be adopted fruitfully outside Francophone geography. Both also demonstrate *Hérodote's* radical departures from conventional geographical thought.

Why **Hérodote?**

The choice of Herodotus, the ancient Greek historian, for the title of the journal is an unexpected one, yet although the journal's subtitle has changed from 'Stratégies, géographies, idéologies' to 'Revue de géographie et de géopolitique', the main title has been maintained. Throughout its issues, the title pages of *Hérodote* have kept the cartoon of Herodotus by Wiaz (see Figure 11.1). As Lacoste notes, it is:

> the symbol of the journal seen through the cheeky talent of Wiaz, the good-natured Herodotus. He holds an anachronistic and somewhat absurd tool: a revolver fitted with a silencer, the World, and Herodotus' gaze is worrying, for he observes things others do not see.
>
> (Lacoste 1985a: 8)

Figure 11.1 The Wiaz cartoon of Herodotus used on the title page of *Hérodote* since the 1970s

275

Herodotus, author of *The Histories*, is usually seen as the father of the discipline of History, but Lacoste also claims him as the originator of Geography, arguing that Herodotus' work is better titled 'Enquiries', with much of it being the assembly of precisely that political-strategic knowledge that Lacoste wants geography to reclaim. Herodotus thus represents that more encompassing view of geographical enquiry, in contrast to more traditional classical geographers such as Ptolemy or Strabo. Herodotus has remained an important figure for Lacoste. In his recent *La Légende de la Terre*, Lacoste begins with Herodotus:

> Herodotus, greetings! – to you this opening invocation. Historians, traditionally, have christened you the 'Father of History', and several ethnologists place themselves under your patronage. For myself, I assert that you are the first geographer known to the history of mankind, and a geographer of the first rank whose ideas and perception of the world still today stimulate our thoughts.
>
> (Lacoste 1996a: 5)

And Herodotus and his perspectives dominate the first hundred pages of the book. As well as seeing Herodotus as 'an intelligence officer' (*un agent de renseignement*), Lacoste now also sees a connection between Herodotus' (public) enquiries and the (limited) democracy of classical Greece: 'Paradoxically, it allows us to advance the hypothesis that the appearance of geographical reasoning with Herodotus is inseparable from the political conditions of Athenian democracy' (Lacoste 1996a: 43).

Hérodote's geopolitics

The rebirth of *Géopolitique* and the 'rebranding' of **Hérodote**

The work and presentation of Lacoste-*Hérodote* undoubtedly underwent an important change in the early 1980s, centring on use of the term 'géopolitique'. In none of the early analysis was there any positive use of the word: linked to Ratzel, Haushofer, Mahan and Mackinder, it is seen as simply a form of strategic geography in military-political discourse that needs exposing and unmasking. 'Since 1945 it has not been good taste to make reference to geopolitics' (Lacoste 1976a: 9). Yet by 1983, *Hérodote* had become a 'Revue de géographie et de géopolitique', and Lacoste-*Hérodote* were 'branding' or labelling much of their work as geopolitics (Lacoste 1983). The fifth edition of *Géographie du Sous-développement* had the new subtitle 'Géopolitique d'une crise' (Lacoste 1982b), and the revised version of *La géographie, ça sert, d'abord, à faire la guerre* in 1985 (Lacoste 1985a) included an important chapter on 'the ghost or spectre of geopolitics'. Having criticized geopolitics in the 1970s as a partial and ideological form of geographical reasoning, what had happened to prompt this about-turn?

The answer was provided by Lacoste himself (Lacoste 1993a: 14–15; 1998: 27–34): towards the end of the 1970s, the Vietnamese-Cambodian conflict had focused attention on the geographical context of the war and French journalists started using the word 'géopolitique' in a non-pejorative way to describe the geo-political factors and contexts of the conflict. The word became frequently used by the press, even fashionable, and its former connotations with Nazi Germany were very largely forgotten. And although not discussed by Lacoste, another factor was the creation of L'Institut International de Géopolitique in Paris in 1981, together with its quarterly review *Géopolitique*. The Institute, under the direction of General Pierre Galois, was a politically conservative body, with a galaxy of French, American and British defence and diplomatic figures as its 'founding members' (*membres fondateurs*). The revival of the word 'géopolitique' created a rhetorical space in French public life that had not been there before 1978, but the space was being occupied by non-geographers. A similar trajectory has been traced in English-language (and other) forms of the word 'geopolitics' (Hepple 1986). If geographers became excluded from this new rhetorical space and the political dialogues and analysis generated therein, then it would simply have confirmed and reinforced the very depoliticization of geography that Lacoste-*Hérodote* had complained about. Geographers therefore needed to stake their claim to this rhetorical space, and the move by Lacoste and *Hérodote* can only be labelled oppor-tunistic in an approbatory way (though, as we shall see, some Francophone geographers have disagreed).

From the early 1980s then, Lacoste and *Hérodote* started to use the term 'géopolitique' to describe the type of geographical-political analysis they had already been advocating for almost a decade. They argued that the problem with the Ratzelian German geopolitical analysis was not its mixing of the political and geographical but the distorted and ideological way in which they were mingled and deployed.

Radical geopolitics and the roots of geopolitical revival

The popular revival of geopolitics in France through journalism and *Géopolitique* had much in common with the revival in English-language sources: a focus on global and regional issues, with some references to the geostrategic heritage of Mahan and Mackinder, but very little recognition or exploration of the academic or linguistic origins of geopolitics. In Anglophone geopolitics, serious re-examination of the roots of geopolitics only took place as new German historical research emerged in the 1980s; in general the earlier debates accepted the bifurcation between 'good' (Anglo-American) global geopolitics and 'bad' (German, Nazi) geopolitics established in Second World War texts. This has now changed, but this sequence helps differentiate the *Hérodote* group's work.

Lacoste and *Hérodote* certainly did not trace their work back to the English-language global geopolitics of Mahan and Mackinder. Indeed, Lacoste has never had much time for, or given much attention to, the Anglo-American classics of geostrategy. In a 1984 paper, published in English, he noted:

> The theses of Mahan and Mackinder, to which today's geopoliticians attach too much importance, rest more on historical evocations than on rigorous strategic thinking, based as they are on the grandiose geographical metaphors of the Land and the Sea. Although the theses lack scientific value, their lyric function is unquestionable.
>
> (Lacoste 1984b: 214; published translation by S.Kennedy)

Damning praise indeed! Lacoste made the point that in fact the revival of 'géopolitique' in debates on the Vietnam-Cambodian conflict flowed directly from the collapse of grand binary divisions in international politics (Land power versus Sea power, Heartland versus Maritime Crescent, USSR versus USA), and signalled the need for a more regional, geographically-sensitive and inter-national analysis.

Instead, Lacoste and his *Hérodote* colleagues, notably Béatrice Giblin and the German specialist Michel Korinman, traced the origins of geopolitics to Ratzel's work (Korinman 1983), his German precursors (Korinman 1991) and the whole history of German *Geopolitik* (Korinman 1990). In these accounts Kjellén, who actually coined the term 'geopolitics' (in Swedish), gets only brief mention, but Ratzel's geographical work is put centre stage, and its setting of conflicts between states (especially nation-states) in their geographical contexts is seen as the root of geopolitical analysis. Lacoste-*Hérodote* also located other 'geopolitical' precedents in the French work of Reclus, Vidal de la Blache (Lacoste 1979) and other work in the period 1918–1930 before historians imposed an 'unpretentious or modest geography' (Lacoste 1985a: 91–114). Korinman argued that the wholesale rejection of German geopolitics had created a dangerous silence, namely a lack of reflection on the important issues of geography and political power, precisely those issues on which *Hérodote* focused. He argued that the ideological and propagandist parts of Ratzel's work should not be allowed to obscure and erase

> those sections which form an effective means of analysis of certain sorts of problems, particularly those which have strategic significance. To radically challenge the first types does not imply rejecting or being unaware of the other types.
>
> (Korinman 1983: 136)

Lacoste characterized the Europe of the period 1918 to 1933 as one of active geopolitical debate, a debate curtailed by the rise of Nazism in Germany and the gradual exclusion of geopolitical reflection within French geography. He even argued that France would have replicated the German ferment of geopolitical writing after 1918 if the war had gone differently:

> If France had been defeated in 1918 (if the Americans had not intervened), it is likely that the first geopolitical movement would have appeared in Paris, for the characteristics of French society were not so different.
>
> (Lacoste 1993a: 19)

Such views have not met with approbation by all Francophone geographers, several of whom still reject the revival of the term 'géopolitique'. The most strident has been Claude Raffestin, whose history of geopolitics sees the term and its perspectives as irredeemably linked with its fascist, nationalist and expansionist past (Raffestin, Lopreno and Pasteur 1995), a critique examined further below.

Le raisonnement géographique

Lacoste and *Hérodote* place '*le raisonnement géographique*' or 'geographical reasoning' at the heart of their work. Such geographical reasoning is by no means restricted to geopolitical analysis, but the application of this reasoning to geopolitical situations is their contribution to geopolitics. The main features of this geographical reasoning have not changed since Lacoste's 1973 essay, and it is important to examine them. First, it is interesting to note that Lacoste does not argue geography has an *object* to study. Instead, the geographer's contribution lies in the approach and the style of *reasoning* deployed. He does not substitute a new object for the Vidalian region, nor does he make grand synthesizing claims for geography, except in so far as a particular style of reasoning draws upon specific elements from other disciplines. This approach offers the possibility of seeing geographical reasoning as a form of 'thoughtful practice', a linkage I shall try to open up later.

The centre of geographical reasoning lies in the analysis of 'intersections of spatial sets or groups and different levels of spatial analysis' (*intersections d'ensembles spatiaux et differents niveaux d'analyse spatiale*). *Hérodote's* geographical reasoning does not privilege any one scale (that was the error of classical French regional geography), nor does it claim to study any specific phenomena across space (as many disciplines study the spatial distributions of their own phenomena). Instead it examines the complexity of geographical contexts by moving between scales of analysis. Lacoste frequently uses a diagram to explain this (see Figure 11.2). It shows intersecting distributions at four different spatial scales (hypothetically 10km, 100km, 1,000km and 10,000km). He notes:

This diagram illustrates this way of thinking about space based fundamentally on the combination of two methods of spatial analysis – on the one hand, the systematic distinction of different levels of analysis

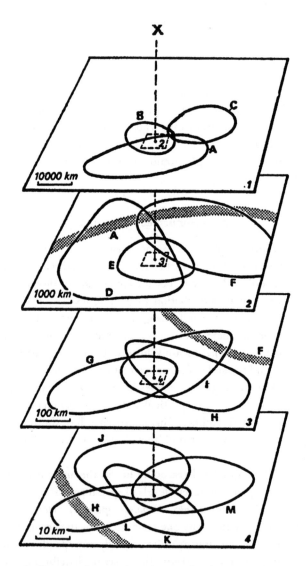

Figure 11.2 Lacoste's diagram showing how 'geographical reasoning' relates different scales and levels of spatial analysis

according to the different orders of magnitude, according to the dimensions that the multiple spatial groups have in reality; – on the other hand, at each of these levels, the systematic examination of the intersections amongst the contours of the diverse spatial sets [groups] of the same order of magnitude.

. . . The geographical characteristics of a particular place or the inter-action of phenomena which one must take account of in order to act in this place – on this sketch it is the point x which can be found at the centre of each of the planes – can only be established by reference to the intersections of different sets at different levels of analysis. Strategically, each set corresponds to a favourable or unfavourable factor for the action being considered.

(Lacoste 1985a: 72)

As a general statement much of this is admirable, though its geometric tone should warn us of internal assumptions. But as a more specific methodology it raises issues of objectivity, perspective and social construction, all issues recognized and discussed (not necessarily under these banners) by Lacoste-*Hérodote*. The assertion of Lacoste-*Hérodote* is that this geographical reasoning provides 'a scientific approach' (*une démarche scientifique*) to geopolitics, one which allows and facilitates a more objective analysis of geopolitical situations:

The bringing into play of geographical reasoning allows one to construct, for each geopolitical situation, a representation more complete and more objective than those provided (in a contradictory manner, and which one must take into account) by the principal actors involved in the territorial rivalry at issue. Indeed, geopolitics as a scientific approach is not limited to the analysis of conflicting repre-sentations; it must itself strive to construct a more global and much more objective representation of the situations, in order to suggest solutions to confrontations in progress but also to attempt to forecast scenarios of development. This particularly hazardous and delicate enterprise requires recourse to the historian's approach.

(Lacoste 1993a: 32)

As Ó Tuathail has noted, this approach of Lacoste-*Hérodote* prioritizes a Cartesian perspectivism, the 'view from nowhere', one that is objective and detached, and he has drawn attention to Lacoste's significant use of Altdorfer's painting of *La Bataille d'Alexandre* on the cover of the 1988 *Questions de Géopolitique* book (Lacoste 1988; see also Ó Tuathail 1994, 1996). Lacoste also uses this picture as the first, full-page illustration in *La légende de la terre*, noting

This scene, painted in the era of conquest of the New World, traces a connection between the completely military explorations of the great Alexander and those of the Spanish conquistadors. It concerns without doubt the first representation of such vast horizons: a confused melee, troop movements around a besieged town, an island with steep slopes, in the distance a mysterious continent and, under the clouds which the setting sun pierces, immense distances which the gaze can take in, whilst at the horizon the curvature of the earth is guessed at.

(Lacoste 1996a: 5)

The grand, strategic perspective makes sense of the complexity of the scene. Indeed, for all their passage into post-Marxism, Lacoste and *Hérodote* remain remarkably untouched by the doubts about objectivity that have exercised post-structuralist and post-modernist writings in both French social theory and Anglophone social science (including human geography). This difference lies behind much of the difficulty of comparing the Anglophone and Francophone approaches to geopolitics.

Yet the Lacoste-*Hérodote* approach is more subtle than the geometrical language of intersecting sets suggests. First there is recognition that the identification of both the relevant scales of analysis and then the associated 'spatial sets' are issues of 'conceptualization'. Second many of these sets are undoubtedly abstractions, especially those at grander spatial scales:

It is important to note that in the majority of cases (and with the exception of deserts) the more the groups (sets) have large dimensions, the greater is their population and the more they are conceived, shaped by a growing degree of abstraction; this is particularly the case for the world-wide set called the 'Third World', which includes more than four billion people. It is a question of an abstraction, but it is useful to debate the relatively specific limits of it on the globe.

It is not easy to articulate scientifically a representation shaped to a significant extent as an abstraction, and a set of much smaller dimensions is, by this fact, much more concrete.

(Lacoste 1993a: 31–2)

However, many of the sets operative at smaller scales, such as ethnic identities and cultural groupings, are also abstractions and are certainly social constructions. Moreover, even with more concrete phenomena, actually mapping and delimiting a spatial range involves abstraction in the forms of representation, measurement and even 'contour definition and cut-off'. Nor should the geometric picture be taken too literally as end-product: phenomena need to be 'mappable (*cartographiables*), that is

282

to say one may be able to recognize significant differences across the surface of the globe' (Lacoste 1985a: 108). Thus one may be able to recognize and study spatial variation without being able to map it to any degree of precision or representation, yet it can be (and may be) an integral part of geographical reasoning.

In these ways, the geographical reasoning of Lacoste-*Hérodote* is more flexible and encompassing than it may appear initially. However, there is much less flexibility or accommodation towards an abstract notion of 'theory' in such geographical reasoning. The method outlined provides a way of understanding the real complexity of concrete geopolitical situations, and Lacoste-*Hérodote* provide no discussion of (or encouragement to) a further role for theory or formal method. Here the alignment with history as an academic practice and the antipathy towards social science theory stands out markedly.

Back in 1976 Lacoste had drawn parallels between scales of analysis in geographical analysis and those in history, citing Althusser's *Lire le Capital*: 'there is, for each mode of production, an appropriate time and history . . . an appropriate history for the political superstructure. . . . The specificity of these times and histories is thus differentiated, since it is based on the differentiated relationships that exist in the whole between the different levels' (Althusser 1965: vol. 2, 47). Lacoste (1976a: 69) provides a longer extract, and he later admires Fernand Braudel's writings on the different 'durées' or periods of history (Lacoste 1986a). Lacoste thus sees linkages between 'les longues durées' and the stability of larger geopolitical spaces (linkages that could be paralleled in Peter Taylor's English-language geopolitical interpretations; see Taylor 1990).

Here it may be useful to consider 'geographical reasoning' as a 'thoughtful practice'. When Lacoste writes of the development of professional geopolitical skills, he argues:

> On the other hand, one must train researchers of great skill through the methodical and thorough analysis of a large number of cases, including in the field, through a structure of interdisciplinary research; it is one of the tasks of the producers of this dictionary, in the framework of the 'doctoral programme in geopolitics' (University de Paris-VIII).
>
> (Lacoste 1993a: 35)

The skills of the geopolitical analyst cannot be defined by theories or formal methodologies; they are thus similar to the American philosopher Dewey's 'thoughtful practices' or the feminist philosopher Lisa Heldke's 'recipes' (Heldke 1992). They are practices with reflexive, thoughtful guidelines, but processes guided additionally by experience, judgement and actual practice. Experience and judgement are accumulated during the study of 'cases', and this experience cannot be reduced to theoretical models or rules. Medicine is a 'thoughtful practice' of this type. These skills can be very real, and Lacoste's expression 'geographical reasoning' is undoubtedly worth further investigation in this light.

The tendency to eschew theory is undoubtedly linked in a further way to the discipline of history: the desire to make the writing (geographical or historical) accessible to a wide public of both citizens and political actors. But the elevation of the concrete, complex, specific and particular (in both disciplines) has the corollary of largely dismissing the abstract, general, ordered and theorized conceptions and analyses of the social and human sciences. Here I believe *Hérodote* tends to limit its own potential. The insights of political science (such as the 'models' derived in comparative politics of the varying roles of the military in authoritarian regimes) are valuable for geopolitical analysis (such as in Latin America). Such theories and models are built from mixtures of comparative empirical generalizations (the basis for much of Ratzel's *Politische Geographie* in 1897) and deductive and discursive reasoning. The closer links of much Anglophone geopolitical analysis to such work is one of its strengths, and the lack of such insight is one of *Hérodote's* weaknesses. These questions of the role (and nature) of theory in geographical and geopolitical analysis will be explored further below.

Themes, theory and politics

Geopolitical themes

Themes and coverage

It is time to look at the actual geopolitical practices as set out in *Hérodote*. When the journal changed its subtitle, Giblin noted that it was the intention to devote approximately two issues a year to reflection on geography and two to geopolitics (Giblin 1985). However, the geopolitical soon came to dominate. In recent years only a few issues have been devoted to 'internal' geographical debates, such as that on the contentious issues of '*chorèmes*'. Most theme issues are regionally based, and there is little explicit general (still less, theoretical) analysis. Claval (1999) has detailed the evolution of topics covered, and the most recent sequence includes:

La question serbe (67, 1992)
La question allemande (68, 1993)
Démocratie et géopolitique en France (69/70, 1993)
L'Inde et la question nationale (71, 1993)
Nation, nations, nationalisme (72/73, 1994)
Géographie historique (74/75, 1994)
Les géographes, la science et l'illusion (76, 1995)
Maîtriser ou accepter les Islamistes (77, 1995)

Japon et géopolitique (78/79, 1995)
Périls géopolitiques en France (80, 1996)
Géopolitique du Caucase (81, 1996)
La nouvelle Afrique du Sud (82/83, 1996)
Le Cercle du Samarcande (84, 1997)
États-Unis: Le racisme contre la nation (85, 1997)
Géopolitique d'une Afrique Médiane (86/87, 1997)
Indonésie, l'Orient de l'Islam (88, 1998)
Italie, la question nationale (89, 1998)
Méditerranée: nations en conflit (90, 1998)
La question de l'Espagne (91, 1998)
La santé publique et sa géopolitique (92, 1999)

The editorial organization of *Hérodote* has ensured a tightness and coherence to the accumulating body of work that cannot be rivalled by a standard journal: the editorial group builds each issue around a theme, invites the contributions and ensures a degree of uniformity of both approach and style. Lacoste writes most of the editorial essays himself. Of course, with a changing set of authors, there is variety in both analytical approach and literary style, but much less than in a standard, contributor-led journal such as *Annales de Géographie* or *L'Espace Géographique*. Over the last twenty years the cumulative set of *Hérodote* issues adds up to a veritable *Géographie Universelle* for today (and a *Géographie Universelle* much more in Reclus' footsteps than the version being produced by GIP-RECLUS!). This is a major achievement for Lacoste and the *Hérodote* group. However, the central control of the journal inevitably makes it less receptive to new, unexpected, conflicting and serendipitous trends than a standard journal might be.

In any exploration of *Hérodote's* contents there is a very real embarrassment of riches, and this essay can only select one or two examples from the treasure-house for brief consideration. Tempting as the geopolitics of Islam and the 'Circle of Samarkand' are – and *Hérodote* has provided an examination of the whole range of geopolitical issues raised by Islamic fundamentalism that is certainly unrivalled in any Anglophone writing – I will discuss two items only here: Lacoste-*Hérodote's* perspective on global and geoeconomic themes, and Lacoste-*Hérodote* on the nation and its role as a geopolitical concept.

The global and the national

Throughout the more than ninety issues of *Hérodote* over more than twenty years, very little direct attention has been paid to geostrategic or geoeconomic analyses at the global scale. This stands in very stark contrast to Anglophone work in critical geopolitics, where global strategy and rivalry between the USA and USSR, and the post-1990 attempts at a 'New International Order' have been central themes and form almost the entire contents of the recent *The Geopolitics Reader* (Ó Tuathail, Dalby and Routledge 1998). Indeed, it was not until 1997, after twenty years, that an issue of *Hérodote* was directly focused on the USA, and that was on the theme of 'Racism against the Nation' rather than on issues of US global geopolitical strategy. Why this choice – apparently a deliberate one – to reject global analysis?

The choice primarily flows from Lacoste-*Hérodote's* interpretation of the nature and history of geopolitics. Lacoste is no stranger to the literature on global analysis. His earlier work on underdevelopment, replete with (critical) references to the grand narratives and theories of dependency of Amin, Frank and others, his polemic on *Contre les anti-Tiers-Mondistes*, and his membership of the editorial board of the new *Revue Française de Géoéconomie* (which began publication in 1997), all indicate his knowledge of these issues. Despite this he has rejected them as the foundations for geopolitical analysis or as the key current geopolitical issues. Lacoste-*Hérodote's* antipathies to globalization perspectives, and to the 'geopolitics-to-geoeoconomics' theme popular in American writings, are made evident in Lacoste's critique of Jacques Lévy's work, which emphasizes geoeconomic view-points. Lévy asserted that, after the end of the Cold War 'the world is becoming degeopoliticized' (*le monde se degéopolitise*) and that 'the progress of political construction and degeopoliticization of the world seem incontestable' (Durand, Lévy and Retaille 1992: 189). Against this, Lacoste places the growth of geopolitical movements (separatist and nationalist) in the aftermath of the Cold War, as well as in states such as Spain, Italy and even France, to say nothing of the Islamic and non-European world (Lacoste 1993c). Lacoste and *Hérodote* thus reject much of the globalization and demise of the nation-state themes that are strong in current Anglophone geopolitical work. They do not deny the spread of global economic influences, but argue that these often generate (or reinvigorate) national and local political and cultural responses that limit any globalization perspective on geopolitics.[2]

The Lacoste-*Hérodote* perspective, building on their interpretation of the geopolitical tradition from Ratzel, Reclus and Vidal de la Blache, focuses on geographical contexts (at varying scales) below the global level, and takes the level of the state as the *primary* scale. This is certainly not to say that this is in any way the exclusive level of analysis – such would entirely contradict the method of 'geographical reasoning' – but it does reflect the centrality of the state as political actor and the geopolitical centrality of external conflicts between states and internal conflicts over control of states. Thus Giblin-Delvallet, in her editorial essay to the issue on the United States, argues that the geopolitics of US foreign policy has been discussed in numerous issues of *Hérodote* in terms of its regional impacts on particular states and regions of the world (Giblin-Delvallet 1997).

At *Hérodote's* favoured scale of the nation-state, the main concerns have been the geopolitical significance of culture and nationhood, and especially the pivotal roles played by language and religion. These factors have been concerns of *Hérodote* since the 1970s, but the political changes since 1989 have reinforced this perspective. *Hérodote* has devoted considerable attention to the different forms – separatist, aggrandizing, democratic – of nationalism. *Hérodote's* examination of nationalisms has reached world-wide, but the most detailed studies are probably those that debate nationalisms and tensions in the new Germany, the former Yugoslavia, and within France itself (e.g. Lacoste's own essays 1992, 1993b, 1994). Tensions based on religious, linguistic and other ethnic cleavages have become very prominent in recent years, and, in the light of the attention that *Hérodote* has focused on these geopolitical tensions in many parts of the world (including France itself), it should be no surprise that the United States issue (85, 1997) is devoted to examining these aspects of the USA: racial tensions, ethnic minorities, migration, the crisis of American identity.

Lacoste-*Hérodote* see the (nation-)state as sandwiched between external, supra-national and global forces and internal, regionalist/localist forces of fragmentation. At this level of generality, most Anglophone geopolitical writers would not dissent from this analysis. However, not only do Lacoste-*Hérodote* not foresee the imminent demise of the (nation-)state, but they would also resist it. This is somewhat dangerous political and intellectual territory: in contrast to most leftist writing in both France and Anglo-America, Lacoste-*Hérodote* do not view the (nation-)state as necessarily politically regressive, a dinosaur whose end is to be welcomed. For Lacoste 'the nation is, to my thinking, the fundamental geopolitical concept, not only on the theoretical level, but also by reason of the very great political assets which are associated with it' (Lacoste 1996b: 207). This difference over the state creates a gulf between Lacoste and Raffestin. For Lacoste the state, and a tradition of geopolitical analysis built upon it, can have political legitimacy and value, whereas for Raffestin such state-based geopolitical analysis is inexorably linked to (and tainted by) its fascist and imperialist-capitalist past and only an internationalist-proletarian programme can be legitimate (see Raffestin, Lopreno and Pasteur 1995, and Lacoste 1996b).

Whilst the essays in *Hérodote* develop the complicated and interwoven strands on nation and nationalism in geopolitics, it is Lacoste's own recent book *Vive la Nation* which expresses the active political role such analysis can play. The book is a contribution to current French debate over citizenship, immigration and race, a debate in which 'La Nation' is often a rallying symbol captures and deployed by the extreme right. The title of Lacoste's book itself is provocative in French national debates; it would be even more so in the UK or USA and it is hard to imagine an Anglophone geographer writing such a book. The book takes its title from the cry of the French troops at Valmy in 1789, and Lacoste argues that France needs to reconstruct itself around a much clearer view of the nation, a nation united by both language and civic tolerance. It is an argument that he has also taken into the pages of the rival journal *Géopolitique* (Lacoste 1997) in their issue devoted to the problems of 'La République'. Reflecting the earlier analyses in *Hérodote*, Lacoste sees the idea of the French nation threatened by the xenophobic nationalism of Le Pen, by the minority separatisms of some regionalist groups, and also by a multiculturalism (espoused by both Islamic groups and by many radicals) that will not assimilate Arabs into French language and society. Lacoste perceives a particular risk in groups speaking not French but 'a sort of identity slang' (*une sorte de jargon identitaire*), reinforcing their isolation from French society. Lacoste also identifies another threat: the potential loss of French identity in a European Union dominated by Anglo-American global interests. As he portrays it: 'only the reinforcement of the idea of the nation in France, in Germany and in the countries of Latin Europe can check this Anglo-Saxon hegemonic tendency' (Lacoste 1998: 152–3). In Lacoste's opinion, only the nation is substantial enough to resist these globalizing pressures; and any move towards smaller regional and local identities would, in the absence of the nation, simply collapse before these pressures. *Vive la Nation* stands as provocative and challenging study, using the more dispassionate and detached analyses developed in *Hérodote* for polemic and debate in French political life, a polemic that had attracted considerable attention in the French media (Cassen 1998; Oliva 1998).

Comparisons with Anglophone geopolitics

How do *Hérodote's* concerns and geopolitical analyses relate to Anglophone research? Are they replicated in Anglophone work, even if under somewhat different banners? Here it is important to make differentiations *within* Anglophone geopolitical writing, and especially between 'conventional' and strongly empiricist work and the much more theorized work in critical geopolitics and critical human geography. This is a crude and oversimplified division, but one useful for present purposes.

Within the first category there is a considerable amount of work on similar topics to those covered by *Hérodote*, written by both geographers and international relations specialists. In geographical journals (and most notably *Political Geography*) one can find many regional political studies, and in international affairs journals (such as *International Affairs* or *Strategic Studies*) and regional political-cultural journals (e.g. *Slavic Review, Journal of Latin American Studies*) equivalent studies by other specialists. Some of this would be self-consciously geopolitical, for example the studies produced by the Geopolitics Research Centre (SOAS/GRC) at London University, as in their study of the Balkans problems (Carter and Norris 1996), or the Oslo study edited by Tunander, Baev and Einagel (1997). However, much of the relevant literature, notably work on nationalisms and ethnic and regionalist separatisms, would not sail under a 'geopolitical' flag.

This work is disparate, and scattered through different journals and outlets. Much of it is 'regional-political', rather than explicitly geographical, and only a small proportion would draw out the interplay of different scale and contexts in the manner that characterizes the 'geographical reasoning' of Lacoste-*Hérodote*. In contrast, a great strength of *Hérodote* lies in the structuring of its regional-theme issues and the cumulative development of a large, coherent and related body of geopolitical analysis. This has been reinforced by the growing facility of *Hérodote* to provide 'geopolitical' (as against other specialist) expertise to write the essays, a facility particularly evident since the creation of CRAG, so that there is more consistency to the analysis. On numerous occasions, *Hérodote* has provided insights and analyses that are hard to find in the English-language literature (certainly at the date they appeared in *Hérodote*). Any list can only be very selective and personal. During the Russian-Afghan war, when there were fears of a 'drive to the Gulf', it was Dresch's paper in *'Points Chaudes'* (Hot Spots) (Issue 18, 1980) that showed the physical geography context and the way the Lout desert, impassable to tanks, stood in the way (Dresch 1980), precisely the sort of dimension ignored by conventional studies. *Hérodote* also provided very early studies of the geopolitics of Kosovo (in Roux 1982), the geopolitical perspectives of the new Russian right (Vichnevski 1994), and the geopolitics of deforestation and the environment in Indonesia (Durand 1998), amongst many other possible examples.

The differences between Lacoste-*Hérodote's* geopolitics and the perspectives of Anglo-American critical geopolitics are intriguing and substantial. There are the differences already noted in terms of their relative attention to, and assessments of, the nation-state and global geopolitics. These are very important, but at least as significant are the differences in terms of attitudes to the role of theory in geographical and geopolitical analysis. From the perspective of Anglo-American critical geopolitics, *Hérodote's* stance is very anti-theoretical, and this generates a gulf between the two 'schools'. Yet the basic starting positions of the two schools, in terms of both radical politics and intellectual analysis were very similar: the critique produced by Lacoste-*Hérodote* in the 1970s and its development of a radical geopolitics in the 1980s was replicated quite closely by Anglo-American work in critical geopolitics and critical human geography. The very different attitudes to theory therefore need some examination, a theme of the following section.

Both Lacoste-*Hérodote* and Anglophone geopolitics argue for a closer engagement between the political and the geographical, a breaking down of 'mystifying' barriers to intellectual inquiry and political analysis. However, in this engagement Lacoste-*Hérodote* have a conventional view of the political: the geopolitical is expanded to encompass a much wider (and internal) perspective on the state, but the focus remains primarily on territorial and state policies, differences and control. There has been very little engagement with the strands of 'cultural politics' popular in Anglophone geography (and also within Parisian social and cultural analysis). Lacoste's early rejection of Foucault's micro-power (discussed in the following section) is the pointer here, and *Hérodote* has not focused on issues of gender/feminist or identity politics, issues central to much current Anglophone work. The 'politics of difference' as expressed in multiculturalism, has been examined, notably in *Vive la Nation*, where it is seen as a problem to be countered, not a direction to be embraced.

Politics and theory

Two intertwined issues emerge from this analysis and critique of Lacoste-*Hérodote's* geopolitics. First, the question of its politics: to what extent does *Hérodote* remain a radical, leftist perspective on geography and geopolitics? Second, the question of the role of theory: why do Lacoste-*Hérodote* and Anglo-American critical geopolitics differ so much on the value of 'theory', and why does *Hérodote* often appear (to Anglo-American eyes) geographically conservative rather than radical? These issues, together with the important dimension of the audience and popular impact of Lacoste-*Hérodote's* work, now need examination.

The politics of Hérodote

Hérodote began in 1976 as the '*enfant terrible*' of French radical geography, post-Marxist but undoubtedly anti-establishment both in geography and politics. What of the development of the geopolitics of *Hérodote* over more than twenty years: is it still an undoubted '*géopolitique de gauche*'? That the question needs asking, and then that it needs some care in suggesting an answer, shows the evolution and perhaps depoliticization of the group. Broc identified a softening of view quite early: the review of *Hérodote*, Issue 1 (1976), was entitled '*Hérodote* à la Sauce tartare' (Broc 1976), but seven years on the reappraisal, examining *Hérodote*, issue 23 (1982), became '*Hérodote* à l'eau de rose' (Broc 1983). '*Hérodote* has put water into his wine, the lion has retracted its claws, the Cossack has left his blade in the cloakroom. Where are the insults, the curses, the condemnations?' (Broc 1983: 708). And it is likely that this trajectory has been reinforced by the emphasis on the 'scientific approach' to geopolitical analysis provided by 'geographical reasoning' (see also Boyer 1986; Giblin 1985, on the evolution of *Hérodote*).

There is some truth in this perspective. Many of the contributors to *Hérodote* would not be subscribers to any particular political outlook, and a clear political tone in *Hérodote* is harder to discern than in the (in some ways) parallel Anglo-American journal *Antipode*, which retains a more explicit leftist agenda. Yet many of the core group of *Hérodote* (including of course Lacoste himself) retain their leftist politics, but these are expressed in a context that has changed dramatically since the 1970s, both internationally and within France. The 'revolutionary' position has gone, but a leftist position within French politics is now expressed in terms of the socialist party within France. Within the pages of *Hérodote*, with its emphasis on the 'scientific approach' to geopolitical analysis, explicit politics are today much less obvious than formerly, and Lacoste's '*géopolitique de gauche*' is most explicitly found in his recent *Vive La Nation*.

The leftist politics can also be traced in two other elements of Lacoste-*Hérodote*: the linkage of geopolitical analysis and democracy, and the context of the emphasis on the role of nation(s) as geopolitical identities and representations. Lacoste-*Hérodote* emphasize the significance of geopolitical analysis for citizens in a democracy, for in a democracy, unlike pre-1900 societies, the power/knowledge linking geography and the state should be understood by all. The critical unmasking of 'geo-power', to use Ó Tuathail's phrase (Ó Tuathail 1996), emphasized in the early Lacoste-*Hérodote* writing is now represented by an emancipatory faith in an informed democracy. This – apart from the specific positions adopted by some of the group on French political issues – is the main leftist inheritance. In a wider context of post-modernist doubt, Lacoste-*Hérodote's* faith here can be seen as part of its (rejected) Marxist heritage, and the group retains a clear enlightenment and emancipatory position.

The second element is more contentious. Lacoste-*Hérodote's* fiercest critic, Raffestin, has attacked what he sees as the reactionary emphasis on nation and nationalism in *Hérodote*, concluding 'there can be no doubt that *Hérodote's* geopolitics are, in themselves, a French nationalist science' (Raffestin, Lopreno and Pasteur 1995: 294). There is much that is misleading about this comment, but in one sense – not perhaps one intended by Raffestin – it has some force. Lacoste-*Hérodote* see the emergence of certain forms of nationalism (and certain definitions of 'nation') as the best-available forms of political expression and community available to groups in many parts of the world. These representations can be very misused and misappropriated, but they are also ways to organize and resist wider dominance. Lacoste-*Hérodote* interpret 'national identity' as a resistance to globalization (and to internationalist, fundamentalist Islam), and, as an immediate reflection of this, they are quite prepared to promote the idea of 'the French nation' against what they term 'Anglo-Yankee hegemony' in Europe. They would, undoubtedly, deny that *Hérodote* presented a 'French nationalist science', but might accept that the 'science' was sometimes reflected through a French view of the world. But, again, they would undoubtedly see any French-bias in this as being also a way of promoting similar 'resistances' internationally. In that sense, Lacoste-*Hérodote* do promote an 'anti-(American)globalization' geopolitical strategy, which again would be part of *Hérodote's* leftist roots.

One might (perhaps fancifully, perhaps not) see the whole intellectual project of *Hérodote* as a geopolitical strategy in itself, a Francophone act of resistance against the tide of Anglophone intellectual imperialism – with the neglect of Anglophone perspectives a deliberate piece of resistance – and the building of an alternative international coalition of 'national perspectives' against globalizing perspectives. And perhaps this very description – with its assumption of Anglophone perspectives as the 'global norm' and Francophone work as the act of (local) resistance – would be seen as an example of the very intellectual imperialism to be challenged! Geopolitics is inevitably situated and political, and this applies to geopolitical analyses as well as practices.

The radical and the conservative: the role of theory

The very first issue of *Hérodote* included an interview with the Michel Foucault on 'Questions sur géographie' (Foucault 1976a). This was Foucault's only direct contact with geography and, because it has been translated and included in collected essays (Foucault 1980), it may be the best-known *Hérodote* essay in the English-language world. In 1976 Foucault had recently published *Discipline and Punish*, and the interview related both to this (and the role of a geopolitics of power) and to *The Order of Things* and the place of geography in an archaeology of knowledge. The interview was in many ways a disappointment: Foucault wanted geographers to do their own 'archaeology', though he did admit, somewhat reluctantly, to the importance of space (and hence geography) in power/knowledge. Much less well-known in the Anglophone world is the fact that Foucault addressed some questions back to *Hérodote* (Foucault 1976b), and elicited extensive responses (Bernard *et al.* 1977).[3] This engagement does highlight the convergence between Foucauldian thought and the geopolitical perspectives of the *Hérodote* group well before Foucault's impact upon the construction of an Anglophone critical geopolitics by Dalby, Ó Tuathail and other in later years. Lacoste (1976a) and *Hérodote* both emphasized one dimension of Foucault's power/knowledge, that of the pervading role of state power (including class power) and its influence on intellectual, academic and political structures. Both Lacoste-*Hérodote* and Foucault draw on the concept of 'le pouvoir'.

But Lacoste-*Hérodote* saw Foucault's wider ideas of power (including micro-power) as less relevant to their political concerns, for in their analysis power is always instrumental and exercised by politicians, the military or business interests. These divergences emerged when Lacoste reviewed Raffestin's *Pour une géographie du pouvoir* (1980). Raffestin deployed both geometric diagrams and Foucauldian concepts in his micro-analysis of power and space, including man–woman–family relationships, citing Foucault's 'power comes from below' (Foucault 1978: 127). Lacoste had no time for 'sophisms which rely on the confusion between very different types of power (sexual power and state power) and on the confusion between levels of analysis (relationships between two individuals and the role of the state apparatus for thousands or millions of individuals)' (Lacoste 1981b: 157). This divide between politics at the traditional scale and those of the politics of identity, sexuality and objectivities characterizes *Hérodote's* work quite markedly.

In many ways the early interchange with Foucault did not reach deeply into Lacoste-*Hérodote's* methodological or conceptual structure. This type of interchange with social theory was not repeated, and *Hérodote* has seldom engaged with the many other forms of French social theory. In fact, after the initial process of 'epistemological reflection' which gave rise to *Hérodote*, Lacoste-*Hérodote* have been sceptical of the need for any further methodological reflection and quite hostile to the role of theory in geographical analysis. Each of these aspects deserves discussion.

Lacoste's initial 'epistemological reflection' was the essential key to questioning and then demolishing the artificial (and ideological) barriers between the 'political' and the 'geographical'. Paul Claval, in one of his studies of the history of geography (Claval 1976), uses the expression 'the enlargement of the past' (*l'enlargissement du passé*), arguing for a more comprehensive and contextual history of geography. Lacoste's epistemological reflection and his case for a wider '*géographicité*' do precisely this. Yet Lacoste and his *Hérodote* colleagues (see the analysis in Giblin 1985) always present this as a task that has been completed: this task is done, the job is now to apply the method. Their reflection on method is not an ongoing, continuous process of critique, yet they nowhere really justify this standpoint.

Lacoste-*Hérodote's* attitude to 'theory' is based on (and scarred by) the experience of Marxism and by the forms of (positivistic) theory presented by the 'New (quantitative) Geography'. This view of theory has undoubtedly been reinforced by recent debates with Brunet and colleagues over the political use of the geometric 'models' known as *chorèmes* (see Giblin-Delvallet 1995; Lacoste 1993c, 1995; and the summary in chapter ten of this volume). When theory and models are discussed, *Hérodote* always represents these in a reductionist manner as attempting to reduce social life to scientific laws and order, contrasting them with the complexity, diversity and uniqueness of geographical analysis. Yet Lacoste-*Hérodote* still assert that the method of geographical reasoning provides 'a scientific approach': 'all this operative reasoning is the matter for a perfectly scientific approach, without needing Order and models' (*d'une démarche parfaitement scientifique, sans pour autant Ordre et modèles*) (Lacoste 1995: 18). To many Anglophone geographers, this faith in the 'scientific approach' to geopolitical and geographical analysis needs, at the very least, some re-examination. Such 'geographical reasoning' not only builds on the entirely sensible verities of careful (and self-critical) collection and appraisal of evidence, and on an entirely commendable practice of interweaving different spatial contexts and scales, but it also suggests that persistent practice can 'step out' of our own positions and contexts to achieve 'an objective analysis'. Where this is phrased in the language of a 'more objective analysis', this is more acceptable, and to some extent this may just be a question of rhetoric. But rhetoric is rarely 'mere rhetoric'. Claims to science and objectivity may be 'promotional' in academic terms, but they may also limit, hinder and distort the analysis. There is a real risk of once again filtering out or hiding much of the political, including the political in one's own perspective and method. Here Lacoste-*Hérodote's* original claims to a '*savoir penser l'espace*' seem much more useful.

Almost the lone social theorist discussed (and commended) by Lacoste is Robert Fossaert, whose multi-volume *La Société* (Fossaert 1977–84) was reviewed by Lacoste in *Hérodote* (Lacoste 1978, 1982a). Fossaert also wrote for *Hérodote* (Fossaert 1979), and contributed to the *Dictionnaire* on the linkages between the social sciences and geopolitics. The appeal of Fossaert's work has been his pluralist analysis. Like Lacoste, Fossaert's analysis grows out of a Marxist background, but Fossaert rejects the economicist base of Marxism, arguing for the trinity of politics, economics and ideology in the social formation, and for a nuanced account of different societies that takes account of variations in space and scale. His unusual emphasis on these latter two makes his ideas particularly attractive to Lacoste. However, the enthusiasm for Fossaert's social science finds little expression in the actual practices of Lacoste-*Hérodote*.

Why is this? It flows from the stance taken by Lacoste-*Hérodote* on theory, seen through Marxist or positivistic lenses as highly reductionist. Yet theory in the social sciences need not be of this form. Most social theories are much more modest affairs, attempts to generalize, gain insights by limited abstraction, nuanced and partial in applicability. If the world is highly diverse and complex, any analysis can only be partial – not just theoretical analyses – and we have to make use of abstractions and theoretical constructs to see through the mass of detail. 'Theory', of these sorts, is needed to navigate these complicated waters or, to change the metaphor, to see the wood through the trees. Most theory in the social sciences today is of this type, and does not make the grand claims of classical Marxism or positivistic social science.

This role of 'theory' is the central divide between Anglophone critical geopolitics and Lacoste-*Hérodote*, a divide that seems hard to bridge. For the Anglophone, the French geographers' neglect of their local post-structuralist social theorists seems perverse; on the other side, one suspects Lacoste-*Hérodote* are quite amused by this appeal of French social theorists, which they may well see as an ivory-tower distraction from serious geopolitical analysis (and possibly from serious politics too). But there are possible linkages at lower theoretical levels in the social sciences, such as the very extensive literature on forms of nationalism, theories of separatism, representations of cultural and ethnic identity, which would be very relevant to Lacoste-*Hérodote's* concerns, and also relate to recent Anglophone critical perspectives (e.g. the essays in Ó Tuathail and Dalby 1998, or Dijkink 1996). These issues need much fuller consideration elsewhere, but they would provide a starting point.

Hérodote, and Lacoste's writings, remain an intriguing blend of the radical and the conservative, and it may be that this blending reflects not only the intellectual stance of the Lacoste-*Hérodote* group, but also a response to the specific context of French geography and academic life in the 1970s. To the outsider the sweep of Lacoste-*Hérodote's* geographicité and 'geography as a political knowledge' is radical, but the geographical reasoning is conservative and defensive. Perhaps one was the practical price of the other: in the French universities and schools of the 1970s and 1980s geography had to defend its academic turf (e.g. Lacoste 1986c), and a clearly defined, and differentiated, geographical approach, fending off the social and environmental sciences in particular, was a necessary strategy. Lacoste-*Hérodote* would certainly not see an instrumental trade-off in this way, but it may be part of the explanation. In the contrasting contexts of Anglo-American academia, geographers have perhaps had less need to be defensive about the role of geography, and have taken a less essentialist stance on the nature of geography, being content (and confident enough) to explore the intersections of space with both the social and natural sciences: Lacoste would certainly identify a fragmentation between human and physical geography as the price being paid for this, a price that might threaten the institutional future of geography.

The audience of Hérodote and the scope of Lacoste's project

Any interpretation of Lacoste's (and *Hérodote's*) intellectual project must take account of its scope and ambition. Although substantially targeted at the Francophone academic geographical world of both the universities and the *lycées*, it has also looked outwards to a wider general public and to the political culture in France. *Hérodote* itself, through its strong theme-issue format and its sales in bookshops rather than through standard journal-subscriptions, reaches a general public, much broader than the usual readership for an academic journal, and its appeal is reinforced by being marketed as a geopolitics and geography journal. Books by Lacoste and other past and present members of the *Hérodote* group also spread the message of careful geopolitical analysis (Foucher 1986, 1988; Lacoste (ed.) 1986b; Loyer 1997). In addition Lacoste and his *Hérodote* collaborators have engaged in a number of original publishing ventures. With Francois Gèze and Alfredo Valladão, Lacoste inaugurated the La Découverte/Maspero series of *L'État du Monde 1981, Annuaire économique et géopolitique mondial* (Geze, Lacoste and Valladão 1981). As its title suggests this is an annual review of world events (economic, political and cultural), with up-to-date facts, maps and contexts. With the Lacoste-*Hérodote* involvement, and again as the subtitle suggests, *L'État du Monde* has a strong geographical and geopolitical emphasis. From out of this venture the annual *L'État de la France* has also emerged, and this also has *Junior* and *Cadet* editions and now a CD-ROM version. Less regular versions have also been produced on many other regions, such as *L'État du Maghreb* by Camille and Yves Lacoste. *Hérodote's* ideas and approach are indeed disseminated far beyond the ivory towers of the academy.

Lacoste has also written 'polemical' works as contributions to political debate, and, of all present-day French geographers, he probably has the most claim on the term 'public intellectual'. He contributes to different journals and papers, he is given interviews and profiles by the press, and he appears alongside leading artists and intellectuals in the series of videos 'Ateliers de rencontres' produced by the École Nationale Supérieure des Arts Décoratifs (ENSAD) (in their listing Lacoste appears under 1991, next to Jacques Derrida). Two books in particular have been written for such public debates. His *Contre les anti-tiers-mondistes et contre certains tiers-mondistes* (Lacoste 1985b) was written in direct response to Pascal Bruckner's *Le sanglot de l'homme blanc* and other studies that attacked the notion of the Third World and its need for economic assistance, and his recent *Vive la nation, destin d'une idée géopolitique* (Lacoste 1998) is a contribution to current debates in French politics. Lacoste's other recent book *La légende de la terre* (Lacoste 1996a) should also be noted: it appears to be a 'coffee-table' book, with magnificent illustrations, and in the same Flammarion series as Michel Serres' *La légende des anges*. However, like Serres' work, it is much more than a coffee-table book for the general reader, and it argues powerfully for an encompassing view of geography and the earth, developing the geopolitical perspective from Herodotus' time onwards but also arguing for the role of 'geographical sensibility or viewpoint' (*le sentiment géographique*) in its broadest sense, notably the growth of environmental knowledge and concern.

The wide scope of Lacoste's writings reflects the aims and ambition of his project. The analyses of Lacoste and *Hérodote* are not just aimed at a University and educational readership (as much Anglophone writing on critical geopolitics is), nor at the world of the public and foreign policy professional (which much 'orthodox' Anglophone and Francophone geopolitical analysis is), but also at a wider educated public and at the arena of political debate. Writing for such wider groups is often, at least in English-language geography, derogated as 'mere popularization'. But this is to confuse categories (and to elide such writing into a narrow model of 'scientific popularization'): such writing is different, but it is often only in such writing that important intellectual (and political) arguments can be developed. And any attempt to alter our conceptions of 'the geopolitical' must embrace the popular, the academic and the policy professional, as Lacoste has recognized.

If *Hérodote's* dismissal of 'theory' as a component in geographical reasoning provides the major barrier between its approach and that of Anglophone geographers, the sheer scale of the Lacoste-*Hérodote* contribution, its coherence, its regional detail within a world-wide sweep, its wide-reach readership – beyond both the bounds of geography within academe, and beyond academe itself – and its direct engagement in political debates, all generate great appeal as a model. Does this mean that Anglophone geography should create an equivalent to *Hérodote*, with 'imitation as the sincerest form of flattery'? To be successful such a venture would have to identify and creatively exploit the specific context of Anglo-America today, just as Lacoste and *Hérodote* were created and grew in a specific French context. Undoubtedly such an Anglophone venture would want to explore theory more directly and positively than *Hérodote* has, but it would also have to address issues of a wider readership: scepticism towards the abstractions of social theory has helped keep *Hérodote* accessible to that wider public. It would also be interesting to see how such a venture would express its radical political stance, and how coherent its project could be.

Conclusions

The work of Yves Lacoste and the *Hérodote* group comprises the most substantial and coherent body of contemporary geopolitical analysis. The scale of the achievements of the *Hérodote* school since the 1970s has been immense, and in this essay only a few aspects have been examined in any detail. The *Hérodote* school deserves much more study, particularly by those outside the Francophone world. There are many differences between *Hérodote's* geopolitical approach and those of Anglophone geographers and critical geopolitical analysts, and both groups have much to learn from the other. This is not to argue for merger, but for a critical and hopefully constructive dialogue. The present essay is offered in that spirit. In a period where respect for difference is a value both Francophone and Anglophone geographers would advocate, both sides need to pay more attention to the work of the other group. It is likely that Lacoste-*Hérodote* will not agree with many of the arguments made here, but opening up a debate would be helpful, and it would be interesting to see *Hérodote* review recent work in Anglophone critical geopolitics. Certainly Anglophone geopolitical analysis needs to set a trajectory towards the concerns of the Lacoste-*Hérodote* tradition, addressing the differences between them and learning from them. Difference is to be celebrated for itself, but also because it reveals to all parties their individualities, characters and directions for change and growth.

Notes

All translations from the French are my own, and I am grateful to Jean Hepple for checking them.

1 CRAG has an internet website, for which the current URL address is www.univ-Paris8.fr/geopolitique/ or www.multimania.com/geopolitique/, and the web pages

contain details of the CRAG doctoral programme, the contents of *Hérodote*, details of CRAG theses and publications, together with downloadable copies of some recent CRAG research papers.

2 Douzet's recent study on how the 'Internet geopoliticises the world' (Douzet 1997) is a particularly good example. He examines how access to the internet is highly geographical and geopolitical, how Anglophone domination poses problems for Francophone and other language groups, but also how the net can provide effective linkages for national and local resistance, using the case of anti-Milosevic groups in Serbia as his prime example. His further reference to the Zapatista revolt in the Mexican state of Chiapas and the very effective presentation of the Zapatista case on the internet shows some convergence with recent work in Anglophone critical geopolitics, notably the papers by Ó Tuathail (1997) and Routledge (1998) on this same case-study and role of electronic geopolitics.

3 Foucault's questions concerned the nature of *Hérodote's* project: how did this group of geographers plan to analyse the notion of power, strategy, the relationship between strategy and war, power and domination? Was it possible to construct geographies of medicine? (an intriguing question, given Foucault's developing interests, and one only recently tackled in geopolitical terms in a CRAG thesis by Olivier Lacoste, published in book form (O.Lacoste 1995)). Foucault also asked 'if I understand you correctly, you are searching to construct a knowledge of spaces (savoir des espaces). Is it important for you to construct it as a science?' (Foucault 1976b). The responses came from a total of thirteen geographers. Although diverse, the general tenor was that presented by Lacoste (1976a) and the early issues of *Hérodote*.

Bibliography

Agnew, J. (1998a) 'Territoire et politique dans l'Italie de l'après-guerre', *Hérodote* 89: 105–6.
—— (1998b) *Geopolitics. Re-visioning World Politics*, London: Routledge.
Agnew, J. A. and Corbridge, S. (1995) *Mastering Space: Hegemony, Territory and International Political Economy*, London: Routledge.
Althusser, L. (1965) *Lire le Capital*, Paris: Maspero.
Baker, A. R. H. (1994) 'Evolution de la géographie historique en Grand-Bretagne et en Amérique du Nord', *Hérodote* 74/75: 70–86.
Bernard, O. *et al.* (1977) 'Des réponses aux questions de Michel Foucault', *Hérodote* 6: 5–30.
Boyer, J.-C. (1986) '*Hérodote*: dix ans, l'âge de raison?', *L'Espace Géographique* 4: 297–301.
Broc, N. (1976) '"*Hérodote*" à la sauce tartare', *Annales de Géographie* 85: 503–6.
—— (1983) '*Hérodote* à l'eau de rose', *Annales de Géographie* 92: 708.
Buleon, P. (1992) 'The state of political geography in France in the 1970s and 1980s', *Progress in Human Geography* 16 (1): 24–40.
Carter, F .W. and Norris, H. T. (eds) (1996) *The Changing Shape of the Balkans*, London: UCL Press (SOAS/GRC Geopolitics series).
Cassen, B. (1998) 'La nation contre le nationalisme', *Le Monde Diplomatique*, March 1998 (Internet edition).
Claval, P. (1976) *Essai sur l'evolution de la géographie humaine*, Paris: Les Belles Lettres.
Clout, H. (1985) 'French geography in the 1980s', *Progress in Human Geography* 9: 473–90.

—— (1994) 'La reconstruction de la campagne de la France, 1918–30', *Hérodote* 74/75: 111–26.

Dijkink, G. (1996) *National Identity and Geopolitical Visions, Maps of Pride and Pain*, London: Routledge.

Douzet, F.(1997) 'Internet géopolitise le monde', *Hérodote* 86/87: 222–33.

Dresch, J. (1980) 'Le desert du Lout en Iran. Un desert absolut est-il franchissable?', *Hérodote* 18: 46–56.

Durand, F. (1998) 'Les forêts indonésiennes à l'orée de l'an 2000, un capital en péril', *Hérodote*, 88: 62–75.

Durand, M.-F., Lévy, J. and Retaille, D. (1992) *Le Monde, Espaces et Systemes*, Paris: Dalloz et les Presses de la Fondation nationale des sciences politiques.

Duroy, L. (1998) 'Yves Lacoste, combattant de la géographie', *La Recherche* 306 (Feb. 1998).

Eribon, D. (1991) *Michel Foucault*, London and Boston: Faber and Faber.

Fossaert, R. (1977–84) *La Société*, Paris: Editions du Seuil, 6 volumes.

—— (1979) 'À propos de l'Europe, de Marx et de la géographie', *Hérodote* 14–15: 171–211.

Foucault, M. (1976a) 'Questions à Michel Foucault sur la géographie', *Hérodote* 1: 71–85.

—— (1976b) 'Des questions de Michel Foucault à *Hérodote*', *Hérodote* 3: 9–10.

—— (1978) *Histoire de la Sexualité, tome 1 La Volonté de Savoir*, Paris: Éditions Gallimard.

—— (1980) *Power/Knowledge. Selected Interviews and Other Writings 1972–1977* (edited by C. Gordon), New York and London: Harvester Wheatsheaf.

Foucher, M. (1986) *L'Invention des Frontières*, Paris: Fondation pour les Études de Défense Nationale.

—— (1988) *Fronts et Frontières. Un Tour du Monde Géopolitique*, Paris: Fayard.

George, P., Guglielmo, R., Kayser, B. and Lacoste, Y. (1964) *La Géographie Active*, Paris: Presses Universitaires de France.

Geze, F., Lacoste, Y. and Valladão, A. (1981) *L'État du Monde 1981, Annuaire économique et géopolitique mondial*, Paris: Éditions La Découverte.

Giblin, B. (1981) Elisée Reclus et les colonisations, *Hérodote* 22: 56–79.

—— (1985) '*Hérodote*, une géographie géopolitique', *Cahiers de Géographie du Quebec* 29: 283–94.

Giblin-Delvallet, B. (1995) 'Les effets de discours du grand chorémateur et leurs conséquences politiques', *Hérodote* 76: 22–38.

—— (1997) 'Le racisme contre la nation', *Hérodote* 85: 3–8.

Girot, P. and Kofman, E. (eds) (1987) *International Geopolitical Analysis. A Selection from Hérodote*, London: Croom Helm.

Heldke, L. M. (1992) 'Foodmaking as a thoughtful practice', in Curtin, D. W.and Heldke, L. M. (eds) *Cooking, Eating, Thinking. Transformative philosophies of food*, 203–29, Bloomington and Indianapolis, Ind.: Indiana University Press.

Hepple, L. W. (1986) 'The revival of geopolitics', *Political Geography Quarterly* vol. 5 supplement: 21–36.

Korinman, M. (1983) 'Friedrich Ratzel et la Politische Geographie', *Hérodote*, 28: 128–40.

—— (1990) *Quand l'Allemagne Pensait Le Monde*, Paris: Fayard.

—— (1991) *Continents Perdus: Les Précurseurs de la Géopolitique Allemande*, Paris: Economica.

Lacoste, O. (1995) *Géopolitique de la Santé. Le cas du Nord-Pas du Calais*, Paris: Recherches.

Lacoste, Y. (1959) *Les Pays Sous-Développée*, Paris: PUF.

—— (1965) *Géographie du Sous-développement*, Paris: PUF.

—— (1966) *Ibn Khaldoun. Naissance de l'Histoire, Passé du Tiers Monde*, Paris: Collection 'Textes a L'Appui' (2nd edition, 1981).

—— (1973a) 'La Géographie', 242–302 in F. Chatelet (ed.) *La Philosophie des Sciences Sociales (Histoire de la Philosophie, tome 7)*, Paris: Hachette-Litterature.

——. (1973b) 'An illustration of geographical warfare: bombing of the dikes on the Red River, North Vietnam', *Antipode* 5 (2), 1–13.

—— (1976a) *La Géographie, ça sert, d'abord, à faire la guerre*, Paris: Maspero (1st edition).

—— (1976b) 'Enquête sur le bombardement des digues du fleuve Rouge (Vietnam, été 1972). Méthode d'analyse et reflexions d'ensemble', *Hérodote* 1: 86–117.

—— (1976c: 3rd edition) *Géographie du Sous-développement*, Paris: PUF.

—— (1977a) 'An illustration of geographical warfare: bombing of the dikes on the Red River, North Vietnam', 244–61 in R. Peet (ed.) *Radical Geography*, London: Methuen.

—— (1977b) 'Self-critical reflections and critique of "A Geography of Underdevelopment"', *Antipode* 9 (3), 117–24.

—— (1978) 'Robert Fossaert, *La Société*', *Hérodote* 10: 155–9.

—— (1979) 'À bas Vidal. . . Viva Vidal!', *Hérodote* 16: 68–81.

——(1980) *Unité et Diversité du Tiers Monde. Des Représentations Planétaires aux Stratégies sur le Terrain*, Paris: Hérodote (Maspero).

—— (1981a) 'Géographicité et géopolitique: Elisée Reclus', *Hérodote* 22: 14–55.

—— (1981b) '*Hérodote* à lu: . . . Pour une géographie du pouvoir', *Hérodote* 22: 149–57.

—— (1982a) '*Hérodote* à lu: Robert Fossaert: les tomes 3–4–5 de *La Société*', *Hérodote*, 25: 152–6.

—— (1982b: 5th edition) *Géographie du Sous-développement*, Paris: PUF.

—— (1983) 'Editorial', *Hérodote* 28: 3–5.

—— (1984a) 'Les géographes, l'action et le politique', *Hérodote* 33–4: 3–32.

—— (1984b) 'Geography and foreign policy', *SAIS Review* 4: 213–27.

—— (1985a) *La Géographie, ça sert, d'abord, à faire la guerre*, Paris: Éditions La Découverte (3rd edition).

—— (1985b) *Contre les anti-Tiers-mondistes et contre certains Tiers-mondistes*, Paris: Éditions La Découverte.

—— (1986a) 'Braudel géographe', *Hérodote* 40: 161–5.

—— (ed.) (1986b) *Géopolitiques de la France* (3 volumes), Paris: Fayard.

—— (1986c) 'Penser et enseigner la geographie', *L'Espace Géographique* 1: 24–7.

—— (1988) *Questions de Géopolitique*, Paris: Le Livre de Poche.

—— (1990) *Geographie und politisches Handeln. Perspektiven einer neuen Geopolitik*, Berlin: Verlag Klaus Wagenbach.

—— (1992) 'Editorial: la question serbe et la question allemande', *Hérodote*, 67: 3–48.

—— (ed.) (1993a) *Dictionnaire de Géopolitique*, Paris: Flammarion.

—— (1993b) 'La question allemande', *Hérodote* 68: 3–17.

—— (1993c) 'Debat: chorématique et géopolitique', *Hérodote* 69/70: 224–59.

—— (1994) 'Nation, nations, nationalistes', *Hérodote* 72/73: 3–8.

—— (1995) 'Les géographes, la Science et l'illusion', *Hérodote*, 76: 3–21.

—— (1996a) *La Légende de la Terre*, Paris: Flammarion.

—— (1996b) '*Hérodote* à lu: Claude Raffestin, Dario Lopreno, Yvan Pasteur, *Géopol-itique et Histoire*', *Hérodote* 80: 204–8.

—— (1997) 'La République et la Nation: quelques réflexions géopolitiques', *Géopol-itique* 60: 60–5.

—— (1998) *Vive la Nation: Destin d'une Idée Géopolitique*, Paris: Fayard.

——, Nouschi, A. and Prenant, A. (1960) *L'Algerie, Passé Present*, Paris: Editions Sociales.

Loyer, B. (1997) *Géopolitique du Pay basque. Nations et nationalismes en Espagne*, Paris: L'Harmattan.

Oliva, J.-C. (1998) 'Nation: une idée neuve pour un vieux monde', *Regards*, April 1998 (Internet edition).

O'Loughlin, J. (ed.) (1994) *Dictionary of Geopolitics*, Westport, Conn.: Greenwood Press.

Ó Tuathail, G. (1994) 'The critical reading/writing of geopolitics: re-reading/writing Wittfogel, Bowman, Lacoste', *Progress in Human Geography* 18,3: 313–32.

——. (1996) *Critical Geopolitics*, London: Routledge.

——. (1997) 'Emerging markets and other simulations: Mexico, the Chiapas revolt and the geofinancial panopticon', *Ecumene* 4: 300–17.

Ó Tuathail, G. and Dalby, S. (eds) (1998) *Rethinking Geopolitics*, London: Routledge.

Ó Tuathail, G., Dalby, S. and Routledge, P. (eds) (1998) *The Geopolitics Reader*, London: Routledge.

Parker, G. (1998) *Geopolitics. Past, Present and Future*, London: Pinter.

Raffestin, C. (1980) *Pour une Géographie du Pouvoir*, Paris: Litec.

Raffestin, C., Lopreno, D. and Pasteur, Y. (1995) *Géopolitique et Histoire*, Paris: Payot.

Reclus, E. (1982) *L'homme et la terre*, Paris: Maspero/La Découverte. 2 volumes (Edited with introductory essay by B. Giblin).

Routledge, P. (1998) 'Going global. Spatiality, embodiment, and mediation in the Zapatista emergency', in G. Ó Tuathail and S. Dalby (eds) *Rethinking Geopolitics*, 240–60, London: Routledge.

Roux, M. (1982) 'Le Kosovo: développement regional et integration nationale en Yougo-slavie', *Hérodote* 25: 10–48.

Sanguin, A.-L. (1993) *Vidal De La Blache. Un genie de la géographie*, Paris: Editions Belin.

Taylor, P. J. (1990) *Britain and the Cold War. 1945 as Geopolitical Transition*, London: Pinter.

Tunander, O., Baev, P. and Einagel, V. I. (eds) (1997) *Geopolitics in Post-Wall Europe. Security, Territory and Identity*, Oslo and London: PRIO (International Peace Research Institute) and Sage.

Vichnevski, A. (1994) 'Le nationalisme russe: à la recherche du totalitarisme perdu', *Hérodote* 72/73, 101–18.

12

CITIZENSHIP, IDENTITY AND LOCATION

The changing discourse of Israeli geopolitics

David Newman

Introduction

States do not occupy a single place within an unchanging geopolitical structure. The geopolitical imagination of the political elites, the residents and citizens, and other groups whose fate is tied up with that of the state, reflect alternative locations within the regional and global setting. The collective imagination of the state, to the extent that it represents the aggregate collective identity of its diverse components, is itself a composition of the individual imaginations of the residents and citizens of that state. The degree to which an individual identifies with the state ethos, sees him/herself as an equal citizen, as a member of the majority or minority groups and/or as a member of the global village, will determine the way in which he/she perceives the location of the state as part of the changing global community (Soysal 1996; Yuval-Davis 1997). The more internally homogeneous is the composition of a state's population and its alternative identities, the less diverse the geopolitical imaginations. The more heterogeneous a population, the more diverse the varied forms of local, national and regional identities and, hence, the positioning within the global system. This becomes all the more diverse as boundaries – both social and spatial – are opened up, as information is disseminated through cyberspace and satellites, as travel restrictions are eased, as diaspora populations become closely linked to 'homeland' populations, and as increasing numbers of migrant workers arrive to take their place within the socio-economic system (Brunn, Jones and Purcell 1994; Morley and Robins 1995; Soysal 1996).

While the geopolitical imagination of the state may be determined from within, its actual positioning within the regional and global systems is largely determined from without. The international community and regional superpowers perceive other component parts of the state system according to their own geopolitical imaginations and determine the extent to which they are prepared to include other states within economic and/or political unions. General perceptions of the world and the 'other' are the result of generations of national socialization in which the 'self' society views itself as the superior culture, the civilizer of others, and by whose definition all others are defined and located within a global hierarchical system. The strength of national identity is something which is taken for granted, an imagined community, and which seeks to reproduce itself elsewhere through a series of ideological practices (Anderson 1983; Billig 1995). Notions of Orientalism, Afrocentrism and the spatial constructs of east and west are good examples of how specific perceptions of the world, for which the point of departure is the national identity of the self, feed into the historical formation of the geographical imagination and are responsible for the ordering of the global political system at any one point in time (Said 1979; Lewis and Wigen 1997). Thus boundaries of inclusion and exclusion, the definition of the 'in' group and the 'other' are as relevant to states within the world system as they are to ethnic and social groups residing within the state (Newman and Paasi 1998). The geopolitical imagination of a country's population or political elites may often contrast with the geopolitical positioning of that state by other states within the system, resulting in inter-state tension on the one hand, and attempts to become accepted on the other. Peer power is important in this respect, with powerful states or regional groupings, such as the United States or the European Union, determining the extent to which other states can gain entry, partial or full, to their own exclusive geopolitical clubs.

Critical geopolitics is concerned with the discursive nature of geopolitical manoeuvring and the way in which global positioning is represented and contested. Agnew and Corbridge (1995) note that the identity and interests of states are formed in interaction with one another. But this interaction itself derives from a combination of realist outlooks by the political elites and leaders concerning the world system as well as, if not more than, the domestic interests of the population (Telhami 1996). One of the main arguments of this chapter is that in order to understand the geopolitical positioning of any country, it is essential to understand the internal discourse of identity of that country's citizens. It is an approach which looks at geopolitics from below, from within the state, rather than the from the perspective of the global system.

It is also an approach which reconciles the geographies of political scale, in each of which the state continues to constitute the central institutional locus (Cox 1998). It allows us to understand the role of the state in the global system through a discursive analysis of the internal components which go to making up that state. Local politics is increasingly influenced by processes of economic and information globalization, while equally global positioning is influenced by local and domestic interests (Smith 1998). Thus the social production of a geopolitics discourse takes place at the levels of local, regional and national identity formation, such that it is impossible to understand the former without the latter. It is the politics of identity which creates and produces the geopolitical knowledge and/or perceptions of the state's elites (Morley and Robins 1995; Dalby and Ó Tuathail 1996).

The geopolitical discourse of any country will vary over time as both the internal identities of the population and the global positioning of the State – the latter representing some form of aggregate collective identity – undergo change. The two are related inasmuch as the imagined national identities of the individuals will influence the way in which the political elites view the role of the state in regional and global affairs. Notwithstanding, the geopolitical positioning of the state *vis-à-vis* the rest of the world may only represent some of the constituent parts of the population, those who are closest to the decision-makers and political elites. Thus, some geopolitical representations, professed by minority groups, are not reflected in the wider geopolitical locus of the state. This assumes that the foreign policy of the state is based, at least partially, on the impact of domestic interests, rather than on a realist perspective which would assume an objective analysis of the state's function in the international system (Telhami 1996).

In this respect, Israel provides an interesting case-study of a country whose geopolitical positioning is diverse and has undergone change over time, but always displays a strong component of domestic interests: a country, recently formed, established by European immigrants in a region in which an Islamic culture dominates, drawing on its links with the global Jewish diaspora as a basic construct of collective identity and as a means of support. The ideology of state formation is viewed by its adherents as an ideology of national renaissance and emancipation after centuries of minority status and geographic dispersion, while at the same time it is seen by its opponents as a form of colonialism through which indigenous peoples have been uprooted and dislocated as a result of European immigration. The traditional geopolitical discourse in Israel has been based around two interrelated themes. First, the State of Israel is a Jewish state and, as such, retains cultural and religious characteristics that are unique to the normative behaviour of the state and its Jewish citizens. It also accords preferential behaviour to members of the Jewish people, world-wide, in terms of immigration, citizenship, the rights to purchase land and build new settlements. Second, Israel sees itself as being isolated and beleaguered in a region of anti-Israel animosity. The existential threat facing the state and its collective citizenry is part of the geopolitical imagination in which the state has to remain militarily strong and to persuade the major world powers, especially the United States, to continue to support the 'only democracy' in the Middle East. It is this traditional discourse which is now being questioned, both in terms of the way the State is defined from within by its citizenry, and from without by the other members of the international community.

In a world in which many boundaries are being opened and some ethno-territorial conflicts are being resolved, Israel finds itself at one and the same time becoming part of a global economic and information system but still tied up in a cycle of conflict and ethnic particularism. While the notion of an imagined national community remains strong, due in no small part to processes of social construction (Anderson 1983; Doty 1996), the changing nature of collective identity within Israel and amongst the Jewish diaspora has meant that alternative and diverse visions of Israel's geopolitical position are emerging, relating to both notions of citizenship and identity and the question of *who is an Israeli* on the one hand, and equally to issues of relative location and the question of *Where is Israel* on the other. Each of these two issues is complex and has to be understood within the context of overlapping and multiple layers, in which there is no single collective identity or location for the country, despite its self-presentation as an internally unified Jewish state. Questions of space and time are inter-linked as issues of individual and collective identity respond to the social and political changes overtaking Israel and transforming it into an increasingly heterogeneous society. The boundaries separating the 'in' group and the 'other' are not as clear-cut as in the past, while notions of inclusion and exclusion have become multi- rather than uni- dimensional.

Who is an Israeli? Issues of citizenship and identity

Who and what is an Israeli is a complex question of identity. At the simplest level, an Israeli is a person who holds Israeli citizenship, by virtue of having been born in the country or having immigrated from elsewhere and taken on citizenship. But the question of Israeli identity is also tied up with Jewish–Arab relations and majority–minority status, as well as with a variety of ideological and cultural/religious considerations (Yuval-Davis 1997). The geopolitical imagination and positioning of the country is, to a great extent, dependent on the way in which the individual identities are defined and understood, both internally (by the residents of the country) and externally (by other countries in the global system). The fact that a decreasing number of its citizens identify with the single, socially constructed, national ethos of Zionism is a testament to the fact that Israel has become a far more heterogeneous society on the one hand, and increasingly critical in its search for alternative forms of meaning and identity (Shefer 1996; Ram 1998a, 1998b). This, in turn, will change the nature of the collective, geopolitical, identity of the State as a player within the global system.

A Jewish state or a state of its citizens: the post-Zionist discourse

Israel as a state is self defined as both a nation (Jewish) state, a homeland for the Jewish people, and a democratic state in which all citizens are, on paper, equal under the law. In reality, this dual definition has proved, over fifty years, to be a

ADA.D VID N N EWMAN

structural contradiction. By giving special status to the Jewish majority and all
Jewish immigrants, the 20 per cent Arab minority of the country have not fully
partaken in the fruits of democracy and equality. This is displayed in the lower
socio-economic levels of development, the unequal allocation of scarce public
resources to Arab and Jewish sectors, and the fact that Arab leaders and representatives do not enjoy the same access to the corridors of power as do their Jewish
counterparts, nor have they ever become full members of Israel cabinets.

The *raison d'être* of the State, as expressed through Zionism as an ideology of
state formation, has been the necessity for an independent and strong Jewish
homeland. This, in turn, was tied in with utopian concepts of 'returning to the
land', the creation of egalitarian rural co-operatives and the renaissance of a
dormant national life. The symbols of statehood were uniquely Jewish and Zionist,
while notions of territorial attachment and spatial exclusivity were an important
part of the educational and socialization process promoted by the State during its
fifty years of existence.

Jewish residents of Israel have had to define themselves in relation to three
identities: an Israeli, a Jew and/or a Zionist. What makes the identity problem
even more complex is the fact that none of these components of identity have a
single meaning. They are interpreted differently by diverse groups within Israeli
society. For some, being Jewish means adhering to the strict letter of the laws of
orthodoxy, while for others being Jewish is a form of cultural attachment which
does not require the unquestioning belief in a Divine role or adherence to stringent customs and ritual. Equally, some groups define a Zionist as being a member
of an exclusive club who has chosen to live in Israel, serves in its army and does
not question the political decisions made by the state concerning the nature of the
Arab–Israel conflict. For some, attempts to reach conflict resolution with Arab
neighbours or with the Palestinians which would entail territorial withdrawal are
nothing short of 'anti-Zionist' and treacherous, while for others the desire to
retain control of every last centimetre of occupied territory is a betrayal of the
Zionist tradition of compromise for the sake of a secure future.

Even amongst those who define themselves as Zionist – the vast majority of the
population – new sub-identities have emerged. These include the separate
Ashkenazi (Jews of European extraction) and *Mizrahi* (Jews of North African and
Middle Eastern extraction) identities and the feelings of social and economic
discrimination felt by the latter (Yiftachel 1997a, 1997b). Gender identities and
empowerment are only just beginning to emerge, although these are not
expressed through the same form of political mobilization as has recently been
practised by both *Mizrahi* and religious Jewish groups (Fogiel-Bijaoui 1997;
Dahan-Kalev 1997; Fenster 1997).

For many religious Jews, the Jewish identity is all encompassing, with Zionism
being no more than one sub-construct of this cultural/religious identity, and

306

Israeli simply being a coincidence of place and time. Equally, for ultra-orthodox groups, the very notion of a Jewish state which is not a fundamentalist theocracy is unacceptable, leading to their total rejection of Zionism as an ideology of state formation which is relevant to the Jewish people. Other religious groups, notably the West Bank settlers, have adopted a neo-Zionist position, focusing on an ideology which places territorial irredentism as a supreme value. The paradox is, therefore, that for some religious groups, Zionism is sacred, while for others it is profane. Their respective subjective interpretations of a common history and culture and the way in which these are imbued with contemporary significance are entirely different.

For some secular Israelis, still the large majority of the country's residents, Zionism is no less than an attempt to create and mould the new Jewish identity, one which is not tied to the traditions of old, but which is an integral part of a modern, emancipated, world. But at the same time they face their own identity dilemma when rejecting the historic and religious link on the one hand, but are not prepared to accept the alternative version of being no more than another example of European colonialists settling in a far-off location and bringing about the partial dislocation of the indigenous population. Their search for identity is tied up with a search for a rationale justifying their presence in this region (Newman in press).

These diverse forms of multi-identity overlap as Israelis spend much of their time trying to work out just who they are. Their relationship with the wider world, their geopolitical positioning, is itself an outcome of this identity game. For some, Israel's unique identity is dependent on it remaining different and isolated from all other countries, while for others State normalcy within an international system can only be achieved by becoming part of that system. Both Jewish and Zionist identities are tied up with this struggle between their respective particularistic and universal characteristics, reflecting diverse forms of citizenship on the one hand, and geopolitical imagination on the other.

Minority identities which do not conform with the single state-constructed Zionist ethos abound within Israel (Peled 1992; Kook 1996; Yiftachel 1997a, 1997b). These range from the Arab-Palestinian residents who make up approximately 20 per cent of the country's population, ultra-religious populations who do not recognize the legitimacy of a secular Zionist state, non-Jewish identities, particularly amongst the hundreds of thousands of Russian immigrants who arrived in Israel during the early part of the 1990s, and most recently over a quarter of a million migrant workers – from places as far afield as the Philippines, Rumania, Africa – who have come to fill many of the menial jobs vacated by the Palestinian population of the West Bank and Gaza Strip (Peled 1992). These groups have begun to form the cores of small, but vibrant, ethnic concentrations in South Tel Aviv which may well signal the growth of future ethnic neighbourhoods in this city, an element previously unknown in the Israeli urban landscape.

The importance of these multi-layered identities for the geopolitical imagination are such that some of these groups identify first and foremost with the state, while others identify primarily with non-state groupings. The post-Zionist discourse which has emerged in recent years argues that for the state to function under conditions of normalcy, then the state itself has to undergo a process of redefinition, from one which places its role as a Jewish state as its primary form of identity, and hence excluding large groups of residents from this single *raison d'être*, to one which is defined as a state for all its citizens (Cohen 1989; Peled and Shafir 1996; Ram 1998a, 1998b; Silberstein 1999). Only the latter can, it is argued, be a true participatory democracy. For the political elites, this is seen as a negation of the state formation process and as being anti, rather than post, Zionist in its orientation. Such a transformation would, by necessity, result in changed geopolitical orientations on the part of the political elites.

One of the paradoxes of recent Israeli history is the fact that the right-wing government which came to power in 1996 and which, like right-wing governments throughout the world, perceived itself as being more loyal to the notions of nationalism and patriotism, is the closest there has ever been to a post-Zionist government in fifty years of State history. It relies greatly for its support on ultra-orthodox non-(some would say anti-) Zionist, and right-wing neo-Zionist, parties, for its support. While not prepared to co-opt politicians from Arab parties (neither was the previous Labour government prepared to), the Netanyahu government, until its fall from power in 1999, had more representatives from parties who did not accept the single socially constructed state ethos of secular Zionism, than any previous Israeli administration.

The notion of Israel as a state of all its citizens rather than a Zionist state could also bring about a redefinition of the basic core of the Arab–Israel conflict. Public surveys continually show that the least desired solution to the conflict for both Arabs and Jews is that of a single binational democratic entity. Since the aspiration for statehood is defined only in terms of national dominance, both Jews and Arabs prefer the creation of separate nation-states rather than a single binational or multi-national entity. The perception by each of the national groups of the 'other' as constituting the basic existential threat facing the 'self' is at the heart of the security fears thrown up by the political elites, and which enables the creation of a socially constructed form of national unity which forms the lowest common denominator of the collective identity, namely the fear of the outsider. Since the outsider constitutes a threat, it therefore follows that he/she cannot live as an equal because he/she cannot be trusted. The solution is therefore to create separate, ethnically homogeneous states, in which citizenship is defined by national identity, or to maintain a situation in which one national group retains military and economic dominance over the other, and in which *de facto* citizenship is not expressed through equal rights to all groups. While a redefinition of the state in terms of all of its citizens is aimed at creating a non-ideological and non-exclusive form of identification with the state on the part of all of its citizens, it is also a means by which the Israel–Palestine conflict could conceivably be solved within the existing territorial frame as a binational entity. This, however, is a solution which is rejected by the vast majority of Israelis and Palestinians alike, each preferring a separate national entity.

Security as national identity

The geopolitical discourse in Israel has, for the past fifty years, evolved around notions of security and collective safety. Focusing on the existential threat, real or perceived, facing Israel from hostile neighbours, the internal discourse has used this as a means of creating national consensus and in socializing generations of Israeli youth to be prepared to fight, and even lay down their lives, for the defence of the homeland. With the growth in internal diversity and ideological polarization within Israeli-Jewish society, it is the Israel–Palestine conflict and the security threat which has remained the single main cement holding these diverse strands together (Shain, 1997).

The notion that that Israel is the only safe haven against a second Holocaust and that all of its (Jewish) citizens have to display unquestioning loyalty to the state's fight for survival in the face of a hostile region have been central to the socialization processes of the state during the past fifty years. There are two major themes in this discourse. The first is that of the Holocaust and its remembrance is an indication of what happened, and could happen again, when the Jewish people were stateless and were unable to defend themselves. The second is that of the need for an independent Jewish state with a strong military deterrence to ensure that 'never again' will such an event be allowed to take place. Visits by thousands of Israeli schoolchildren to the sites of the Nazi atrocities in eastern Europe, sponsored by the Israeli Ministry of Education, and accompanied with the raising of the Israeli flag and the prominence of additional national symbols, has become a powerful means through which this dual message is transmitted. In many cases, it has become the single most powerful factor bonding the child with the fate of the state and impressing upon him/her the need to defend the state against all external threat, real or perceived.

Traditionally, Israel has perceived itself as being a lone and isolated player on the world stage. In terms of the historical discourse, this is partly due to the persecution experiences of Diaspora Jewish communities over centuries of dispersion, culminating in the evil of the Holocaust during the Second World War. This conception has also evolved as part of the reality of statehood, during which period the country has been involved in five major wars, most of which have been perceived as defensive wars against external aggression and which guarantee the continued existence of the State in a hostile region. Israel also sees itself isolated within the major international forum, the United Nations, where vote after vote condemns the country for its continued occupation of the West Bank and the security zone in Southern Lebanon. The 'Zionism = racism' vote which was passed in the mid-1970s, later to be rescinded in the early 1990s, only served to strengthen this feeling of isolation, despite the fact that Israel draws its major legal justification to sovereignty from the United Nations partition resolution of November 1947. The major philosophical context behind this form of geopolitical isolationism is best summed up in a book by former Israeli diplomat, Jacob Herzog, entitled 'Behold a People who Dwell Alone', a title taken from the Biblical description of the Jewish people by the prophet Balaam (Herzog 1975). The contemporary political manifestations of this discourse are themselves summed up in a book written by Herzog's brother, later to be Sixth President of the State of Israel, Chaim Herzog. As ambassador to the United Nations at the time of the Zionism=Racism vote, he publicly tore up the piece of paper on which the resolution was printed, thus demonstrating his disdain for an international community which did not recognize the 'noble' and 'state formation' characteristics of Zionism, as compared to the 'colonial' and exploitative characteristics emphasized by its detractors (Herzog 1978).

The policy implications of this self-perceived isolation is that the country can only rely on itself, through a strong military posture, and that it must maintain an independent foreign policy without external intervention (including that of the USA) in its security decision-making. Defending the homeland against all comers became the ultimate form of heroism, with the ultimate sacrifice being the willingness to die for the defence of the country, a phenomenon which became known as the Masada complex, after the 'heroic' suicides of the warriors at Masada in the face of the Roman attempts to put down their rebellion in the first century AD. Young Israeli soldiers are often taken to the Masada hilltop overlooking the Dead Sea for their swearing-in ceremony, where they declare in unison that 'Masada will not fall again' (Zerubavel 1996). It is also a place where Jews from throughout the world will come to have family occasions, notably a bar-mitzvah ceremony, due to its association with Jewish heroism and, hence, the responsibility incumbent upon the youth to live up to these standards of defending one's people and one's country. In contemporary Israeli history, this has been equated with the mythical declaration of a Jewish pioneer soldier, Joseph Trumpeldour, who was killed while defending a new settlement outpost, Tel Hai, in the 1920s. As he lay dying, he is reputed to have cried out: 'It is good to die for my country', a slogan which has been used, together with that of Masada, to focus on military heroism and bravery, often at the expense of other social, cultural and moral messages.

Loyalty to the common cause is expressed through the role of the army, or as it is known the 'defence forces', which has traditionally been viewed as the all encompassing and unifying factor within Israeli society (Popper 1998). The notion that there were wars in which Israel took part during its fifty years of history which were not forced upon the country as an act of self defence is a subject of heated debate. While the 1948 War of Independence, the 1967 Six Day War and the 1973 Yom Kippur War are commonly seen as legitimate acts of self-defence, the same is not the case regarding the invasion of Lebanon in 1982 or the continued occupation of the West Bank through the 1980s and 1990s. The army itself has become more politicized over time, although it still remains the major institution of consensus. Young adults who finish their army service have, tradi-tionally, been eligible for a range of financial benefits, notably better mortgages, than those who have not served. These policies serve to exclude, amongst others, the Arab citizens of the country, the majority of whom do not serve in the army, but neither have been requested to do so by the state which fears the influx of such a large group of people whose basic loyalty to the state remains under question.

Events of the past decade, notably the Palestinian *intifada* (civil uprising against the Israeli occupation), the 1991 Gulf War which brought Iraqi missiles into the heart of Israel's cities, and the attempts at conflict resolution with Egypt, Jordan and the Palestinians respectively, have only served to strengthen this traditional security discourse (Newman 1997, 1998a). Each is used as part of a socially constructed message to show, both to Israeli citizens at home and the international community, that Israel is threatened and must take measures to defend itself (Falah and Newman 1995; Bar-Tal, Jacobson and Klieman 1998).

The major paradox in this continued sense of security identity is the fact that Israel displays its obvious military strength and superiority while, at the same time, emphasizing the security threat as part of a national discourse aimed at justifying actions and policies which would not normally be supported by the global community. Thus Israel's military strengths, its continued occupation of much of the West Bank and the Golan Heights and the security zone in Southern Lebanon, its' requests for international aid and assistance are interpreted as an outcome of its structural weakness, rather than as the basis of a strong country, as much of the world perceives it. To be strong, one must portray oneself as weak, both internally and externally. By presenting itself as threatened and isolated, the state is able to create a strong military deterrent with the direct assistance of the majority of the population who identify with the need to collec-tively combat the perceived threat. The fact that the perceived threat is based on the objective reality of wars ever since the establishment of the state in 1948 means that the subjective interpretation of that reality continues to be one in which the state will forever face a threat to its very existence. This, despite the fact that the threat facing the collective, as contrasted with the threat facing the individual in the case of a terrorist attack, virtually ceased to exist after the Six

Day War in 1967, and certainly after the implementation of the Israeli–Egyptian peace accords in the early 1980s. While the scud missiles of 1990, and the terrorist bombing of civilian centres within Israel, have not removed the threat altogether, these do not represent the sort of threat which can wipe the state out of existence. But they serve to remind the Israeli population of the threat environment within which they live, making them suspicious of any moves towards a peace process which could, but at the same time may not, bring about the end to an atmosphere of continual violence. So much of the daily discourse is tied up with conflict, threat and security that it is almost impossible for most Israelis to come to grips with the concept of a post-conflict Israel, one in which the main social and political agenda switches away from issues of defence and security and turns its attention to pressing welfare, economic and cultural problems facing society from within.

Contextually, the Israeli government came under strong pressure from within, in 1991, to retaliate for the Iraqi missiles which were fired into the heart of Israel's metropolitan centres, rather than acquiesce to the United States pressure not to take any action. This was seen as much as an ideological issue as it was a purely military one. On the one hand, Israel was seen to bow to external pressure rather than make its own independent decisions concerning its defence priorities. In addition, many saw the country as having professed a weakness to the outside world in the sense that it had been attacked by missiles and had not been able to retaliate for fear, perhaps, that it would not have succeeded in its mission. When, for a short time, in early 1998, there was renewed tension between the United States and Iraq over the issue of the United Nations armaments inspectors, there was a strong feeling within Israel that, if fired upon by Iraq, the government would not hesitate in retaliating, if only to prove that the country was able to make its own decisions and that it was capable of launching an attack, whether or not such an attack would have served the long-term security concerns of the country.

Both geography and demography play an important role in the social construction of the security discourse. The small size of the country in relation to the rest of the Middle East is an important geo-regional component of this discourse. The notion of a minute Israel, in terms of both territory and population, is used as a means of portraying the image of the isolated and besieged nation. For propaganda purposes, the map of Israel is often over-laid on any one of the individual states or provinces of North America as a means of stressing this message (Newman 1997, 1998a). Internally too, the scale dimensions of a small territory and the location of antagonistic populations in upland areas overlooking the Israeli towns and population centres is another important means through which the discourse of fear and threat is used in an attempt to create a single national unifying identity around notions of security.

Demographically, Israel and the Jewish people are portrayed as a small nation surrounded by a hostile Arab population undergoing much faster rates of natural growth. Within Israel and the Occupied Territories, the Jewish–Arab ratio is approximately 65:35, changing to 81:19 respectively within the pre-1967 boundaries of the State of Israel. Israeli non-annexation of the West Bank and Gaza Strip prior to the Oslo Agreements was traditionally explained as being due to the implications of granting full citizenship to the Palestinian population of these areas and thus endangering the long-term Jewish majority in the Jewish state. Even within the pre-1967 boundaries, the demographic ratios play an important part in the security discourse, with Jewish immigration to the country being encouraged and extra child benefits being granted to all families with four or more children (Newman 1998b).

The issue of water geopolitics also plays an important part in the security discourse. As a basic existential, but increasingly scarce, resource, water is viewed in political terms (Kliot 1994; Shapland 1997). Any attempt by a neighbouring country to tamper unilaterally with Israel's water supply is, automatically, seen as being a legitimate *casus belli*. Past conflagrations between Israel and Syria have erupted over issues of water diversion, while Israel's invasion of Lebanon in the early 1980s was interpreted, wrongly as it turns out, by some as having been intended to gain control over some of the Litani River headwaters. Israel's negotiations with both Syria and the Palestinians focus strongly on issues relating to the control of water resources, with neither party prepared to give up ultimate control to the benefit of the 'other' side (Elmusa 1994; Shuval 1996), while nearly half the Israel–Jordan peace agreement deals with issues relating to cooperation in the search and utilization of this important resource.

Paradoxically, the implementation of the peace process has brought with it, if anything, a decreased rather than increased sense of security. The fact that suicide bombings occurred after the initial implementation of the Oslo Accords resulted in the strengthening of fears that the ultimate intention of the 'other' side was to destroy the State of Israel, and that territorial withdrawal from parts of the West Bank as part of the peace agreements enabled the creation of a stronger base from which to carry out these threats, as witnessed by the continued attacks. The transfer, albeit partial, of some territory to Palestinian autonomy is seen, by detractors of the peace process, as providing a territorial base for the continued assault on Israel, rather than as a basis for territorial and ethnic separation which could bring about regional peace and stability. It is this continued security threat which was continually thrown up by the post-1996 Netanyahu administration in its attempt to slow down the peace process and to minimize the continued transfer of territory to the Palestinian Authority.

Alternative discourses to those of perpetual conflict are difficult for the Israeli populace to accept. Isolated, besieged, threatened, having to go it alone in the face of a hostile and anti-Semitic world, these are concepts which are easy to grasp. The notion that peace is a concrete and objective reality, rather than some form of metaphysical aspiration, makes it difficult to come to terms with the government attempts to move towards meaningful conflict resolution. At the same time, the futility of continuous warfare and the occupation of another people has started to erode away the single remaining consensus, that of the army. During the past decade there have been instances of soldiers refusing to obey orders, while a decreasing percentage of Israeli youth undertake their obligatory army service (Linn 1994; Helman 1998). The social ethos of army service is no longer the only card of entry into Israeli society, and while the army continues to occupy a prominent place in daily life, it has ceased to constitute the single frame around which Israeli society automatically rallies and gives its unquestioning loyalty (Shain 1997). The decision by a young rock singer, Aviv Gefen, to refuse to undertake military service for no other reason than self-defined incompatibility, and the fact that, despite this 'unpatriotic' behaviour, he was nevertheless invited to appear together with former Prime Minister Rabin – one of Israel's military heroes – at the mass pro-peace rally which took place only minutes before the latter was assassinated, this also gave credence to the notion that the role of the army in Israeli society was beginning to undergo change.

Thus, inasmuch as notions and perceptions of security are central to the conflict, they also play an important role in the formation of separate national identities. This is in direct contrast to the notions of heterogeneity discussed above. The focus on the conflict and the existential threat facing the national collective has covered up much of the internal heterogeneity which is located within the collective. National identity remains unique and particularistic, while ethnic and group identities are much more diverse and, to a certain extent, can even cross national group boundaries, although this is rare. Emphasis on national identity leads, in turn, to certain types of geopolitical imagination, focusing on Israel's isolated and unique position in the global order, while a focus on inter-group differentiation and the fragmentation of the single Zionist state ethos will, by definition, throw up alternative and more diverse geopolitical imaginations.

Where is Israel?

Israel does not have a single geopolitical imagination. It is located, at one and the same time, in a number of diverse locations, not all of which are geographically contiguous. Both the internal and external political discourse which takes place is characteristic of a sort of geopolitical schizophrenia, if not quadrophenia, as this relatively young country attempts to draw together its contrasting global, national and regional identities. Five separate geopolitical imaginations are discussed here, all of which are relevant to different groups within Israel and are partly an outcome of the extent to which the individual identities of the diverse population groups are translated into collective identities and are expressed in terms of the perceived relationship and positioning of the country *vis-à-vis* neighbouring countries, the region as a whole, and the global system. The five geopolitical imaginations are: the Middle East; Europe; the Jewish Diaspora; the United States, and, finally, the centre of the world. These locations are not exclusive. They overlap to the extent that some groups imagine themselves to be part of more than one geopolitical location, although there is a tendency to prioritize certain locations over others.

In the Middle East

Geographically, Israel is located at the western margins of the Middle East. It is perhaps not surprising that the historic centre for two of the world's monotheistic regions – Christianity and Judaism – and an important focus for the third – Islam – should lie at the confluence of Europe, Asia and Africa. Israel's state-building ideology, Zionism, has at its core the 'return' to the ancient homeland of the Jewish people, but not necessarily a 'return' to the regional culture which is characteristic of that particular location. Political Zionism was a national ideology, created in Central and Eastern Europe, implemented by immigrants and settlers from Europe. The desire to be part of the Middle East, often declared by the country's leaders, may be politically correct but does not necessarily coincide with the cultural origins or aspirations of the political elites.

As a state, Israel has always been ambivalent with respect to its standing in the wider region. On the one hand, the desire to be accepted as a political reality and to be recognized by its neighbours, has always been uppermost in much foreign policy thinking in Israel (Susser, 1997). It remains politically correct for Israeli leaders to state their hopes that the country will one day be truly and fully integrated into their geographic region. At the same time, Israel's political and economic elites have always seen themselves as being part of the western world, with a highly technological, post-industrial economy and with a highly educated and literate workforce. Western Europe and North America have proved to be much more attractive, culturally and economically, to the Israeli population than the Middle East.

The regional vision espoused in the post-Oslo period was that of the 'New Middle East'. Drawing on processes of globalization and the expansion of regional economics, former Israeli Prime Minister, Shimon Peres, described the new regional order which was, in his view, about to dawn and in which Israel would finally be accepted and integrated into the geographic region within which it is located (Peres 1994). Prior to this period, Israel had been unable to play any role in the region owing to Arab opposition to the existence of the state. The Peres vision is based around the economic benefits to be attained from peace and regional integration, similar to much of the discourse surrounding the 'end of the nation-state' and the 'disappearance of boundaries' theories which are part of the post-modern discourse but which focus almost entirely on economic integration in Western Europe and North America, without reflecting the very real ethnic and cultural obstacles in the way of full regional integration, especially in those regions (such as the Middle East) in which ethno-territorial identities and conflicts continue to play a major role in everyday life.

Israel's main contribution to the region is in the field of economics. Given its special status with both the European Community and the United States, the only country on the face of the globe to enjoy preferential status with both western economic superpowers, Israel finds itself in a unique situation. Many countries in Asia and Africa see Israel as a country which can provide a backdoor into the world's major trading blocks, in some cases even using Israel's leaders to argue their case before the American political elites. Israeli businessmen themselves are aware of the enormous economic potential that Israel possesses in future dealings with the Arab world if, and when, real peace is to be achieved. At the same time, the economic benefits which could accrue to the Israeli business community from becoming more fully integrated into the region, are not necessarily welcomed by all Arab countries. For many, it is seen as a modern form of neo-colonialism, through which Israel is able to impose its economic advantages on the region, and achieve what they were unable to achieve during forty years of conflict. Thus, as much as Israeli perceptions of its own role within the region remain ambivalent, so too do the perceptions of the neighbouring countries concerning their willing-ness, even under conditions of conflict resolution, to have Israel participate fully in the region's cultural and economic activities.

The extent to which Israel sees itself as part of the Middle East or as the eastern extension of Europe (see discussion below) has to be seen within the context of the battle for political hegemony within Israel between the *Ashkenazi* (European) and *Mizrahi* (Asian and African) Jewish populations. The dominance of the *Ashkenazi*-European elites is now being questioned as the *Mizrahi* populations are undergoing empowerment. The fact that over half of the Jewish population of the country are of North African and Eastern origin – *mizrahim* – would suggest that there is a desire to become more fully integrated in the Middle East. As a growing number of representatives of these communities attain positions of power, partic-ularly with the emergence of strong ethno-based political parties in recent years, it has been argued that these groups are a more integral part of the geographical region. This translates into two contrary political positions based on the notion of: 'we lived amongst the Arabs for hundreds of years, therefore we know them, their mentality, their negotiating culture and the way to deal with them'. Some groups argue that the experience of living within a Middle Eastern culture will enable them, rather than the European based political elites, to negotiate with the Arab leaders, based on mutual respect for their common regional heritage, while other groups argue that these same experiences of living as a minority amongst Arab nations for hundreds of years have taught them that there is no possibility of ever reaching an agreement with the Arab world. Regardless of which position is adopted, the notion that Israel must become part of its geographical region, rather than a European colonial implantation within a different cultural *milieu*, is common to both perspectives

The fact that an immigrant from Morocco, David Levy, became Israel's foreign minister was an important indication of the shift which is taking place, although he was belittled by many of his political colleagues for not being able to represent Israel adequately in world fora due to his lack of English language skills. The *Ashkenazi–Mizrahi* discourse is not only played out in formal political institutions but also through popular culture. It took until the late 1960s for it to become acceptable to play Oriental music on the Israeli radio, and even then it commenced as part of the Arabic department. By contrast, classical music played a major role in the broad-casting services from their inception. These discourses of music, art and culture equally reflect the geopolitical tensions between a desire to integrate into the region as contrasted with a desire to be part of Europe (Izenberg 1998; Waterman 1998).

In contemporary Israel, the non-integration into the region is perhaps best reflected in the linguistic skills of the inhabitants. English is the preferred second language taught in all schools, while Arabic is compulsory for one year only and is chosen by few students. Despite daily contact with Arab citizens of Israel and Palestinian residents of the Occupied Territories, only a small percentage of Israelis are able to converse with them in Arabic. This is typical of majority–minority relations in general, and in situations of conflict in particular, where the minority population is dependent on the language of the majority popu-lation in order to undertake bureaucratic and economic transactions. This creates an interesting paradox in that while a decreasing number of Jewish citizens speak Arab, a large proportion – if not most – of the Palestinian population understand and speak Hebrew, while the majority of Diaspora Jewry are unable to speak, read or write the language of their Israeli compatriots.

It was, perhaps, Palestinian Authority Chairman Yasir Arafat's analogy of Israel with the twelfth-century Crusader states which characterized much of the Arab thinking on Israel's role in the region. The Crusaders were early colonizers, controlling the 'holy land' through a series of well-fortified army bases, eventually being vanquished by the Islamic warriors under the leadership of Saladin. The Crusaders have always been seen as being external and alien intruders to this region, whose presence proved to be no more than a temporary phenomenon in the long and tumultuous history of the Middle East. At the same time, this state-ment was seen by many Israelis as being an indication of the ultimate Arab intentions concerning Israel and that, by definition, there was no point in attempting to enter into a peace agreement with the Palestinians. This highlights the essentially pragmatic, rather than ideological, underpinnings of the Israel–Arab peace processes, namely the fact that the agreements serve the *Realpolitik* interests of all sides involved, rather than signal any real desire by the Arab countries of the Middle East to accept Israel as one of them (Susser 1997).

In Europe

In terms of cultural aspirations, Israel sees itself firmly rooted in Europe. In a well publicized article some years ago, Meron Benveniste argued that while Israel's geographical location is in the Levant, its preferred cultural location is somewhere between Paris and Prague. Israel's founder elites have their roots in Central and Eastern Europe, and the systems of state government and a host of other state institutions are strongly rooted in European culture. During its first fifty years, all Israel's Prime Ministers, from Ben Gurion to Netanyahu, and all but one of its Presidents, were of European extraction, despite the fact that over half the Jewish population of the country are *Mizrahi* (North African and Asian) in origin. Even the leaders of the right wing *Likud* party, which has traditionally enjoyed the support of the *Mizrahi* population in elections, have been European. This has been strengthened in recent years with the arrival of over 800,000 immigrants from the former Soviet Union. For some, this signals a re-balancing of the delicate ethnic mosaic, swinging the scales back in favour of the European-*Ashkenazi* cultural hegemony, at a time when *Mizrahi* empowerment is coming to the fore for the first time in Israel's political history.

Israel enjoys favoured trading status with the European Community and partic-ipates in many European cultural activities, such as the Eurovision song contest and the European football (soccer) competitions. While this has largely come about as a result of Israel's exclusion from Middle Eastern political and cultural activities, it is unlikely that the stated desire to become part of the natural region within which the country is located actually extends as far as opting to replace those European links with those of the Middle East. Yet for many Israelis, links with Europe are problematic. The collective memory of the country's Jewish citi-zens still finds it difficult to come to terms with the normalcy of relations between Israel and Germany (Timm 1997). Despite the growing cultural and economic ties between the two countries, the idea that Germany could, or should, provide mili-tary or other guarantees to a Middle East peace solution is still unacceptable. Many Israelis still refuse to visit the country, despite the fact that formal relations have been established for over thirty years and Israel has received substantial repara-tions from successive German governments during this same period (Levy 1996).

Israel's relations with the other two major European powers, France and Britain, have been ambivalent. Britain is traditionally perceived as a country whose foreign policy is dictated by a pro-Arab lobby, tracing its roots back to the traditional affinity of past British governments with the romance of the desert and the nomadic culture. In addition, Britain's mandate rule of Palestine, the 1939 White paper limiting the number of Jewish immigrants, and the period following the end of the Second World War is remembered for its attempts to prevent the establishment of an independent Jewish State and its refusal to allow holocaust survivors entry into Palestine. Successive Israeli governments have viewed British interests in the region with great suspicion, despite the interests of the latter in playing a more prominent role in Middle East peacemaking. This was highlighted as recently as 1998, following a disastrous visit by British Foreign Secretary, Robin Cook, who, at the time, also represented the European Community. His decision to ignore much diplomatic protocol in his visit to East Jerusalem was interpreted by most Israelis as being a clear indication of a pro-Arab bias on the part of the European Community. Israel's relationship with France was the warmest of any of the three European powers. During the 1950s, France assisted in the development of Israel's defence industries, particularly in the establishment of Israel's nuclear reactor in Dimona. But this relationship has cooled in recent years as France has opposed many of the Israeli policies concerning the Palestinians and the West Bank occupation.

The notion of a single European entity, as contrasted with the individual member countries of the Union, presents new opportunities of forging stronger links with Europe. The emergence of the European Union has enabled Israel to reassess its regional position *vis-à-vis* Europe. Economic and political ties with the EU are linked to each other, with the Union demanding a more active role in the peace process if Israel is to enjoy even further integration into the regional economic markets. The dissatisfaction of the European Community with the progress of the peace process under the Netanyahu administration, together with the Israeli reticence to allow European leaders to take an active role as third party arbitrators alongside that of the United States, has made it difficult for Israel to gain any further concessions within the European Community.

Somewhere between a Middle Eastern or European affiliation lies the Mediterranean location. In this respect, Israel is similar to additional countries within the Eastern Mediterranean, such as Cyprus and Turkey. The geographical and cultural location at the Europe–Middle East, Christian–Islamic interface, has resulted in the emergence of a common Mediterranean identity, together with other European countries such as Italy and Spain and the North African countries of the Maghreb. But this has not resulted in any significant political positioning on the part of Israel or the other countries involved.

In the Jewish Diaspora

The State of Israel maintains strong links with Jewish communities throughout the world. It also sees itself as the self-appointed protector of these communities in times of persecution or crisis. In his study of Israel as an ethnocracy, Yiftachel (1997a) has described the country as one 'without borders', or a state within which the boundaries of identity do not correspond to the limited territorial boundaries of the country. While some people residing outside the territorial boundaries are included within the national collective, others who reside within the territorial boundaries are excluded. Automatic 'right of return' and instant citizenship is granted to Jews throughout the world whether they have ever actually previously set foot in Israel or not, while local Arab-Palestinian residents who have lived uninterrupted in this region for hundreds of years do not enjoy the same rights. This is particularly the case with respect to the right to purchase land, a right which has been granted to Jewish residents of the Diaspora but not to Palestinian residents of the Occupied Territories. Much of the state-owned land was purchased by global Jewish organizations, notably the Jewish National Fund, while the Jewish Agency remains a quasi public agency which operates on behalf of the global Jewish community and the State of Israel at one and the same time.

The notion that Israel is located within the Jewish Diaspora is, in many ways, an antithesis to the territorial notion of Zionism as an ideology around which state formation took place. The establishment of an independent Jewish state was seen by the state founders as constituting a solution to the condition of Diaspora, a negation of the situation which had lasted for approximately two thousand years during which period the Jewish people had been a dispersed and, in many instances, a persecuted minority throughout the world. The very fact that Zionism, rather than Jewish nationalism, was chosen as the name of the national movement was itself an indication of the strong territorial focus on a particular piece of territory, a territory which had remained a central part of Jewish prayer and ritual throughout the Diaspora period as a result of strong processes of cultural socialization (Davies 1982; Newman 1998c, 1998d).

The relationship between the State of Israel and the Jewish Diaspora has been, and remains, a complex one (Kimmerling 1989; Shefer 1996). On the one hand, an underlying ideological foundation of the State has been to persuade Jews throughout the world to immigrate to Israel. This is not referred to simply as a process of immigration, but one of 'aliyah', a process of 'going up', through which the person achieves a form of national self-fulfilment by choosing to live within the ancient homeland. The process through which emissaries representing the State have traditionally raised funds from the Diaspora Jewish communities has often been described as one in which those who have not chosen to live in Israel pay a form of tax, by assisting those that have made this choice. But at the same time, Jewish criticism of Israeli government policies is met with the response that they should not interfere in the decisions of a state in which they have not chosen to live. Thus it is an asymmetrical relationship in which the Diaspora Jewish communities are told to 'pay up, but shut up'.

In recent years, this complex relationship has began to change, as many Diaspora communities have become disenchanted with Israeli government policies concerning the peace process and the growing influence of fundamental religious groups within government circles. Right-wing Diaspora groups criticized the Rabin government for daring to give up 'holy' territory in the name of peace, while left-wing groups have criticized the Netanyahu administration for slowing down the peace process and giving in to religious and irredentist demands from his coalition partners. Both groups have chosen to publicly show their disaffection for the policies of the Israeli governments and have even demonstrated outside the Israeli embassies in the major capital cities of the world, an action which would have been considered inconceivable some years ago.

Israel perceives itself as being a partner with the Diaspora Jewish communities, but a senior partner who occupies the moral high ground. For their part, the Diaspora communities are no longer willing to accept this secondary role, not least because the very existence of the State has made them more self-assured and culturally independent in their own countries of residence. Whereas much of the money raised by these communities in the past was unquestioningly transferred to the coffers of the State and its institutions, the current trend is for a greater portion of these resources to remain within the communities themselves as a means of increasing religious and cultural awareness and identity. While the Israel-Diaspora partnership remains intact, the role of each has changed, with neither feeling as dependent on the other as in the past. It is by no means ensured that the Diaspora community will continue to represent Israel's interests before foreign governments, especially in the USA, if current political and social trends continue to take hold of the Israeli body politic.

The notion that the Diaspora could, and should, be negated is itself an experiment in inculcating a false consciousness amongst the Jewish communities of the world. With the mass immigration of Russian and Ethiopian Jews to Israel during the 1990s, and the final exodus of the small remaining communities in both Syria and Iraq, the State no longer has a function as a place to which oppressed Jews can come, simply because the number of oppressed communities has virtually disappeared. Within the western world, the Jewish communities are politically and financially secure and do not perceive modern, materialistic, Israel as representing the sort of ideological challenge that it did during the first three decades of statehood. Israel is no longer perceived as occupying the moral high ground in this relationship and, as such, the two constituent parts of this global geopolitical relationship are in the process of taking on new roles. In addition, while the Diaspora community desires to see a strong and independent Israel, its very strength and self-assuredness is a factor which, paradoxically, reduces the feeling of commitment on the part of the Diaspora who no longer feel needed in the same way as before.

The geopolitical imagination of Israel as part of the global Jewish community is also coming into conflict with the very basic notions of Jewish identity. Orthodox Jewry refuses to recognize alternative trends within Judaism – notably the Reform and Conservative communities – as legitimate Jews since they do not always answer to the strict orthodox interpretation of 'who is a Jew?'. Within Israel, Jewish identity is determined by the orthodox groups, not least because they have strong representation in the Israeli parliament through their political parties. The 'status quo' on religious affairs, instituted soon after the establishment of the State in 1948, placed all matters of personal and religious status in the hands of the religious (orthodox) establishment. While the Reform and Conservative movements do not have demographic or political clout within Israel, they nevertheless make up the vast majority of Diaspora Jewry, especially in the United States. These communities have traditionally been strong supporters and lobbyists on behalf of Israel. But as these communities become increasingly estranged from the growing tide of religious fundamentalism within Israel, coupled with attempts by some orthodox groups to deny their Jewishness, so too their direct support for, and automatic identity with, Israel is diminished.

Finally, this process of estrangement is also seen with respect to Israeli emigrants. During the past two decades, the process of Israeli out-migration has grown, in some years exceeding the numbers of Jewish immigrants to the country. Not only is this seen as a bad advert for encouraging further Jewish immigration, but the Israeli communities which have grown up in major cities such as New York, Los Angeles, Toronto and London, function separately from the local Jewish communities. The relationship between the two communities has often been tense, with the former perceiving themselves as just one more amongst a group of migrant communities, while the latter prefer to stress their religious and cultural affinities. Thus, the nature of the Israel–Diaspora relationship is becoming more complex and tenuous than in the past, while Israel is unable to count automatically on the blind support of World Jewry for all its policies.

The fifty-first state of the United States

Israel has always had a special relationship with the United States. Without American recognition of Israel in 1948, it is questionable whether the Zionist leadership of the time would have been as quick to establish the new state following the departure of the British Mandate forces. Israel has relied on United States support, both political and fiscal, during the past fifty years, even to the extent that that support has, on some occasions, been contrary to the weighted opinion of the entire international community. As recently as 1997, there was the weird spectre of the United States and Micronesia as the only two countries voting against a United Nations motion to condemn Israel for establishing further settlements in East Jerusalem.

The United States support for Israel has been based on two factors. The first of these is, undoubtedly, the weight of the pro-Israel lobby within the American body politic. America's Jewish community is well established, both financially and politically. Members of the Jewish community have been elected to important political positions, while they are also an important source of party funding for both Democrats and Republicans. The notion of the 'Jewish vote' is less established, given the fact that the entire community numbers fewer than six million, of which a large proportion are not necessarily affiliated to the Jewish community and do not automatically support Jewish or Israeli causes. Thus the influence of the Jewish community on American foreign policy-making in the Middle East is disproportionate to their actual numbers.

Israeli leaders have traditionally had an ambivalent attitude towards the United States (Reich 1994). On the one hand, there has always been a strong lobby aimed at maintaining, if not increasing, the amount of foreign aid, both military and financial. Appeals to the American conscience in supporting the beleaguered Jewish state have not been offset by Israel's military superiority. At the same time, American pressure on the Israeli government to make political and territorial concessions is displayed within Israel as intervention in the internal affairs of an independent sovereign entity. This has particularly been the case under Israel's right-wing governments, with notable tension having emerged between the Shamir and Bush administrations in the early 1990s, and between the Netanyahu and Clinton administrations seven years later. In both cases, some of the cracks in USA–Israel relations were conveniently papered over by the Iraqi crises, during which periods the United States immediately offered unquestioning assistance to Israel in its preparation of the home front against possible missile attacks.

Right-wing groups in Israel often demonstrate outside the hotels of visiting American leaders and policy-makers when they are directly involved in shuttle negotiations between Israel and her Arab neighbours, in extreme cases crudely displaying the visiting diplomats as being anti-Semitic and damaging Israel's interests. This reached a peak following the Yom Kippur War of October 1973, with United States Secretary Kissinger mediating an Israeli partial withdrawal from the Golan Heights. Despite his own Jewish refugee background, he was displayed as being an American 'self-defined Jew hater' in vocal, and in some cases violent, anti-American demonstrations. Visits by American mediator Denis Ross and Secretary of State Madeleine Albright have, in recent years, been greeted with similar anti-American sentiments on the part of some right-wing groups, often expressed through public advertisements in the daily press. The fact that these diplomats are themselves Jewish places them in an awkward position. Right-wing Israelis see them as traitors to their own people for bringing American pressure to bear, while Arab countries view them as pro-Israel and, hence, too soft in dealing with Israeli settlement policies.

323

Despite these contradictions, America has remained Israel's strongest backer over the past fifty years. Without United States support, Israel would have been unable to maintain its political and financial position in the region. American criticism of Israeli policy and the occasional tensions between the two countries is normally referred to as being no more than an 'in-house feud', to be expected in even the best of family relationships. Perhaps the only situation in which the United States has unilaterally forced Israel to make a major policy change was the Eisenhower insistence on an immediate Israeli withdrawal from the Sinai peninsula, following Israel's capture of the region in 1956 in collusion with Britain and France.

Another factor underlying American support for Israel has been the oft stated notion that the leader of the free world must support the only democracy in the Middle East. This argument is used by Israel in arguing that true peace can only be made between democracies (as defined by the normative western models of participatory democracy) (Garnham and Tessler 1997; Elman 1997) and that the United States must insist on internal political transformations within Arab countries before it can offer them equal support. For as long as Egypt was under the direct influence of the Soviet Union, the United States had a major interest in defending, and even strengthening, Israel as a bulwark against Soviet expansion in the Middle East. But Egypt's decision to move out of the Soviet sphere of influence, the Israel–Egypt peace treaty at Camp David, and the eventual collapse of the Soviet bloc has changed the regional balance of power and has brought much of the region under both direct and indirect American influence. Israel and Egypt are currently seen as being joint allies of the United States in their fight against what is perceived as the new regional threat, namely that of Islamic fundamentalism, and the expansionist policies of Iraq which threaten America's oil interests in the Persian Gulf. After Israel, Egypt is now the second largest recipient of United States foreign aid, with these two countries together accounting for well over half the entire foreign aid package.

Two further factors are likely to change the nature of the US – Israel relationship in the foreseeable future. In the first place, there is a growing division within the American Jewish community concerning the unquestionable support for Israeli governments who do not necessarily represent the interests of the entire community. Second, the growth of the American Islamic community and their entry into local and national politics is likely to create a strong counter force amongst American policy-makers. This is likely to be the case particularly when Israeli governments hold back on the peace process. In the case of the latter, continued and unqualified support for Israel is seen by the growing Arab lobby as biased and as a sign that the United States is unable to be an honest broker for both sides of the conflict, an argument which the Americans are keen to counter. Despite their occasional threats to the contrary, America is unlikely to ever withdraw its involvement from the Middle East, not least because they are unwilling to allow Europe to insert a significant foothold in the region. But overall, Israel has successfully managed to retain American support for the country, even where its policies have not always been in the best global interests of the United States. Israel insists on American, rather than European, guarantees for any peace agreements, or promises of demilitarization. The notion of Israel as the 51st State of America continues to be a source of major geopolitical strength for Israel.

The centre of the world

The most metaphysical of the five geopolitical locations is the notion of Israel being located at the very centre of the world. Israel is the geographical location of the 'holy' land, the birthplace of both Judaism and Christianity, and the third most important centre of Islam. As such, this piece of territory takes on symbolic importance in the sense that it possesses some form of abstract notion of sanctity. It is elevated and more important than other territories, owing to the events, some real, many mythical, which are supposed to have taken place here. Bloody religious wars have been fought over the right to control the holy sites. The Crusaders attempted to recapture the old city of Jerusalem for Christianity, the Muslims built the Mosque of Omar on the site of the ancient Jewish temple, while for religious Jews the capture of the old city of Jerusalem in the Six Day War of 1967 was no less than a miraculous 'liberation' of sites which belonged to them by right through the Divine promise to the Jewish forefathers, outlined in the Biblical narratives (Davies 1982; Newman 1998c).

In this sense, the modern state of Israel is no more than a temporary objective reality within whose boundaries the holy sites happen to be located. As a metaphysical location, the precise boundaries of the modern state are unimportant, although the contemporary State is able to emphasize the fact that they exercise political control of the symbolic spaces and sites. This is reflected in the pilgrimage of Jews and Christians from throughout the world, especially around the time of major festival periods such as Christmas or the Jewish New Year, with a boost of hundreds of thousands of Christian pilgrims expected to take place around the millennium. It is increasingly common for Jewish residents of the Diaspora to purchase holiday apartments in Israel and to spend their vacation periods in the country during the period of the major Jewish festivals. Paradoxically, while the whole world watches the Christmas Mass live from the 'holy land', the neighbouring towns of Bethlehem and Jerusalem are places where there is a normal working day on this most important day in the Christian calendar. Most Jews and Muslims in the State of Israel and the West Bank are virtually unaware of the existence of this religious festival as they go about their normal weekday activities.

As the centre of the world, this particular piece of territory takes on metaphysical dimensions. The notion of the 'Jerusalem of above' as contrasted with that on the ground is perceived as being a place of utopia and perfection, as contrasted with the realities of the bitter conflict which takes place between Arabs and Jews in this city. The idea that Jerusalem can exist in other places, such as in 'England's green and pleasant land' is also part of this perceived a-spatial location of a particular place. This metaphor of place constitutes a powerful component in the geopolitical imagination which moves the place from the realm of a defined territory to a space which is both movable and transient. This explains the use of Biblical place names in locations throughout the world, notably Succot, Shiloh, Nazareth, Lebanon, Jordan River, to name but a few. At the same time, it throws up the paradox of transforming a meta-physical sense of a-spatiality into a concrete political notion of space which has a single physical location, which then takes on greater importance than all other locations and which ends up being fought over. This also explains why the West Bank settlers have been so insistent on naming their settlements after the Biblical places which were located on, or nearby, their settlement sites (Cohen and Kliot 1981, 1992)

As such, the notion of being at the 'centre of the world' is an exclusive, and hence contested, discourse of place. Just as the 'promised land' and Jerusalem are central to Jewish religious and geographical thinking (Davies 1982), so too is it central in Muslim and Christian thought. The Vatican view these same places as central to its own political interpretations of theology and history and, as such, display a great interest and desire to be involved in the political events taking place therein (Perko 1997). But as a contested theological discourse of place, the notion of centrality is exclusive, rather than shared. For each 'centre', the religions of the 'other' are relegated to the periphery, thus enhancing the sense of zero-sum conflict, the sort which takes place at Armageddon (also geographically located within Israel) rather than on a human battlefield.

The symbolic dimension of this territory and its global significance also mean that it is perceived as being much larger than it really is. For a country occupying no more than 20,000 square kilometres (25,000 including the Occupied Territories), with a distance of no more than seventy-five kilometres between the Mediterranean Sea in the west and the Jordan River in the east, and encompassing no more than eight million people (Israelis and Palestinians together), the country and its conflicts take on mega-state proportions in the world media and the general interest expressed by virtually the whole of the international community. Violent conflict elsewhere in the world may only receive passing mention, while the smallest of stone-throwing events in and around Jerusalem or Bethlehem will automatically become headline news. The amount of attention accorded Israel in world affairs is over and beyond its demographic or economic contribution to the changing dynamics of the world political map. While other conflicts, such as Bosnia or Northern Ireland, also take prominence, in terms of ongoing coverage over an uninterrupted period of fifty years there is probably no other single conflict which has continually occupied such a central place in world attention.

At the same time, there is a sense amongst many Israelis that it is fitting to be at the centre of world attention, if only to show the rest of the world that this small state composed of so many Jewish refugees cannot be dismissed as insignificant. This is also part of the post-holocaust syndrome, a need to categorically state to the world that 'we are here' and are in no hurry to disappear from the stage of world history. A joke going around Israel in the early 1980s, partially plagiarized from a Peter Sellers film '*The Mouse that Roared*' plays up this sense of global self-importance. The early 1980s was a period of stagflation in the Israeli economy with the national debt growing to unmanageable proportions. The Israeli cabinet met in emergency session to discuss ways to solve the economic crisis. On observing that two of the world's most successful countries at the time, Germany and Japan, had been the vanquished nations of the Second World War one cabinet minister suggested declaring war on the United States. Having been vanquished in such a war, American would then create a modern version of the Marshall Plan to restructure and strengthen the Israeli economy. As this novel plan was being discussed, the defence minister, warrior general Ariel Sharon stood up, and dismissed the plan out of hand. 'What', he argued, would happen to Israel's economy, 'if and when we win the war'?

The sense of being in the centre of the world and at the centre of world attention therefore has a dual dimension. On the one hand, many Israelis complain at the undue attention given to Israel and its problems when contrasted with other states. But, at the same time, the collective need to be an important player on the world stage, in disproportion to the actual size of either population or territory, continues to play a role in the thinking of most Israelis. The subjective interpretation of Israel's global role as an international player, despite its objective reality of being a small country located at the eastern edge of the Mediterranean Sea, explains this fifth geopolitical dimension of the collective imagination.

Conclusions

The objective of this chapter has been to show the diversity of the Israeli geopolitical imagination. This varies according to the extent to which the objective realities of the country's geographic location and size, or the subjective interpretation of the symbolic significance of the place, is taken into account. These varied interpretations of location, both spatial and a-spatial, are themselves a function of the way in which residents of Israel self identify, as part of an exclusive national club or as citizens of a plural democracy which is expanding its ties within a global economy. As Israel becomes more heterogeneous in its internal composition, as attempts are made to achieve conflict resolution between Israel and her neighbours, and as schisms occur between different branches of Judaism worldwide, so too the nature of both individual and collective identity becomes more diverse, resulting in the formation, and overlapping, of numerous geopolitical imaginations. These can only be understood by reference to the discursive narratives of the different groups themselves and the extent to which they are redefined at local, regional and global levels

Bibliography

Agnew, J. and Corbridge, S. (1995) *Mastering Space: Hegemony, Territory and International Political Economy*, London: Routledge.

Anderson, B. (1983) *Imagined Communities: Reflections on the Origin and Spread of Nationalism*, London and New York: Verso Press.

Bar-Tal, D., Jacobson, D. and Klieman, A. (1998) (eds) *Security Concerns: Learning from the Israeli Experience*, Greenwich, Conn.: JAI Press.

Billig, M.(1995) *Banal Nationalism*, London: Sage Publications.

Brunn, S. D., Jones, J. A. and Purcell, D. (1994) 'Ethnic communities in the evolving "electronic state": cyberplaces in cyberpspaces', 415–24 in W. A. Gallusser (ed.) *Political Boundaries and Coexistence*, Bern: Peter Lang.

Cohen, E. (1989) 'Israel as a Post-Zionist society', 203–14 in R. Wistrich and D. Ohana (eds) *The Shaping of Israeli Identity: Myth Memory and Trauma*, London: Frank Cass.

Cohen, S. B., and Kliot, N. (1981) 'Israel's place names as reflection of continuity and change in nation building', *Names* 29: 227–46.

—— (1992) 'Place names in Israel's ideological struggle over the administered territories', *Annals of the Association of American Geographers* 82: 653–80.

Cox, K. R. (1998) 'Spaces of dependence, spaces of engagement and the politics of scale, or: looking for local politics', *Political Geography* 17 (1): 1–25.

Dalby, S. and Ó Tuathail, G. (1996) (eds) *Critical Geopolitics*, special issue of *Political Geography* 15 (6/7).

Dahan-Kalev, H. (1997) 'The oppression of women by other women: relations and struggle between Mizrahi and Ashkenazi women in Israel', *Israel Social Science Research* 12 (1): 31–44.

Davies. W. D. (1982) *The Territorial Dimension of Judaism*, Berkeley: University of California Press.

Doty, R. L. (1996) 'Sovereignty and the nation: constructing the boundaries of national identity', 121–47 in T. J. Biersteker and C. Weber (eds), *State Sovereignty as Social Construct*, Cambridge, England: Cambridge Studies in International Relations. Cambridge University Press.

Elman, M. F. (1997) (ed.) *Paths to Peace: Is Democracy the Answer?*, Cambridge, Mass.: CSIA Studies in International Security, MIT Press.

Elmusa, S. (1994) 'The Israeli–Palestinian water dispute can be resolved', *Palestine–Israel Journal* 3: 18–26.

Falah, G. and Newman, D. (1995) 'The spatial manifestation of threat: Israelis and Palestinians seek a "good" border', *Political Geography* 14 (8): 689–706.

Fenster, T. (1997) 'Relativism vs. universalism in planning for minority women in Israel', *Israel Social Science Research* 12 (2): 75–95.

Fogiel-Bijaoui, S. (1997) 'Women in Israel: the politics of citizenship as a non-issue', *Israel Social Science Research* 12 (1): 1–30.

Garnham, D. and Tessler, M. (1995) (eds) *Democracy, War and Peace in the Middle East*, Bloomington: Indiana University Press.

Helman, S. (1999) 'Redefining obligations, creating rights: conscientious objection and the redefinition of citizenship in Israel', *Citizenship Studies* 3: 45–70.

Herzog, C. (1978) *Who stands accused? Israel answers its critics*, New York: Random House.

Herzog, J. (1975) *A People that Dwells Alone: Speeches and Writings of Yaacov Herzog*, London: Weidenfeld and Nicholson.

Izenberg, D. (1998) 'A not-so-cultured argument', *The Jerusalem Post*, 24 April: 16.

Kimmerling, B. (1989) 'Between "Alexandria on the Hudson" and Zion', 265–84 in B. Kimmerling (ed.) *The Israeli State and Society: Boundaries and Frontiers*, Albany, N.Y.: SUNY Press.

Kliot, N. (1994) *Water Resources and Conflict in the Middle East*, London: Routledge.

Kook, R. (1996) 'Between uniqueness and exclusion: the politics of identity in Israel', 199–226 in M. N. Barnett (ed.) *Israel in Comparative Perspective: Challenging the Conventional Wisdom*, Albany, N.Y.: SUNY Press.

Levy, G. (1996) *Germany and Israel: Moral Debt and National Interest*, London: Frank Cass.

Lewis, M. W. and Wigen, K. E. (1997) *The Myth of Continents: A Critique of Metageography*, Berkeley: University of California Press.

Linn, R. (1994) *Conscience at War: The Israeli Soldier as Moral Critic*, Albany, N.Y.: SUNY Press.

Morley, D. and Robins, K. (1995) *Spaces of Identity: Global Media, Electronic Landscapes and Cultural Boundaries*, London: Routledge.

Newman, D. (1997) 'Israeli security: reality and myth', *Palestine–Israel Journal* 4 (2): 17–24.

——. (1998a) 'The geographic and territorial imprint on the security discourse', 73–94 in D. Bar-Tal, D. Jacobson and A. Klieman (eds) *Concerned with Security: Learning from the Experience of Israeli Society*, Stamford, Conn.: JAI Press.

—— (1998b) 'Population as security: the Arab–Israeli struggle for demographic hegemony', 163–86 in N. Poku and D. Graham (eds), *Redefining Security: Population Movements and National Security*, Westport, Conn.: Greenwood Publishing Group.

——. (1998c) 'Concrete and metaphysical landscapes: the geopiety of homeland socialization in the Land of Israel', 153–84 in R. W. Mitchell and H. Brodsky (eds) *Visions of Land and Community: Geography in Jewish Studies*, Maryland: University of Maryland Press.

——. (1998d) 'Real spaces – Symbolic spaces: interrelated notions of territory in the Arab–Israel conflict', 3–36 in P. Diehl (ed.) *A Road Map to War: Territorial Dimensions of International Conflict*, Nashville: Vanderbilt University Press.

—— (in press) 'From national to post-national territorial identities in Israel/ Palestine', in A. Kemp, et al. (eds) Israelis in *Conflict: Identities, Challenges, Hegemonies*, Albany, N.Y.: SUNY Press.

Newman, D. and Paasi, A. (1998) 'Fences and neighbours in the postmodern world: boundary studies in political geography', *Progress in Human Geography* 22 (2): 186–207.

Peled, Y. (1992) 'Ethnic democracy and the legal construction of citizenship: Arab citizens of the Jewish state', *American Political Science Review* 86 (2): 432–42.

Peled, Y and Shafir, G. (1996) 'The roots of peacemaking: the dynamics of citizenship in Israel 1948–1993', *International Journal of Middle East Studies* 28 (3): 391–413.

Peres, S. (1994) *The New Middle East*, New York: Henry Ling.

Perko, F. M. (1997) 'Toward a "sound and lasting basis": relations between the Holy See, the Zionist movement, and Israel, 1896–1996', *Israel Studies* 2 (1): 1–21.

Popper, M. (1998) 'The Israel defense forces as a socializing agent', 167–80 in D. Bar-Tal, D. Jacobson and A. Klieman (eds) *Concerned with Security: Learning from the Experience of Israeli Society*, Greenwich, Conn.: JAI Press.

Ram, U. (1998a) 'Post-nationalist historiographies: the case of Israel', *Social Science History* 22: 513–45.

—— (1998b) 'Citizens, consumers and believers: the Israeli public sphere between fundamentalism and capitalism', *Israel Studies* 3 (1): 24–44.

Reich, B. (1994) 'Reassessing the US–Israeli special relationship', *Israel Affairs* 1 (1): 64–83.

Said, E. (1979) *Orientalism*, New York: Vintage Press.

Shain, Y. (1997) 'Israel's state and civil society after fifty years of independence', 224–31 in PASSIA (ed.) *Palestine, Israel, Jordan: Building a Base for Common Scholarship and Understanding in the New Era of the Middle East*, Jerusalem: Passia Publications.

Shapland, G. (1997) *Rivers of Discord: International Water Disputes in the Middle East*, London: Hurst.

Shefer, G. (1996) 'Israel Diaspora relations in comparative perspective', 53–84 in M. N. Barnett (ed.) *Israel in Comparative Perspective: Challenging the Conventional Wisdom*, Albany, N.Y.: SUNY Press.

Shuval, H. (1996) 'Towards resolving conflicts over water: the case of the mountain aquifer', 215–38 in E. Karsh (ed), *Between War and Peace: Dilemmas of Israeli Security*, London: Frank Cass.

Silberstein, L. (1999) *Postzionism Debates: Knowledge and Power in Israeli Culture*, New York: Routledge.

Smith, M. P. (1998) 'Looking for the global spaces in local politics', *Political Geography* 17 (1): 35–40.

Soysal, Y. N. (1996) 'Changing citizenship in Europe: remarks on postnational membership and the national state', 17–29 in D. Cesarani and M. Fulbrook (eds) *Citizenship, Nationality and Migration in Europe*, London and New York: Routledge.

Susser, A. (1997) 'Israel's place in the region', 201–11 in PASSIA (ed.) *Palestine, Israel, Jordan: Building a Base for Common Scholarship and Understanding in the New Era of the Middle East*, Jerusalem: Passia Publications.

Telhami, S. (1996) 'Israeli foreign policy: A realist-ideal type or a breed of its own?', 29–52 in M.N. Barnett (ed.) *Israel in Comparative Perspective: Challenging the Conventional Wisdom*, Albany, N.Y.: SUNY Press.

Timm, A. (1997) 'The burdened relationship between the GDR and Israel', *Israel Studies* 2 (1): 22–49.

Waterman, S. (1998) 'Carnivals for elites? The cultural politics of arts festivals', *Progress in Human Geography* 22 (1): 54–74.

Yiftachel, O. (1997a) 'Israeli society and Jewish–Palestinian reconciliation: ethnocracy and its territorial contradictions', *Middle East Journal* 51 (4): 505–19.

—— (1997b) 'The political geography of ethnic protest: nationalism, deprivation and regionalism among Arabs in Israel', *Transactions of the Institute of British Geographers*, 22 (1): 91–110.

Yuval-Davis, N. (1997) 'National spaces and collective identities: borders, boundaries, citizenship and gender relations', inaugural lecture, delivered at the University of Greenwich, London 22 May 1997.

Zerubavel, Y. (1995) *Recovered Roots: Collective Memory and the Making of Israeli National Tradition*, Chicago: University of Chicago Press.

13

REFIGURING GEOPOLITICS

The *Reader's Digest* and popular geographies of danger at the end of the cold war

Joanne Sharp

Introduction

It has now become something of a cliché that with the decline of a communist threat at the end of the Cold War, conservative American culture has entered a period of crisis that has raised profound questions about both national identity and purpose (Engelhardt 1995). America's Cold War goal of containing the USSR might be understood in terms of a moving frontier between the US and this 'foreign' threat, not unlike the frontier narratives that characterized the nation's initial western colonial expansion. This narrative of frontier containing difference and protecting the American nation from foreign incursion, is key to American national identity. Its absence at the close of the Cold War has made the operation of traditional identity politics problematic. With the demise of the 'Evil Empire' lurking beyond the frontier, arguments for consensus and discipline in the face of this accepted enemy have become less urgent. This has led to an almost tangible fear in conservative political culture that once the recognized enemy has gone, the American populace will not be able to tell right from wrong, nor good from evil.

Many theorists have suggested that at the end of the Cold War, politicians would seek other foreign threats to America to try to reinvigorate the political culture of exclusion. The most prominent alternatives to communism have included terrorists, drug dealers and economic threats. This argument assumes that a foreign Other is required in the construction of territorial nation-state identity. However, this chapter will argue that for mainstream US culture at the end of the Cold War, it is a domestic enemy which is seen to most threaten the country's values, identity and destiny. To explain this contention I will provide a thorough engagement with one source of American culture, the magazine the *Reader's Digest*, at the close of the Cold War.

Popular culture and geopolitics

In one sense, geopolitics, the spatializing of international politics, is inherent to any representation of political process whether of global, regional, national or local scale. The goal of 'critical geopolitics' is to highlight the use of geographical language and to underline the fact that rather than being an apolitical and natural aspect of international politics, geography is a discourse and as such is a form of power/knowledge itself (Ó Tuathail and Agnew 1992: 192; Ó Tuathail 1996). Critical geopolitical approaches seek to examine how it is that international politics are imagined spatially or geographically and in so doing to uncover the politics involved in writing the geography of global space.

The majority of critical geopolitics focuses at the state level where policy is enacted. Here the language used in the formal 'scripting' of international poli- tics is examined. Although recognizing a hierarchy of knowledge producers, the location of geopolitical agency in such a restricted figure is problematic. There are different authorities of knowledge production within any society and the geographical imaginations that they project may not coincide. Dominant images of the world and its workings do not emerge from a single source, but from the complex – and fragile – workings of hegemony. Understanding the wider cultural context of geopolitical models is important for two primary reasons. First, it is through institutions such as the media and education that people are drawn into the political process as subjects of various political discourses. The media and education explain the linkages between their audiences and what is being explained in order to provide a context of interpretation. People are told what various changes and occurrences both at home and around the globe mean to them personally. Popular culture presents imagined geographies to their audiences and explain where individuals fit into these political models.

Second, intellectuals of statecraft are not somehow beyond or outside of hegemonic national culture, nor are their pronouncements somehow unaffected by the circulation of ideas and beliefs therein. In order to make their arguments sensible, they must refer to concepts and values that have consonance for the population at large, if their support is to be assured. As Ó Tuathail and Agnew argue, 'geopolitics is not a discrete and relatively contained activity confined only to a small group of 'wise men' who speak in the language of classical geopolitics' (Ó Tuathail and Agnew 1992: 194). Simply to describe a foreign policy is to engage in geopolitics and so normalize particular world-views. If this is the case, then surely the media – both high and low culture alike – is inti- mately bound up with geographing the world, as are a range of activities normally described as occurring outside the sphere of international politics. By forcing apart the 'geo-politics' of the everyday and the 'geopolitics' of statecraft, some commentators too readily accept a neo-realist view of state actors as the primary agents in world politics, rather than accepting the fluid and contested nature of hegemonic values and norms.

The scripting of global politics in popular culture – popular geopolitics – is also significant, in that it is within the sphere of popular culture that national cultures are formed and reinforced. A national culture represents a common source of narratives and understandings which attempt to produce a sense of belonging. These narratives and beliefs are drawn upon to define and explain new situations and their importance to individuals in the community. Both politicians and the media are storytellers, and in order for their stories to be accepted by their audience they have to resonate with meta-level hegemonic cultural values. Those values that flow between sectors of hegemonic culture are those which facilitate the narration of events and processes in an acceptable or meaningful way in the context of national self-identification. These narratives and beliefs are drawn upon to define and explain new situations and their importance to individuals in the community. For example, the values that John Wayne epitomized, celebrated and reinforced, act to strengthen – and in many ways make possible – the decisions and actions made by political leaders. Richard Nixon can refer to Wayne's film *Chisum* and the values of the Western, and the populace will understand the reference, and understand its origin in relation to an imagined geography of America (see Wills 1997).

As a result of the influence of cultural context, different countries' geopolitical traditions draw upon specific metaphors to create images of international geography. Political elites must use stories and images that are central to their citizens' daily lives and experiences. By reducing complex processes to simple images with which their audiences would be familiar, geopoliticians could render political decisions quite natural, or could make the result of the process appear predetermined. Sport metaphors have been particularly prominent in the USA. Such language points to the 'essential' differences between national potentials for world-class performance and naturalizes a global arena in which the rules of the game are understood, and within which there are clear (unequivocal) winners and losers. Agnew (1998) argues that in so doing, the ambiguities of conflict are reduced to technicalities in game play. Michael Shapiro (1989: 70) points out that comparing world politics to sporting contests serves the geopolitical purpose of emptying world space of any particular content: places lose their uniqueness and world politics becomes a strategy played out on a familiar sports field. American presidents (particularly Presidents Nixon and Bush) have been particularly fond of sports metaphors as applied to world politics. In a Vietnam bombing campaign, Nixon adopted the code-name 'quarterback' (Shapiro 1989: 87) and used terms such as 'end-run' and 'play selection' in foreign policy. As Agnew (1998: 71) has suggested, such metaphors 'allowed a notoriously socially awkward man to appear as "one of the boys"', engaging in dialogue with other sports-loving men. This not only explained the condition of conflict, but also why those hearing about it should accept the interpretation that Nixon offered: it provided a taken-for-granted cultural referent that the majority of his American audience would accept.

This wider context of interpretation is important in that geopolitical descriptions and arguments often rely upon accepted models, metaphors and images. These are naturalized – made into 'common-sense' statements – through reproduction in education and popular culture. Through these institutions, people learn about different places whether this is a list of 'factual' data or a more metaphorical narration. Such context allows geopolitics to function because of the naturalization of certain political assumptions and relationships. As a result, popular geopolitics have a special significance in reproducing the values and beliefs upon which more formal geopolitical statements must draw in order to resonate with various audiences.

In the next section I explain why the US magazine the *Reader's Digest* is of particular importance in the popular reconstruction of imagined geographies of America's place in the world, and in the circulation and normalization of geopolitical models and arguments.

The *Reader's Digest* and the (re)construction of 'America'

DeWitt and Lila Acheson Wallace launched the *Reader's Digest* in 1922. Their goal was to publish a digest of what they considered to be the best articles from the variety of magazines and journals that were proliferating in America at the time. This digest was intended to present essential common-sense advice to its readers so that they could make sense of the conflicting information they faced on a day-to-day basis from various sources. This common-sense world-view would allow *Digest* readers to act as good citizens: they could make informed decisions regarding current issues by understanding them within the context of the 'lessons of history' and American Destiny, so that they could see through the rhetoric of politicians and other figures.

Initially, twenty-eight articles were published each month, reduced to their bare elements for busy readers through the editors' 'art of condensation'. The print run was initially small but grew exponentially. Now the magazine is the world's most widely read publication, and a cultural influence in its own right. This means that although prominent people publish or are reprinted in the *Reader's Digest*, through the rigorous editorial process the magazine itself has become recognized as an author of importance.

The *Reader's Digest* emerged out of a period of reorganization of the American magazine industry at the turn of this century (Wilson 1983). The intellectual literary and political magazines that had been dominant at the end of the nineteenth century went into decline to be replaced by a new type of publication. This new form of publication was aimed at the growing middle classes, written for a middle-brow rather than an intellectual audience, and for the first time encompassing the entire country. This latter point is of significance for the constitution of national identification. Benedict Anderson (1983) has highlighted the central role of print capitalism to modern nationalism. Anderson argues that the imaginings of people across the country are drawn together in a communal bond through the juxtaposition of concerns and issues from around the state territory on the pages of newspapers and magazines. The *Reader's Digest* scripts national concerns and goals and, by explicitly addressing them as Americans, unites its readers into a national community. The magazine is thus a unique source for investigations of the changing nature of mainstream American identity.

The *Reader's Digest* is important to the study of American national identity for a number of more specific reasons. First, the *Digest* has the highest subscription rate of any magazine in the US with the exception of *TV-Guide* and *Modern Maturity*, and a remarkably high resubscription rate of around 70 per cent, suggesting significant reader loyalty. As over sixteen million copies are sold per month in America, the *Digest* is a fairly significant presence in the day-to-day representation of identity and purpose to American people. Second, the *Reader's Digest* presents itself as guardian of American morals. Events are not simply described; there is always a moral to the story. Third, because it juxtaposes articles about global affairs with pieces on issues of personal importance to the individual, the *Reader's Digest* articulates personal and moral concerns with global and national issues. An analysis of the *Digest* can help to present an image of how political issues can be made to resonate with individuals who have little or no direct material or obvious relationship to such issues.

Very much the product of founder DeWitt Wallace who edited the magazine from 1922 until 1973, the *Reader's Digest's* content and style have become so established that it has remained more or less unchanged. Over its seventy-six years of publication, the *Digest* had successfully written itself into the heart of American identity constructed around the geopolitics of the Cold War, in some ways becoming symbolic of US Cold War culture. Like mainstream/conservative America more generally, the magazine is now searching for an understanding of world politics within which to script America's role and destiny. This chapter will consider the magazine's engagement with the end of the Cold War, and the subsequent strategies it has employed as its map of the international realm has been challenged as a credible guide to America's place in the world. This analysis is based upon a detailed reading of all articles published between 1986 and 1994 that dealt with threats to the US and the American way, America's role and mission, and the role that good American citizens should play in maintaining America's international stature.[1]

336

The end of the Cold War

I expect that many people would date the beginning of the end of the Cold War to the rise of Mikhail Gorbachev to General Secretary of the Soviet Communist Party. However, as much of the world was welcoming Gorbachev and his reforms as heralding the end of international danger, the *Reader's Digest* sensed in the new leader an ever more significant threat to American security and the primacy of the democratic world. The *Digest* conveyed this wary interpretation of the Soviet presence through two distinct narratives and here provides an example of its own geopolitical representations of the world that underpinned and reinforced those of American statecraft.

The first narrative was one which refused to accept the ability of the Soviet Union to change; this Cold Warrior mindset could not readily comprehend a world order outside of the binary cartography of opposing superpowers. This narrative continued to write world politics within the structure of the Cold War, acknowledging only cosmetic changes to the Soviet system (for example, see Evans and Novak 1987). Indeed, the *Digest's* insistence upon the superficiality of change actually reinforced its Cold War geopolitics: Gorbachev's expression of progressive change represented nothing more than a continuation of the communist tactic of seducing people and nations by projecting positive appearances that would serve to deflect attention away from true (expansionist) intentions.

Second, in other articles, *Digest* writers recognized the significance of changes in the USSR but challenged the benign intentions said to drive them. Just as in the 1970s when the *Reader's Digest* had feared a weakness arising from détente policies towards the communist world, now it worried about the effect of 'too many Gorbachev boosters' (Richard Nixon, quoted in Barnes 1988: 89) who readily accepted the intentions of the USSR as benevolent. 'Gorbachev's nuclear-weapon-free-world proposal,' stated one author in 1989, 'was only an attempt to woo public opinion in the west' (Adelman 1989: 69). In an interview with Ronald Reagan, the magazine put it to the President that 'It's widely reported that Gorbachev is winning the propaganda war in Europe' (*Reader's Digest* 1990: 54).

It was this widespread public acceptance of Gorbachev's intentions that seemed to unsettle the *Digest* most and was apparently more important to the magazine than military or political power. To the *Digest*, Gorbachev's goal was

> the most ambitious ever sought by a Soviet leader, [and it] has profound meaning for the Western alliance. It is nothing less than achieving, in the eyes of the world, full moral equivalence with the United States. . . . It is hugely important to Moscow that the world believe that there is no great difference between us.
>
> (Rosenthal 1988: 71–2)

337

Although it stands as the sole superpower at the close of the Cold War, America's Manifest Destiny is at stake as there is no longer a commonly accepted 'Evil Empire' against which it must continue to struggle as champion – and paragon – of the 'free world'.

John McClure (1994) has suggested that this loss of defining national role is central to the state of imperial decline. He suggests that embroiled within the structure of identity is a romantic desire for chaos and uncertainty which can be tamed through various heroic trials. McClure's argument is based on an analysis of literature but I would contend that the desire for a realm of alterity can also be read in realist narrations of international relations. For the *Digest* the world is a moral landscape in which various countries provide settings as testing grounds for American moral authority. This version of national identity requires an organizing purpose and a danger over which America can triumph. Just as Edward Said (1978) saw European culture and identity written into the Orientalists' texts describing other places, so the *Digest's* descriptions of threats to America – the magazine's 'others' – tell much about its understanding of American culture, identity and mission. Like much of hegemonic American Cold War culture, the *Digest* represented the disorder of global spaces which lay outside its definition of freedom and democracy. This allowed the magazine to construct a normative agenda in which order could be projected onto this international chaos. This means that when America's enemies are not clearly recognizable, identity becomes difficult to define in this way. The magazine's account of American Destiny and identity depends upon the existence of an identifiable threat somewhere, whatever this threat might be. McClure explains this requirement as follows:

> [w]ithout the unordered spaces, or spaces distorted by war, it is impossible to stage the wanderings and disorientations, the quests and conquests and conversions, the ordeals and sacrifices and triumphs that are the stuff of romance. The ultimate enemies of romance, then, are not the foreign foes confronted on the field of battle in the test itself, but the foes held at bay by these essential antagonists: the banal, quotidian world of calculation and compromise from which the heroes of romance are always in flight.
>
> (McClure 1994: 3)

The fact that the *Reader's Digest* understood the USSR to be constructive of American identity rather than simply as a material threat to the country is perhaps indicated by the fact that just when the region became most unstable – when it was most in flux and it would appear that there should be much to discuss – there was actually a *drop* in coverage (see table 13.1). Once the unity of the communist world finally crumbled for the *Digest*, the magazine did not greatly concern itself about the rise of Yeltsin or Zhirinovsky, or the continued battles for territorial

Table 13.1 *Reader's Digest* articles concerned with themes of danger and American identity, 1986–1994

	1986–88	*1989–91*	*1992–94*
Russia and communism[a]	48	50	34
Terrorism	6	8	1
Drugs	13	19	5
Japan and the economy	7	9	10
Domestic danger[b]	27	51	76
American dream and American values	13	11	25

Notes

a Before the mid-1980s, the figure for Russia and communism was much higher. The average three-year total between 1974 and 1985 was sixty-one, and the figure was considerably larger in years of the two decaades before.

b This category includes articles concerned with bureaucracy, big government, political correctness, victimism and injustice.

sovereignty or weapons ownership. The USSR no longer embodied the Evil Empire standing in the path of America's Manifest Destiny. Instead the magazine sought new locations within which to stage America's historic struggle.

New 'others'

Theories of the spatiality of nation-states insist that for identity to function, an external enemy must be located, contained and marked as different. A number of theorists have argued that the process of defining an external enemy is an integral part of *domestic* policy (Dalby 1990), and as a result, central to the formation of a national identity (Campbell 1992). The majority of writings concerned with the end of the Cold War assumed that as communism declined as an opposing presence to the US, other threats would arise in the international arena. These other threats were most likely to be terrorists (especially Islamic fundamentalists), drug traffickers, and in the realms of international trade, Japan (Campbell 1992; Der Derian 1992). Each of these themes had been covered by the *Digest* during the Cold War but in the late 1980s appeared to take on a new urgency. Moreover, *Reader's Digest* cold warriors – including Eugene Methvin, Ralph Bennett, Rowland Evans and Robert Novak – now turned their attention from communism to these other sources of danger to America. I will now briefly look at each new threat as a further demonstration of the popular dimensions of geopolitical doctrine, and its everyday articulation through popular texts.

Terrorists

Terrorism has the potential to present the US with the same scenario of Total War as did the Cold War: perpetual vigilance and preemptive action are required to combat what is often described as an incessant threat. Anyone could be a terrorist just as anyone in the past might have been a communist. Similarly, in the fight against terrorism, there was no front line: terrorist intervention might be enacted anywhere throughout society. Whereas earlier articles had provided information on how to recongnize a communist, now readers were instructed on how to recognize a terrorist in their midst.

The *Reader's Digest* compared terrorism both directly and indirectly to communism. For example, one author stated in 1993 that 'what Moscow was to world communism, Teheran is to holy revolution and radical international fundamentalism' (Adams 1993: 76–7). As with its treatment of communism, the *Reader's Digest* was quick to expose the acts of terrorists not as being driven by a higher purpose but rather by darker, even psychopathic, motivations. Here the *Digest* implies that readers should not expect terrorists to show remorse or any other trait of humanity. One author, Netanyahu (Israel's former Prime Minister) claimed that 'terrorists are calculating murderers, swaggeringly proud of their acts, and accompanying these acts with clever political propaganda' (Netanyahu 1986: 110–11). One author, describing his kidnapping in Beirut, claims to have expected his captors to be 'fanatical and devout, but,' he insisted, 'I never saw them pray' (Glass 1988).[2] This echoes the magazine's previous assertions of the hypocrisy of communist leaders who claimed to be devoted to the ideology of communist equality and yet enjoyed great wealth at the expense of their people.

An earlier article inscribed the battle lines along a familiar East–West cartography by presenting 'Terrorism: how the west can win' (Netanyahu 1986). A report following the World Trade Center Bombing echoed the Domino Theory by positing the disruption of America once more as the major ambition of evil. 'The United States', claimed the article (Adams 1990), 'remains the single most important target for international terrorists' for mullahs whose 'hatred of the United States reached psychotic proportions.' The effect of Middle Eastern pressure groups within the US might act to reinforce this embattled mentality.

Drugs

As in previous articles on communism, those involved with drugs were seen as being evil, dangerous and perverse. In 'Cocaine king: a study in evil' the *Digest* suggested that the rising power of drug lords had allowed them to 'become a secret government, corrupting entire societies' (Adams 1988: 228). In a similar vein to its frightening descriptions of communists during the Cold War, the *Digest* now accused drug lords of having grand ambition: 'his idol: Adolf Hitler. His goal: to destroy the United States, to rule a kingdom of cocaine. Here is his story' (Adams 1988: 230).

It has been a harder task for the *Digest* to suggest that its readership is in more immediate danger from illegal drugs than from the threat of terrorism. Some articles did attempt to do this by utilizing the disease metaphors that had been prevalent at the height of the Cold War. One suggested that 'no-one is immune from [cocaine's] savage infection' (Hurt 1988). Other articles contained anecdotes of the struggles of normal or respectable people with the drugs that they thought posed no danger to them (e.g. 'champions who chose drugs' in May 1991).

Increasingly the *Reader's Digest* concentrated upon illustrating the dangers of the secondary effects of other peoples' drug abuse to its own readership. It suggested that as with communism, drugs would gradually take over American society if not checked: the process was inevitable so that the *Digest* could predict the outcome. The fight against drugs stated the *Digest*, is 'everyone's fight.' Even in apparently 'idyllic' villages, drug use could rapidly spread if discipline and moral standards were not maintained, particularly among the young. 'Crack invades the country-side' warned an article in 1989. In this tale, a quiet West Virginian village 'was engaged in a life-and-death struggle against drugs, and it was losing' (McConnell 1989). The article promised a lesson to all Americans in this story: citizen outrage and action saved the village. But any areas could potentially be invaded. Just as during the Cold War, readers were advised to act now before the danger was upon them and it was too late: 'if you don't help us now, crack will invade your community too' (Abe Brown, community organizer, quoted in Methvin 1991: 57).

For the *Reader's Digest*, the real problem facing America was not the drugs themselves but societal attitudes toward them. The magazine's writers believed US institutions to be too liberal in their attitude towards drug use. This lax attitude could apparently be attributed to the continuing hold of 1960s and 1970s permissiveness. The magazine held a black and white view of drug culture so that all who were involved were to blame. Only those who actively opposed drug usage were exempt from blame:

> we must never forget that the enemy we are fighting is not a chemical or a country or a social condition. Rather, it is each and every individual who sells, consumes or condones an illegal drug.
>
> (*Reader's Digest* condensation 1989: 88)

The economy and 'Nippophobia'

The *Reader's Digest* has been determined not to admit to American economic decline. Written into the magazine's image of America as world leader is a necessary belief in American economic and industrial superiority, and more importantly that the world hegemon exhibits a culture which most clearly exemplifies the industrial spirit (House 1989). The *Reader's Digest* is underwritten by a belief in the power of optimism, a belief that negative thinking and criticism will lead to decline. The American Dream is, after all, itself a narrative of optimism in the possibility of any individual American 'making it.' The *Reader's Digest* has been scornful of those who have given up on American chances of future economic leadership. For example, a series of articles under the general heading 'America on the rise' in 1989 described American economic buoyancy by way of a series of anecdotes about individuals whose ingenuity, hard work and persistence had made their business successful in the face of 'the prophets of doom' (Gilder 1989).

Often economic articles related conditions in America to those in Japan. Some suggested that Americans should not envy the Japanese because of the low standard of living and hard work that they suffered. One article proclaimed that, 'the Miyakawas do not expect ever to buy a home, traditionally the dream of every Japanese' (Shear 1991: 44). The choice of this example was surely not random. Given the centrality of home ownership to the American Dream, the Japanese system's denial of home-ownership (rather than any other form of consumption) to this hard working couple must have been intended to strike a chord with many readers.

However, other articles considered the economy to be under threat. It was sometimes rendered in terms of a battlefield, in which for example, ex-Cold Warrior Eugene Methvin (1986) could suggest that America utilize 'our new defense weapon – competition'. On the terrain of economic battles, the 'enemy' has clearly been written as Japan. Another Cold War *Digest* regular Fred Barnes complained that in economic issues, Japan, 'won't play fair.' In his article Barnes interspersed the text with subtitles which encoded the economy in military terms: 'Economic warfare,' 'Targeting for dominance,' and 'Hired guns' (Barnes 1990: 34–6).

When taken to an extreme, this representation of Japan renders it in terms previously reserved for the imperialist Soviet state. The *Reader's Digest* warned of the dangers of American inaction. One 1991 article pronounced 'Don't remilitarize Japan', continuing that if 'a warlike nation rises again, we may only have ourselves to blame . . . Asia tempts them, as it did the Japanese generals whose shades they worship' (Rosenthal 1991: 59). In the Cold War, any neutral space was seen to offer irresistible temptation to the Soviets. Now a lapse in the containment of Japan would apparently release the same expansionary desire.

The *Reader's Digest* saw other threats to American identity and values continuing into this period and new fears arose, most prominently of AIDS, but none ever came close to occupying the position of communism as America's alter-ego. Perhaps the *Digest* editors realized that communism was in fact a unique danger against which American identity and mission could be forged. As David Campbell has suggested for American political culture more generally:

The operations of anticommunism as a prominent discourse of danger in the United States throughout the nineteenth and twentieth centuries – with its ability to encompass the entire population, intensively structure the practices of everyday life, and offer a link between internal and external threats in ways that circumscribed the boundaries of legitimacy – is probably the best example of an effective discourse of danger.

(Campbell 1992: 196)

This was not the case for new dangers. In short, despite the evident encoding of terrorism, drug culture, Japan and others in the language of Cold War opposition, the importance of these themes should not be overstated. Table 13.1 shows that the relative importance of these themes when measured by volume of articles, is much less significant than communism ever was. Even in the after-effects of the World Trade Center, Oklahoma and Atlanta bombings, foreign terrorism has not had a particularly high profile in America, drugs have failed to sweep through the nation in the biblical proportions suggested in the 1980s, and perhaps the economy is too intangible a concept around which to forge a sense of national identity.[3]

'The barbarians are not at the gates. They are inside'

The relative unimportance of the threats to the *Digest's* America discussed above reveal myopia in dominant critical theorizing of American national identity. Such theories are contained by (neo)realist theory which assume that it is the structure of the international state system rather than the historically specific characteristics of the individual state societies that characterize and shape political events. For orthodox accounts, the dissolution of bipolarity and the apparent chaos of the multipolar state system shape the character of America and the future world-system. Many theorists and commentators have suggested that multiple regional threats will replace the Soviet threat, and that it is this combination of Others which will produce the greatest Danger in the destabilized and chaotic multipolar international politics that it fosters (see especially Mearsheimer 1990). Some critical analysts are also apparently trapped within this theoretical mold which anticipates that the agents of hegemonic US culture will seek an external source of danger against which conservative American political culture can write its vision of American destiny. Their understanding of the spatial politics of nation-state identity requires that an external threat is required in order to produce a coherent sense of self (e.g. see Campbell 1992). With the end of communism, the US will simply look for another malevolent presence to take over from the Soviet Union.

However what has characterized conservative American political culture in the late 1980s and particularly since the early 1990s is not its concern with a new external threat, but rather with dangerous groups within America. An analysis of the *Reader's Digest* since the end of the Cold War reveals how this magazine had turned to one of its perennial fears for US society: increased power to central government, and the resulting culture of dependency (see Table 13.1). The immediate threat to the *Reader's Digest's* 'America' is weakness in the morality and resolve of the American people. As one author stated: 'the barbarians are not at the gates. They are inside' (Sowell 1994: 180).

Post Cold War 'America'

Two interlinked themes dominated the *Digest's* representation of America in the early 1990s: a concern with big government and bureaucracy, and a fear of moral decline, particularly in what the magazine calls a culture of 'victimism'. Although such concerns have been prominent in the magazine for decades, both the level of coverage and the magnitude of the stated problem have increased in the wake of the Cold War. The magazine suggested that too much government was exacerbating all other problems. Articles about the US government sought to illustrate incompetence ranging from its failed good intentions to scandalous excesses. Most of the former articles provided anecdotes of the government's ineffective attempts to help the poor or disadvantaged. In most cases, the *Reader's Digest* suggested, people should be left to pull themselves up as help from the government would only lead to dependence.

One 1989 article explained the *Digest's* position on the dangers of government 'interference' most clearly through the voice of a successful ex-patriot Peruvian, Hernando de Soto, who returned to his homeland after the fall of the socialist government (Methvin 1989). De Soto found two settlements on his land which had been established at the same time, yet one group was rich, and the other poor. In the article he posed the question of why some people (and, adopting a questionable universalizing logic, also some nations) are rich and others poor. As all the settlers on his land were indigenous he concluded that the difference was not due to cultural factors, nor could it be due to external influences given the smallness of the area in question. Instead, he claimed that the wealthier community was successful because its members had 'badgered' government bureaucrats until they were given the titles to their land. These lucky peasants invested their wages on improvements to their own property and had become wealthy (Methvin 1989: 140). The poor peasants could not improve their lot because of a failure to beat government bureaucracy. Thus, de Soto concludes his report:

> I now know why some countries are poor and others rich. . . . We're a world of 169 countries, and only about twenty-five of them have 'made it' economically. They were able to do so because they stripped governments

344

of the power to deprive the humblest citizens of the fruits of their industry and creativity. The answer boils down to one word: *Freedom*.

(Methvin 1989: 140)

The *Reader's Digest* frequently berates the American government for its attempts to 'limit' personal freedom: 'why are they always trying to take our freedom away?' asked one *Digest* article (Armbrister 1986). Government bureaucracy is also blamed for causing problems as diverse as 'crippling the CIA' (Evans and Novak 1986), increasing spending through taxes (Brookes 1987), and imposing economic sanctions that worked against black interest in apartheid South Africa (Reed 1989).

Running implicitly through the articles described above, and more explicitly through others, is an opposition between the American Dream and a new social condition which the *Reader's Digest* recognized as the 'culture of victimism'. During the Cold War, the individualism of the American Dream was implied in the *Digest's* reporting of unnatural constraints to 'human nature' found under communism. Over the course of the late 1980s and 1990s, the American Dream emerged from the edges of articles on 'victims' which saw American individualism crushed by people who expected everything to be provided for them as a right. An article in 1993 explains that 'stripped of its pretensions, victimism is an ideology of the ego, an impulse to deny personal responsibility' (Sykes 1993). When people think that their lives are not perfect, claims the *Digest*, they can act badly without taking responsibility, an 'it's not my fault!' mentality (Hamill 1991; Reed 1994: 114). Moreover, this is not confined to those who naïvely expect too much but are now being manipulated by 'today's professional victims' which number '[r]adical feminists, ethnic minorities, homosexuals and other activists'. These people, the *Digest* continued, 'are taking *us* on a guilt trip into a mine field – no matter where *we* step, *we're* in trouble' (Epstein 1991: 122, emphasis mine).[4]

The victim mindset offers a profound challenge to the *Digest's* utopian view of American society within which each individual has the potential to 'make it'. Rather than realizing this potential to succeed, Americans seem to the magazine's editors, to be lapsing into an acceptance of disadvantage, an acceptance of a view that inequalities mean that not everyone can make it in America, at least not without the aid of the state. If people do not succeed they are seen to blame 'the system' rather than their own lack of effort and ingenuity. This undermines the ethos of individualism central to the *Digest's* understanding of American identity. A 1992 article explained the decline of the ability of the American Dream to script a national identification:

This isn't what our Founding Fathers had in mind when they enshrined in the Declaration of Independence the inalienable right to 'Life, Liberty and the pursuit of Happiness'. . . . Too many Americans have twisted the sensible right to *pursue* happiness into the delusion that we are *entitled* to a guarantee of happiness. If we don't get exactly what we want, we assume that someone must be violating our rights.

(Jacoby 19.92: 129)

The *Reader's Digest* was correct in a sense. On route to the New World, John Winthrop presented American exceptionalism in his 'A modell of Christian charity'. In Winthrop's view, a striking fact of humanity was inequality. Winthrop explained that God's will was for difference and inequality, and although this might seem to provoke antagonism, it made His creation all the more wondrous because the variety of people would be compelled to depend upon each other. This, claimed Winthrop, would strengthen community in the New World (Dolan 1994: chapter one). The *Reader's Digest* has simply secularized this view (although often still couched in quasi-religious overtones) to suggest that equality of opportunity should be at the heart of American democracy and not equality *per se*, which the magazine regarded as quite unnatural (Sharp 1996).

The creation of structures in society – which the notion of victims requires – is inimical to the *Digest's* desire to promote individual potential. Moreover, it runs contrary to the magazine's understanding of society as constructed of nothing more than the sum of its autonomous individual parts. Articles asked 'what *really* ails America?' (Bennett 1994). The answer is repeated many times over: a dissolving of morality and personal responsibility had led to a perception of barriers and hopelessness (*Reader's Digest* editorial review 1987). For the *Digest*, America had 'lost sight of the moral truths that give meaning to our lives' (Bennett 1987). 'Moral truths' have given way to an 'anything goes' mentality. Acceptance of difference – manifested to the magazine's writers as tolerance or promotion of multiculturalism, the presentation of 'alternative' histories, religious values and literary canons in schools, and multi-lingual US state bodies – suggest to the *Digest* that America's 'moral truths' are under attack.[5] A 1994 article quoted an author's greatest fear as:

seeing America, with all its great strength and beauty and freedom . . . gradually subside into decay through default and be defeated, not by the communist movement, but from within, from weariness, boredom, cynicism, greed and in the end helplessness before its great problems.

(Bennett 1994:201).

The *Reader's Digest* blames much of America's cultural decline on the school system which, it claims, threatens to leave students 'morally adrift' (Bauer 1987). The *Digest* insists that children need lessons in patriotism to appreciate freedom and is horrified that textbooks do not teach about, for example, the failings of the Soviet system. The magazine reprinted part of E. D. Hirsch's 'Cultural literacy' in 1987 to argue the point that schools were not fulfilling their societal role because they failed to teach 'the basic information necessary to maintain our democratic society' (Hirsch 1987). Cultural literacy is very much in synch with the *Reader's Digest* production of knowledge: it is not complex, abstract or theoretical academic knowledge, but instead a kind of informed common sense.

Inserted into Hirsch's article was an article about standards of 'cultural literacy' among students at the University of Southern California. This report presented to readers the fact that 'only a few [of the students] could articulate in any way at all why life in a free country is different from life in a non-free country' and concluded that '[i]n a state of such astonishing ignorance, young Americans may well not be prepared for even the most basic national responsibility – understanding what the society is about and why it must be preserved' (Stein 1987: 81). Hirsch's article concluded with the full implications of this lack in American society:

> I have in mind the Founding Fathers' idea of an informed citizenry. This is the basic principle that underlies our national system of education in the first place – that people in a democracy can be entrusted to decide all important matters for themselves because they can deliberate and communicate with one another.
>
> (Hirsch 1987: 83)

Evidently this poses a fundamental challenge to the *Digest's* belief that American democracy is upheld by voting and other political involvement undertaken by 'informed' US citizens. The danger lay in the fact that people in this state of 'astonishing ignorance' would not be able to distinguish good from evil and so would derail American destiny.

Despite this imminent sense of danger, the *Reader's Digest* has never reported American life as purely negative. In the post Cold War period it has juxtaposed articles on victimism with tales of triumph over adversity and initial disadvantage. This position is best illustrated by the *Digest's* remark that for one American hero, the 'concept of service to one's country flowed . . . as naturally as the American flag flies outside his family's Maryland home' (Hurt 1989: 66). In a similar vein, an October 1991 article told of Colin Powell's rise to power despite his lowly background (Reed 1994), and a new series called 'my first job' introduced noted Americans who told of humble early employment to claim that 'it's not what you earn, it's what you learn'. A 1992 article, 'From outcast to supercop', illustrated most clearly the *Digest's* belief in the power of individual decisions over disadvantage: 'Raped as a child, pregnant at sixteen,

Jacklean Davis had every reason to fail. Instead, she *chose* success' (Michelmore 1992: 179, emphasis mine). As a consequence, the *Reader's Digest* juxtaposes the new social threats facing the American nation and state with the possibility of moral and intellectual transformation in the 1990's.

Conclusion

The *Digest's* 1988 claim that Gorbachev's rise presented a moral threat to the US appears to be valid from the magazine's point of view. In the post Cold War period, not only has the 'moral void' of the Soviet Other, against which the magazine defined American identity and mission collapsed, but as a result, America's global moral leadership of the 'free world' has also declined from its Cold War heights.

David Campbell (1992: 195) claims that the set of practices comprising the Cold War represented a series of boundaries between 'civilization' and 'barbarity', and as a result, rendered a contingent identity of 'America' secure as the nation's essence. Thus, he argues that containment was not just a historically significant foreign policy strategy. Rather 'containment is a strategy associated with the logic of identity whereby the ethical powers of segregation that make up foreign policy constitute the identity of an agent in whose name they operate, and give rise to a geography of evil' (Campbell 1992: 195).

The territoriality of today's 'geography of evil' is not so easy to define as its Cold War counterpart was. There is no longer an 'evil empire' against which America's moral mission can be cast: terrorists, drug dealers and US government power as often compound the break-up of an overarching moral geography as hold it together. Without popular recognition of a powerful and ideological opponent, the *Reader's Digest's* ability to define, map, and contain the threat, is greatly limited. In the face of this fragmentation of danger it is very difficult for the magazine to present a mirror-image of American identity of any coherence. As the threats to 'America' fragment, so does its own identity. The implications of this are greater than the sales figures of a magazine however. Popular geopolitics are crucial to the cultural legitimation of more traditionally conceived geopolitics of statescraft. The collapse of moral geographies such as that produced by the *Reader's Digest*, undermines the commonly accepted values and narratives from which political leaders can draw in their scripting of world politics.

As a result, a fear at the end of the Cold War is the loss of international recognition of American moral superiority and imperial decline as American culture itself falls from its high ground in the eyes of its own people. Like Francis Fukuyama in his influential polemic on 'The end of history', the *Digest* must regard the end of the Cold War as 'a very sad time.' In Fukuyama's words, the struggle for recognition, the willingness to risk one's life for a purely abstract goal, the worldwide ideological struggle that called forth courage, imagination, and idealism, will be replaced by economic calculation, the endless solving of technical problems, environmental concerns, and the satisfaction of sophisticated consumer demands.

(Fukuyama 1989: 16)

The end of Cold War culture for Fukuyama, the end of Cold War romance in McClure's understanding, is characterized by boredom as the great challenges of previous decades – the challenges which defined 'America' for publications such as the *Digest* – fall away. As one commentator of Fukuyama's work suggested, 'his comment of boredom was meant to indicate that liberal states do not refer their citizens to "higher aims" leaving a vacuum that can be filled with sloth, self-indulgence, banality and the desire for wealth' (Peet 1993:64). The containment of the USSR acted simultaneously to contain 'America': it acted to discipline the myriad possible characterizations of 'America' into a coherent moral agent, and provided that power of authority to those who upheld and espoused these characterizations. Now that the geopolitical containment no longer operates, there is not the same common enemy to hold together Americans into following an agreed story and goal. As a result, the *Reader's Digest*, in its continued effort to promote its version of American identity, first reintegrated containment in narratives of terrorism, drug dealing and the economy, but eventually faced up to the question of the nature of changes to the American character directly.

In addition to offering insight into the ways in which people are drawn into the political process on a day-to-day basis, this form of analysis of popular culture has also challenged the view of American identity as always being organized around the spatial exclusion of an external enemy. American identity is not always wrought through the spatial exclusion of the enemy outside. Rather it is also formed through the identification of an enemy at home, an enemy which potentially poses a greater threat due to its challenge of the politics of inside and outside.

'Moral equality,' with the Soviet Union, wrote a 1988 *Digest* article, 'erodes our values and visions and compassion; that is its greatest danger' (Rosenthal 1988: 72). Morality is perhaps the key concept to understanding the *Reader's Digest's* 'America' and recent mainstream American political culture more generally. Morality structures the magazine's notion of American mission and destiny. But this is a morality which has been drawn through the magazine's lens of 'common sense', into the simple binary logic of right or wrong. As early as 1988 the *Reader's Digest* realized the moral dangers of the end of communism. If the world could no longer distinguish Soviet from American morality, then America's moral role would be lost in the ensuing relativism. The loss of America's high ground in the wake of a new moral geopolitics means that the authority of the *Reader's Digest*, as

349

supporter of the old geopolitics, is also under renewed challenge. The *Digest* has written its way into a particular history and identity of America. With the decay of the moral geopolitics of this old order, the *Reader's Digest*, like the conservative America it has helped to construct, is struggling to redefine its role.

Notes

1 This work is part of a larger project which has involved an examination of the content of the *Reader's Digest* and its role within American culture between 1922 and 1994 (Sharp 2000).
2 This is in stark contrast to 'good muslims': in Robert Kaplan's 'Why the Afghans fight' (1989), we are told that the author's translator prayed five times a day and exhibited no tension when author says he is Jewish: 'in Afghanistan – the Islamic faith has not been poisoned by Middle East politics' p.129.
3 Revelations about the internal base of more recent attacks in Oklhoma and Atlanta makes this issue much more complex, especially as it relates to the reconstruction of US identity.
4 Despite the *Digest's* claim to have reach into the greatest number of subgroups of all popular magazine's (Thomson, editor in chief 1976–1984, *pers. comm.*), this article is evidently addressing an audience which is exclusionary of 'radical feminists, ethnic minorities, homosexuals and other activists'.
5 For example: 'let's hear it [the Constitution] in English' September 1994, 'The supreme court is wrong about religion', December 1994, and 'thought police on campus' May 1991.

Bibliography

Adams, N. (1988) 'Cocaine king: a study in evil', *Reader's Digest*, Dec.: 227–72.
—— (1990) 'Iran's mastermind of world terrorism', *Reader's Digest*, Sept.: 59–65.
—— (1993) 'The terrorists among us', *Reader's Digest*, Dec.: 76–7.
Adelman, K. (1989) 'Arms control: games Soviets play', *Reader's Digest*, March: 65–9.
Agnew, J. (1998) *Geopolitics*, London: Routledge.
Anderson, B. (1983) *Imagined Communities*, London: Verso.
Armbrister, T. (1986) 'Why are they always trying to take our freedom away?', *Reader's Digest*, Oct.: 124–8.
Barnes, F. (1988) 'Can Gorbachev last?', *Reader's Digest*, May: 88–93.
—— (1990) 'The Japan that won't play fair', *Reader's Digest*, Aug.: 33–8.
Bauer, G. (1987) 'What we must teach our children about freedom', *Reader's Digest*, May: 102–4.
Bennett, R. K. (1987) 'The closing of the American mind', *Reader's Digest*, Oct.: 81–4.
Bennett, W. (1994) 'What really ails America?', *Reader's Digest*, April: 197–202.
Brookes, W. (1987) 'Don't raise taxes', *Reader's Digest*, Oct.: 163–6.
Campbell, D. (1992) *Writing Security: United States Foreign Policy and the Politics of Identity*, Minneapolis: University of Minnesota Press.
Dalby, S. (1990) 'American security discourse: the persistence of geopolitics', *Political*

Geography Quarterly 9(2): 171–88.

Der Derian, J. (1992) *Anti-Diplomacy:Spies,Terror, Speed andWar*, Oxford: Blackwell.

Dolan, F. (1994) *Allegories of America: Narratives–Metaphysics–Politics*, Ithaca and London: Cornell University Press.

Engelhardt, T. (1995) *The End ofVictory Culture: ColdWar American and the Disillusioning of a Generation*, Basic Books.

Epstein, J. (1991) 'Today's professional victims', *Reader's Digest,* April: 122–4.

Evans, R. and Novak, R. (1986) 'Congress is crippling the CIA', *Reader's Digest*, Nov.: 99–103.

Evans, R. and Novak, R. (1987) 'Gorbachev: the man with a nice smile and iron teeth', *Reader's Digest*, Oct.: 70–5.

Fukuyama, F. (1989) 'The end of history', *National Interest* supplement, Summer, 16pp.

Gilder, G. (1989) 'A new breed of innovators', *Reader's Digest*, Aug.: 126–8.

Glass, C. (1988) 'Kidnapped in Beirut', *Reader's Digest*, April: 90–7.

Hamill, P. (1991) 'It's not my fault', *Reader's Digest*, Oct.: 11–12.

Hirsch, E. D. (1987) 'Cultural literacy: what every American needs to know', *Reader's Digest*, Dec.:, 79–83.

House, K. (1989) 'Are we underestimating America's future?', *Reader's Digest*, May: 185–92.

Hurt, H. (1988) 'They dared cocaine – and lost', *Reader's Digest*, May: 81–7.

—— (1989) 'Portrait of a patriot', *Reader's Digest*, July: 65–9.

Jacoby, S. (1992) 'When rights run wild', *Reader's Digest*, Aug.: 129–30.

Kaplan, R. (1989) 'Why the Afghans fight', *Reader's Digest*, May: 128–32.

McClure, J. (1994) *Late Imperial Romance*, London and NewYork: Verso.

McConnell, M. (1989) 'Crack invades the countryside', *Reader's Digest*, Feb.: 73–8.

Mearsheimer, J. (1990) 'Why we will soon miss the ColdWar', *Atlantic* 266(2): 35–50.

Methvin, E. (1986) 'Our new defense weapon – competition', *Reader's Digest*, Sept.: 99–103.

—— (1989) 'Crusader for Peru's have-nots', *Reader's Digest*, Jan.: 137–40.

—— (1991) 'Tampa's winning war on drugs', *Reader's Digest*, July: 56–60.

Michelmore, P. (1992) 'From outcast to supercop', *Reader's Digest*, Nov.: 179–86.

Netanyahu, B. (1986) 'Terrorism: how theWest can win', *Reader's Digest*, July: 110–15.

Ó Tuathail, G. (1996) *Critical Geopolitics*, Minnesota: Minnesota University Press.

Ó Tuathail, G. and Agnew, J. (1992) 'Geopolitics and discourse: practical geopolitical reasoning in American foreign policy', *Political Geography* 11(2): 190–204.

Peet, R. (1993) 'Reading Fukuyama: politics at the end of history', *Political Geography* 12(1): 64–78.

Reader's Digest editorial review (1987) 'The closing of the American mind', *Reader's Digest*, Oct.: 81–7.

Reader's Digest Condensation (1989) 'Why we're losing the war on drugs', *Reader's Digest*, Oct.: 83–8.

Reader's Digest (1990) 'Exclusive, Interview with the President', *Reader's Digest*, Jan.: 53–9.

Reed, D. (1989) 'Do South Africa sanctions make sense?', *Reader's Digest*, Feb.: 51–6.

Reed, J. (1994) 'It's not my fault!', *Reader's Digest*, Aug.: 113–14.

Rosenthal, A. (1988) 'Gorbachev's hidden agenda', *Reader's Digest*, March: 71–2.

—— (1991) 'Don't remilitarize Japan', *Reader's Digest*, Feb.: 59–60.

Said, E. (1978) *Orientalism*, New York and London: Vintage.

Sharp, J. (1996) 'Hegemony, popular culture and geopolitics: the *Reader's Digest* and the construction of danger', *Political Geography* 15(6/7): 557–70.

—— (2000) *Condensing Communism: the Reader's Digest and American identity, 1922–1994*, Minneapolis: University of Minnesota Press.

Shapiro, M. (1989) 'Representing world politics: the sport/war intertext', 69–96 in J. Der Derian and M. J. Shapiro (eds) *International/Intertextual Relations: Postmodern Readings of World Politics*, Lexington, Mass.: Lexington Books.

Shear, J. (1991) 'Don't envy the Japanese', *Reader's Digest*, May: 43–6.

Sowell, T. (1994) 'Who say's it's hopeless?', *Reader's Digest*, June: 179–80.

Stein, B. (1987) 'Cultural literacy', *Reader's Digest*, Dec.: 79–83.

Sykes, C. (1993) 'No more victims please', *Reader's Digest*, Feb.: 21–4.

Wills, G. (1997) *John Wayne: The Politics of Celebrity*, London: Faber.

Wilson, C. (1983) 'The rhetoric of consumption: mass market magazines and the demise of the gentle reader, 1880–1920, 39–64 in R. W. Fox and T. J. Jackson Lears (eds) *The Culture of Consumption: Critical Essays in American History 1880–1980*, New York: Pantheon.

14

TOWARD A GREEN
GEOPOLITICS

Politicizing ecology at the Worldwatch institute

Timothy W. Luke

Introduction

This chapter re-frames geopolitics in ecological terms to test the depth and breadth of some perplexing new tendencies. With the end of the Cold War, transnational corporate enterprise now reigns more or less supreme as the planet's most effective bloc of productive forces as well as its most articulated relations of production. This corporate capitalist economy, in turn, legitimizes its operations for many clienteles around the world with measures of how fully, broadly or deeply it satisfies the wants and needs of consumers. One of the few remaining sites of effective resistance to this globalized corporate economy has been environmentalism. Yet, to gain mass support or build new political constituencies, environmentalism, like transnational corporate capitalism, increasingly is forced to pitch its message in consumerist terms (Luke 1993).

In struggling to control the industrial uses of nature, corporate capitalism and organized environmentalism tussle over the conditions of consumption, struggling to determine who should manage the material ends and productive means of global markets (Luke 1997). Plainly, not all businesses are mindless polluters, and not all environmentalists are anti-business. Likewise, not all environmentalism is consumerist, and not all consumption is anti-ecological, but these sets of economic and technological contradictions produce some intriguing new connections between geography and politics that merit future investigation. Therefore, this chapter reconsiders one prominent environmentalized reconstruction of geopolitics, as it is emerging today out of the elective affinities drawing together mainstream environmentalism and modern consumerism. Whether we stand at 'the end of history' or 'the end of Nature,' a green geopolitics shows how many older ideological divisions now seem much less certain as new battles over the control of our environments are being waged all over the world (Fukuyama 1992; and McKibben 1989). The specific example that will be addressed here is a well-known, environmental organization in the United States, namely, the Worldwatch Institute. Its many activities represent very well the operational philosophies, ecological goals, and policy orientations of a new, green geopolitics that seeks to guide human political economy by the stars of its technoscientific political ecology.

Green geopolitics

The widespread interest in talking politically about managing the Earth's ecology first surfaced as a new kind of popular geopolitics in the United States among local and national ecology movements during the 1960s. It has become far more pronounced, however, in the 1990s, following the end of the Cold War. Having won the long twilight struggle against communist totalitarianism, the United States is governed by leaders who now see 'Earth in the balance', arguing that global ecologies should incarnate what is best, and not worst, in the human spirit (Gore 1992). Economists, industrialists, and political leaders represent the strategic terrain of the post-1991 world system as one on which all nations must compete ruthlessly to control the future development of the world economy by developing new technologies, dominating more markets, and exploiting every national economic asset. Indeed, the phenomenon of 'failed states', ranging from virtually dysfunctional jurisdictions, like Rwanda, Somalia or Angola to essentially crippled entities like Ukraine, Afghanistan or Kazakhstan, is attributed to the severe environmental frictions caused by the unwise abuse of Nature following ineffective attempts to create economic growth (Kaplan 1996).

Taking 'ecology' into account, then, creates a series of discourses about 'the environment' that derive not only from morality, but from rationality as well. As humanity has faced 'the limits of growth' and heard 'the population bomb' ticking away, ecologies and environments have become something more than what one must judge morally; they are things that governments must administer. To follow Foucault, ecology has evolved into 'a public potential; it called for management procedures; it had to be taken charge of by analytical discourses,' as it was recognized in these more environmentalized manifestations to be 'a police matter' – 'not the repression of disorder, but an ordered maximization of collective and individual forces' (Foucault 1980: 146).

'Geo-politics', as Ó Tuathail argues, 'does not mark a fixed pressure but an unstable and indeterminate problematic', and it should not be regarded as 'an "is", but as a question' (Ó Tuathail 1996: 67). Destabilizing geopolitics as 'geo-politics' in accord with Ó Tuathail's insight permits one to disconnect geopolitics from its historical engagements in *Realpolitik*, and then search for new connections in the geopolitical amidst its ultimately ambiguous and indeterminate meanings. The unnamed operational practices embedded in 'geo-politics' permit this cluster of conceptual values and practices to assume and acquire new practical deployments elsewhere and at other times much removed from its diplomatic/imperial/military traditions. The full development of a 'closed political system . . . of world-wide scope' described by Halford Mackinder as the 'post-Columbian age' (1904: 422) does not acquire complete quiddity in the collective political imagination until 1968 when Apollo 8 returns with its full-colour photographs of the Earth floating alone in the darkness of space (Cosgrove 1994). While the closure of most terrestrial space to easy territorial acquisition was underscored by the First and Second World Wars, the ecological reimagination of the planet as Spaceship Earth on or around 20 July 1969 – when Neil Armstrong set foot on the moon – finally confirmed Mackinder's belief that what has been understood as geopolitics must divert 'the attention of statesmen in all parts of the world from territorial expansion to the struggle for relative efficiency' (Mackinder, 1904: 422). Mackinder's own uniquely Victorian British anxieties about the future well-being of the empire and nation manifest themselves in his thoughts about 'relative efficiency' in the increasingly competitive and interconnected world of the early 1900s. His conservative imperialist proclivities led him to worry about how capable the British Empire would be at managing its hard-won territorial assets in this new era of comparatively complex interdependence (Ó Tuathail 1996; Ryan 1996).

Interestingly enough, ecology emerges as a geopolitical metaphor at the same time that the millenarian spark in world communism finally died out, leaving the whole world more or less locked into fixed patterns of growing industrialization and urbanization. The socialist critique of capitalism and the capitalist counter-attacks against communism come to an ideological impasse in the 1960s, but both systems could easily be subjected to an ecological critique of their relative efficiencies. By the dawning of Earth Day in 1970, many people in the capitalist west and communist east no longer held out much promise for a real socialist transformation of existing industrial economies and societies. New anxieties about the security and sustainability of new consumerist lifestyles, which were proving much more attractive than those productivist ones provided by neo-Stalinism, bubble into political discourse in the forms of ecological thought. In ecology, there are highly suggestive echoes and reflections of Mackinder's *fin de siècle* world vision (Kearns 1993). Indeed, with ecology, geopolitics for the first time truly 'can perceive something of the real proportion of features and events on the stage of the whole world, and may seek a formula which shall express certain aspects, at any rate, of geographical causation in universal history', and these new ecological constructs 'should have a practical value as setting into perspective some of the competing forces in current international politics' (Mackinder 1904: 422). Mackinder's geopolitical gaze, then, ironically can prepare the ground for a new standard of practical value to the environmentalized perspectives driving contemporary international economics and politics (Ó Tuathail 1992). A green geo-politics would complete the logic of Mackinder's ethnocentric imperialist vision by globalizing its administrative reach all the way down into the last micro-organisms in the remotest biome of the Earth's biosphere. Like Mackinder, a green geopolitics suggests that:

> The space of the world is now, for all intents and purposes, known, occupied and closed. The world has become a single unified globe of occupied space, a system of closed space (a closed spaceship earth) where events in one part inevitably have their consequences in all other parts. It is no longer possible to treat various struggles for space in isolation from one another, for all are part of a single world-wide system of closed space. The world of international interactions is now global.
>
> (Ó Tuathail 1996: 27)

With these operational assumptions in place, 'the struggle for relative efficiency' can emerge as the *leitmotif* of an environmentalistic political economy, which can be well-supported by re-examining the Worldwatch Institute's studies of the Earth's systems of ecological and economic exchange.

The Worldwatch Institute, as its mission statement suggests, 'believes that information is a powerful tool of social change', and, its goals, therefore, are to provide the information needed 'to raise public awareness of global environmental threats to the point where it will support effective policy responses' (Worldwatch 1999). By conducting what it sees as non-partisan interdisciplinary research on major environmental challenges at its headquarters in Washington DC, the Worldwatch Institute helps to set the larger agenda for international regulatory practices pertaining to global climate change, ozone depletion, resource conservation or biodiversity protection. In this mode of operation, Worldwatch Institute experts often participate in ongoing international negotiations over climate, ozone, biodiversity or overpopulation. Yet, it has no formal role in actually monitoring or managing the processes of environmental regulation, because the Worldwatch Institute sees its institutional mission in terms of policy research and information dissemination.

On one level, the Worldwatch Institute worries about chaos in this closed system when it cites with approval scientific ecologists, like Jane Lubchenco, who argues 'we're changing the world in ways that it's never been changed before, at faster rates and over larger scales, and we don't know the consequences. It's a massive experiment, and we don't know the outcome' (Knickerbocker 1998: 1). Yet, on another level, the operational approach of every Worldwatch Institute report is to deal with this chaos managerially and scientifically by generating a preliminary set of hard numbers with considerable accuracy on all the ecological crises currently being caused by the global economy. This operational approach to the Earth as a single world ecological system constitutes a new geo-political stance *vis-à-vis* both the workings of technoscience and the state.

Very little of the Worldwatch Institute's work takes the form of direct action out in the field, because most of its practices instead have been steered toward geopolitical analysis and stocktaking in the administrative registers of ecological intelligence-gathering and dissemination. After fifteen years of growing support and increasing circulation, the annual reports of the Worldwatch Institute now provide key intelligence for a global network of green geopolitical strategists and tacticians. As Brown suggests, the impact of the Worldwatch Institute's work is now felt world-wide:

> When we launched the first *State of the World* report fifteen years ago, we had high hopes for its impact, but did not anticipate that it would become semi-official, widely used by government officials, UN agencies, corporate planners, educators, and environment activists around the world. . . . never in our fondest dreams did we anticipate that it would one day appear in some 30 languages. . . . We did not anticipate that one day the first printing in the United States would be 100,000 copies. Nor did we expect that we would be on the best seller list in Finland and Argentina.
>
> (Brown 1998: xix)

With this self-celebratory review of its first fifteen years of service, the reception of the Worldwatch Institute's *State of the World* reports also reveals the main audiences and major advocates for its unique form of green geopolitics. One can follow Foucault here by seeing how the Worldwatch Institute's environmentalism operates as 'a whole series of different tactics that combined in varying proportions the objective of disciplining the body and that of regulating populations' (Foucault 1980: 146). The political project of 'sustainability' in green geopolitics, whether one addresses either the sustainable development of one national economy or sustainable use of all global resources, embeds this new managerial responsibility for the life processes of Earth's ecologies in many governments' efforts to rationally harmonize their political economy with some vision of world ecology.

From ecology to hyperecology

To preserve the various ecosystems of the earth, as the Worldwatch Institute asserts, the inhabitants of each human community must rethink the entire range of their economic and technological interconnections to local habitats in terms of how they are meshed into the regional, national, and international exchange of goods and services. Beginning this strategic review immediately poses the question of protecting all existing concrete 'bioregions' in first nature, or the larger *biosphere* of

357

the planet, within which the ecologies of any and all human communities are rooted. Bioregions historically have constituted the particular spatial setting of human beings' social connections to specific lands, waters, plants, animals, peoples, and climates from which their communities culturally constitute meaningful places for themselves in the 'first nature' of the natural biosphere (Sale 1985).

The 'domination of nature,' however, is not so much the total control of all natural events in the environment as much as it is a wilful disregard by many people of localized ecological conditions and constraints in building human settlements (Berry, 1989; Commoner 1990). Out of this disregard, humanity constructs many abstract 'technoregions' from fabricated objects which become so pervasive that these domains constitute a 'second nature'. This always emergent *technosphere* of the planet, within which all modernizing human communities also are now mostly embedded, operate by virtue of technoeconomic exchanges that far exceed the carrying capacities of their natural habitats. These transactions, in turn, create new anthropogenic conditions in the natural environment, which generate their own highly artificial hyperecology of an unsustainable type (McKibben 1989).

This managerial reconstruction of the environment by the Worldwatch Institute as an interlaced ensemble of loosely coupled cybernetic systems in the biosphere and artificial technosphere echoes Spykman's reading of geophysical features on the Earth as ontologically fundamental forces. Just as 'geography does not argue; it just is', (Spykman 1938: 237) so too does 'the environment' in every Worldwatch Institute Report not argue; it just is. Yet, these naive assertions already have integrated into their ontological assumptions those postulates of definition needed to determine what thoughts are now legitimate, which actions must be taken, and whose interests shall be served. Presuming simply that the environment 'just is' only submerges their particular normative agendas beneath the general presence of ecological facticity. Approaching the environment as a bundle of inexorable necessities, immanent imperatives, and unalterable truths, however, assumes a rhetorical stance that mobilizes the Earth's constraints and energies to support this over-determined reading of Nature.

As new geopolitical spaces, environments can be monitored to judge their relative success or failure in terms of abstract mathematical measures of consumption, surveying national gains or losses by the density, velocity, intensity, and quantity of goods and services being exchanged in the systems of mass consumption constituting the technosphere (Luke 1995). Here one finds the Worldwatch Institute pushing for wiser uses for all biotic assets from the biosphere. Consumption is outsourced from many different planetary sites by using varying levels of standardized energy, natural resources, food, water and labour inputs drawn from all over the Earth through the mode of production's transnational commodity, energy, and labour markets (Brown *et al.*, 1999, 1998, 1997, 1996, 1995). Green geopolitics works behind state power and/or market clout by providing new measures of the requisite force and/or capital needed to impose these costs on the many for the benefit of the few. By substituting the symbolism of 'Earth Days' for substantial ecological transformation, technospheric hyperecologies of transnational exchange are repackaging themselves successfully in green wrappers of ecological concern; but, they still require the profligate waste of energy, resources, and time in biospheric ecologies to maintain the abstract aggregate satisfaction of 'average consumers' enjoying 'the typical standard of living' in the developed world's cities and suburbs.

Within the existing system of industrial objects, the Worldwatch Institute finds a new fundamental truth:

> *objects now are by no means meant to be owned and used but solely to be produced and bought.* In other words, they are structured as a function neither of needs nor of a more rational organization of the world, but instead constitute a system determined by an ideological regime of production and social integration.
>
> (Baudrillard 1996: 162–3)

The technosphere's system of objects reifies systematic human behaviours in each and every technoregion, which also rest upon inputs from and outputs into the biosphere. The *habitus* in this system of objects now is humanity's most real habitat, and consumer society is the (con)fusion of bioregions with technoregions: 'a social realm, a temporal realm, a realm of things by virtue of which, and by virtue of the strategy that imposes it, objects are able to fulfil their function as accelerators and multipliers of tasks, satisfactions, and expenditures.' (Baudrillard 1996: 102) To save bioregional habitat, a green geopolitics must reshape the technoregional *habitus* (Bourdieu 1984: 170).

Connecting this system of consumption to a new geopolitics is how the contemporary state and economy transform raw consumerism with all of its inherent tasks, satisfactions, and expenditures into 'sustainable development'. Transnational capitalist production's (con)fusion of *habitus* and habitat, economy and ecology, domicile and dominion culminates in a global geopolitics that intermixes carrying capacities with credit cycles as environmentalized biospheres/technospheres: 'everything has to intercommunicate, everything has to be functional – no more secrets, no more mysteries, everything is organized, everything is clear' (Bourdieu 1984: 170). The real cultural contradictions of contemporary capitalism are not those of accumulation versus expenditure or repressiveness versus permissiveness, but rather those of ecology versus exchange as the goals of this new geopolitics require a re-engineering of environments within the oxymoronic practices of sustainable development.

Environmentalism as green geopolitics

Geopolitical logics come from somewhere, and the environmental movement is now where many of their latest axioms arise. As one very high-profile attempt to recast the forces of nature to serve the economic exploitation of advanced technologies, the Worldwatch Institute's rational management of ecological energies provides a significant geopolitical supplement to transnational commercial interests promoting the growth of the global economy.

Seeing the path of untrammelled consumerism as the cause of today's environmental crises, a major Worldwatch Institute study by Brown, Flavin and Postel attributes the prevailing faith in more growth to 'a narrow economic view of the world' (1991: 21). Any constraints on further growth are cast by conventional economics 'in terms of inadequate demand growth rather than limits imposed by the earth's resources' (Brown, Flavin, and Postel 1991: 22). For the Worldwatch Institute, however, ecologists should push outside of the technosphere to study the complex changing relationships of organisms with their environments, and, for them, 'growth is confined by the parameters of the biosphere' (ibid.: 22). Economists ironically regard ecologists' concerns as 'a minor subdiscipline of economics – to be "internalized" in economic models and dealt with at the margins of economic planning', while 'to an ecologist, the economy is a narrow subset of the global ecosystem' (ibid.: 23). To end this division, the discourse of dangers propagated by the Worldwatch Institute pushes to merge ecology with economics. This union, in turn, can both infuse environmental analysis with economic instrumental rationality and defuse economics with ecological systems reasoning. Once this is done, economic growth no longer can be divorced from 'the natural systems and resources from which they ultimately derive,' and any economic process that 'undermines the global ecosystem cannot continue indefinitely', (ibid.: 23) which permits the Worldwatch Institute to give this new geographic and political administration of the planet a green tint.

With such rhetorical manoeuvres, the Worldwatch Institute articulates its designs for consumer economics as the instrumental rationality of resource managerialism. Working on a global scale in transnationalized registers of global administration, the Worldwatch Institute hopes to perfect the wastefulness of overly consumptive societies. Nature is to be reformed by the Worldwatch as a cybernetic system of biophysical systems, whose many geophysical formations appear among today's nation-states in 'four biological systems – forests, grasslands, fisheries, and croplands – which supply all of our food and much of the raw materials for industry, with the notable exceptions of fossil fuels and minerals.' (ibid.:73). The joint performance of such systems must be monitored with analytical spreadsheets written on bioeconomic terms. It then can be judged in managerial equations, which balance constantly increasing human population, constantly running base ecosystem outputs, and highly constrained possibilities for increasing ecosystem output against fairly inflexible limits on more technical inputs. In re-examining these four systems, the Worldwatch Institute recognizes that Nature mostly is a system of energy-conversion systems:

> Each of these systems is fuelled by photosynthesis, the process by which plants use solar energy to combine water and carbon dioxide to form carbohydrates. Indeed, this process for converting solar energy into biochemical energy supports all life on earth, including the 5.4 billion members of our species. Unless we manage these basic biological systems more intelligently than we now are, the earth will never meet the basic needs of 8 billion people.
>
> Photosynthesis is the common currency of biological systems, the yardstick by which their output can be aggregated and changes in their productivity measured. Although the estimated 41 per cent of photosynthetic activity that takes place in the oceans supplies us with seafood, it is the 59 per cent occurring on land that supports the world economy. And it is the loss of terrestrial photosynthesis as a result of environmental degradation that is undermining many national economies.
>
> (Brown, Flavin and Postel 1991: 73–4).

Photosynthetic energy generation and accumulation, then, becomes the new accounting standard for submitting the Earth's ecologies to environmentalizing geopolitical discipline. It imposes upper limits on economic expansion; the earth is only so large. The 41 per cent that is aquatic and marine as well as the 59 per cent that is terrestrial are actually decreasing in magnitude and efficiency due to 'environmental degradation'. Partly localized within national territories as politically bordered destruction, and partly globalized all over the biosphere as biologically unbounded transboundary pollution, the technosphere's imbrication with the biosphere's system of systems needs global management, or some powerful, all-knowing centre of surveillance, like the Worldwatch Institute, to mind its environmental resources.

Such managerial requirements follow from the convergence of dangerous trends, namely, bioeconomic accounting suggests that,

> 40 per cent of the earth's annual net primary production on land now goes directly to meet human needs or is indirectly used or destroyed by human activity – leaving 60 per cent for the millions of other land-based species with which humans share the planet. While it took all of human history to reach this point, the share could double to 80 per cent by 2030

if current rates of population growth continue; rising per capita consumption could shorten the doubling time considerably. Along the way, with people usurping an ever larger share of the earth's life-sustaining energy, natural systems will unravel faster.

(Brown, Flavin, and Postel 1991: 74)

To avoid a collapse of ecological throughput, ever increasing raw levels of mass consumption must end. Human beings must check their increasing levels of population, reorganize their wasteful resource-intensive modes of production, and limit their rising rates of excessive material consumption. All of these ends, in turn, require a measure of geographical surveillance and degree of political administration perhaps beyond the powers of modern nation-states, but not beyond those exercised by some transnational intergovernmental organization following a new globalized green geopolitics.

This geopolitics clearly engages the disciplinary task of policing the Earth by somehow equilibrating the 'net primary production' of solar energy fixed by photosynthesis in the Institute's four biological systems to match human needs in global production and consumption. Natural processes in the planet's total solar economy, first, are reduced rhetorically to energy extraction and distribution systems, such as food stocks, fisheries, forest preserves, and grass lands only to return, second, as geopolitically environmental assets, enveloped in bioeconomic accounting procedures and encircled by green managerial programs. Worldwatch presumes to know all of this, and in knowing it, to have mastered all of its economic/ecological implications. This sort of authoritative technical analysis, in turn, can perfect human material consumption for any would-be wardens of this overall planetary solar economy. By questioning the old truth regime of mere productive growth, environmentalized consumption within a more sophisticated ecological economy stands ready to reintegrate human production and consumption in a balance with the Earth's four biological systems.

The World Commission on Environment and Development admits humanity has been unable to fit 'its doings' into the 'pattern of clouds, oceans, greenery, and soils' that is the Earth. The hazards of this new reality cannot be escaped, but they 'must be recognized – and managed' (World Commision on Environment and Development 1987: 1). Through a green geopolitics, 'we can see and study the Earth as an organism whose health depends on the health of all its parts', which gives us 'the power to reconcile human affairs with natural laws and to thrive in the process' (Grubb et al. 1993: 87). This reconciliation rests upon the work of agencies like the Worldwatch Institute, which provides an understanding of 'natural systems', furthers the expanding of 'the environmental resource base', helps in managing 'environmental decay', and assists the controlling of 'environmental trends'. As the Rio Declaration asserts, Earth's 'integral and interdependent nature' can be redefined and reduced to 'the global environmental and developmental system', and with this rhetorical manoeuvre, what was once 'wild Nature' becomes tame ecosystemic infrastructure for green geopolitics.

Ecology, as green geopolitics, can be recalibrated in discourses of governmentality such as those circulated by the Worldwatch Institute, as an absolutely essential science with the most apt theories and practices for a world whose gross planetary product acquires form and substance in the commodified space of global product flows. Transnational enterprises, in turn, are now the major new life-forms inhabiting the world-wide range of the biosphere, and green geopolitics with its environmentalistic reasoning best represents their struggle for survival in accord with Mackinder's goal of relative efficiency. Indeed, green geopolitics must rationalize social goals with regard to the convenience of their ends and the rightness in their disposition of things. For such world-class traders, workers, and citizens, the Worldwatch Institute can monitor constantly the environment for any signs of dysfunction, disruption, and delegitimation which might resonate negatively for the clients and servers of global business.

No longer nature, not merely ecosystem, the biospheres/technospheres of a world under this kind of watch truly reduce it to geopolitical strategic spaces. The health of global populations as well as the survival of the planet itself compel humanity to draw up a bioeconomic spreadsheet that can be draped over the Earth, generating an elaborate set of accounts for a economical geopolitics of global reach and local scope. Hovering over the world in their scientifically-centred institutes of green surveillance, the disciplinary grids of this Worldwatch now track efficiency and waste, health and disease, poverty and wealth as well as growth and stasis, stability and instability employment and unemployment. Fusing ecology and economics in geo-politics, Brown, Flavin and Postel declare 'the once separate issues of environment and development are now inextricably linked' (1991: 25). They are linked, at least, in the discourses of Worldwatch Institute as its experts survey nature-in-crisis by auditing the Earth's levels of topsoil depletion, air pollution, acid rain, global warming, ozone destruction, water pollution, forest reduction, and species extinction as problems brought about by excessive mass consumption.

The Worldwatch Institute's geopolitics would have all states govern people through their things, and the instrumental ends that things serve, by restructuring today's ecologically unsound system of objects with elaborate managerial designs to create tomorrow's environmentally sustainable economy out of ecologically vetted objects for environmentalized living (Brown 1981). The shape of a new green geopolitics emerges from a re-engineered economy of environmental technologies and practices (wind power, bicycles, vegetarian diet) approved by the Worldwatch Institute. The individual human subject, and all of his or her things tied to their currently unsustainable ecological practices, would be reshaped through this green geopolitics, as the practices, discourses, and ensembles of such geopolitical administration would more efficiently synchronize the bio-powers of populations with the geo-powers of their environments. To police global carrying capacity, the Worldwatch Institute would direct each human subject to assume the much less capacious carriage of disciplinary frugality instead of affluent suburban abundance. All of the world must come under this geopolitical watch, and the Worldwatch Institute would police its human charges to dispose of their things and arrange their ends – in re-engineered spaces using new energies at new jobs and leisures – around these agendas.

Sustainability, like sexuality, now can become another expert discourse about exerting power over life. This fascination with relative efficiency at the Worldwatch Institute is behind its endorsement of 'an economy that does not rely on the one-time pollution of the atmosphere, clear-cutting of forests, and over-pumping of aquifers', which is, at the same time, not only within reach, but 'in the end would be more economical – and productive – than the one that supports us today' (Brown, Flavin, and Postel 1991: xviii). What the bio-power strategies of the eighteenth and nineteenth centuries helped fabricate in terms of human sexuality are now being re-imagined for humanity in worsening global conditions of survival as a more perfect ecological survivalism. How economic development might 'invest life through and through' becomes a new sustainability challenge, once geopolitical relations are established, must make these investments permanently profitable as new consumer systems of objects (Foucault 1980: 13). As a result, the Worldwatch Institute issues pamphlet after monograph after book on the supreme virtues of bicycles, solar power, windmills, urban planning, or organic agriculture to reveal higher forms of consumer goods perfection now fully attainable by the system of objects. Sustainability discourses more or less presume that some level of material and cultural existence has been attained that is indeed worth sustaining. This economic formation, then, constitutes 'a new distribution of pleasures, discourses, truths, and powers; it has to be seen as the self-affirmation of one class rather than the enslavement of another: a defence, a protection, a strengthening, and an exaltation . . . as a means of social control and political subjugation' (Foucault 1980: 123). Sustainable development means developing new consumer powers by defining a new model of green subjectivity organized around sustaining both new object worlds in a more survivable second nature and new consumer systems for their surviving subjects (Luke 1996).

These new geopolitical readings of the environment have brought fresh discourses of social responsibility to public attention, including the light green geo-politics of the Clinton administration with its own intriguing codes of limited ecological reflexivity. The presidential pledge to deploy American power as an environmental protection agency has waxed and waned over the past quarter century, but in 1995 President Clinton made this green geo-politics an integral part of his global doctrine of 'engagement'. 'To reassert America's leadership in the post-Cold War world', and in moving 'from the industrial to the information age, from the Cold War world to the global village', President Clinton asserted

> we know that abroad we have the responsibility to advance freedom and democracy – to advance prosperity and the preservation of our planet . . . in a world where the dividing line between domestic and foreign policy is increasingly blurred. . . . Our personal, family, and national future is affected by our policies on the environment at home and abroad. The common good at home is simply not separate from our efforts to advance the common good around the world. They must be one and the same if we are to be truly secure in the world of the twenty-first century.
>
> (Clinton 1995: 43)

By becoming the key agency of environmental protection on a global level, the Clinton administration sees itself reasserting American world leadership after the Cold War in accord with Mackinder's search for a new relative ecological efficiency among nations. As the world's geopolitical leader, America stipulates that it cannot advance economic prosperity and ecological preservation without erasing more of the dividing lines between domestic and foreign policy. In the blur of the coming Information Age and its global villages, the United States must not separate America's common good from the common goods of the larger world. To be truly secure in the twenty-first century, each American's personal, family, and national stake in their collective future should be served through the workings of national environmental policies. Secretary of State Christopher confirmed President Clinton's engagement with the environment through domestic statecraft and diplomatic action: 'protecting our fragile environment also has profound long-range importance for our country, and in 1996 we will strive to fully integrate our environmental goals into our diplomacy – something that has never been done before' (Christopher 1996: 12).

These efforts to connect economic growth with ecological responsibility are developed most systematically in Vice President Al Gore's environmental musings. To ground his green geo-politics, Gore argues that 'the task of restoring the natural balance of the Earth's ecological system' should also reaffirm America's long-standing 'interest in social justice, democratic government, and free market economics' (1992: 270). The moral authority unlocked by this official ecology might even be seen as bringing 'a renewed dedication to what Jefferson believed were not merely American but universal inalienable rights: life, liberty, and the pursuit of happiness' (Gore 1992: 270). At another level, however, Gore asserts that America's global strategies after the Cold War must also re-establish 'a natural and healthy relationship between human beings and the earth', replacing the brutal exploitation of Nature with an 'environmentalism of the spirit' (Gore 1992: 218, 238).

Gore's programme for earth stewardship takes something of a geopolitical turn when he calls for the creation of a Global Marshall Plan to embed sustainable development at the heart of ecological policy. In that historic post-Second World War programme program, as Gore notes, several nations joined together 'to reorganize an entire region of the world and change its way of life' (Gore 1992: 296). Like the Marshall Plan, his new Global Marshall Plan would 'focus on strategic goals and emphasize actions and programs that are likely to remove the bottlenecks presently inhibiting the healthy functioning of the global economy . . . to serve human needs and promote sustained economic progress' (ibid.: 297). The possible forms of ecological sustainability in mass consumption are re-moulded here into the actual program of a new economic growth ideology. Sustaining Nature by preserving consumption from its ecosystems in green geo-politics becomes now an essential goal of American foreign policy in complete accord with hopes advanced by the Worldwatch Institute. Vice-President Gore says the right things about changing our economic assumptions about mindless consumerism, but his bottom line for sustainable development is to be found in sustaining American business, industry and science through more ecologically mindful forms of consumption. As the world's leading capitalist economy, Gore concludes 'the United States has a special obligation to discover effective ways of using the power of market forces to help save the global environment' (ibid.:274). Vice-President Gore already has mobilized a small cadre of around forty men and women, known informally as the 'Gore-techs', to help him with this unique mind meld of high-tech entrepreneurialism and high-touch environmentalism for the year 2000 presidential race (Simons, 1998: A1, A6). Whatever exists in the natural world that cannot come under such a high-tech Worldwatch simply will be recast on the Internet as a zone suitable for digital simulations.

Vice-President Gore recently confirmed these ideological commitments both to the telematics and the environment by charging NASA to provide images of Earth, spinning in space, on the Internet. Billed as 'all-Earth, all-the-time', these video pictures would be broadcast live over the WWW from a small spacecraft stationed somewhere between Earth and the Sun (Sawyer 1998: A1). Gore still has a photo blow-up of the Earth from Apollo 17 displayed in his West Wing office at the White House, and he now claims this continuous live feed of Spaceship Earth shot from space by NASA would have 'tremendous scientific value' (Sawyer 1998: A13).

While its scientific value is contestable, this space-based system of surveillance has real geopolitical effects. The reaffirmation of environmental vigilance in green geopolitical discourses in the 1980s and 1990s arguably is altering the behaviour of many corporate and state agencies toward Nature. Because Earth, as Al Gore asserts, is in the balance, the raw externalization of many environmental costs to generate some economic benefits is becoming less common in most countries around the world, if not in fact then, at least, in principle. Yet, a more refined internalization of ecological debits and credits, such as that given by the Worldwatch Institute, also implicitly articulates a new geopolitical understanding of the world. One must push past the gratifying green glow emanating from documents like the Brundtland Report or Agenda 21 in which humanity allegedly appears ready to call an end to its war against Nature and to launch a new era of peaceful coexistence with the Earth's wild expanses and untamed creatures. In fact, these mostly diplomatic initiatives, like many other visions of sustainable development, balanced growth or ecological modernization, only underscore the validity of Jameson's take on post-modernity: a situation in which 'the modernization process is complete and Nature is gone for good' (Jameson 1991: ix). The open acceptance of a Worldwatch Institute by many governments only marks how far the infrastructuralization of the planet has advanced in its technoscientific merging of biosphere and technosphere.

Conclusions

When the Agenda 21 principles advanced at the 1992 Rio Environmental Conference have us recognize 'the integral and interdependent Nature of the Earth', they emphasize how the Earth is 'our home' (Grubb et al. 1993: 87). Instituting a worldwatch over the Earth, then, is a form of globalized 'home building', whose processes and progress should be monitored, as the Worldwatch Institute suggests, from two sets of now commonly-denominated books: the registers of economy as well as the ledgers of ecology. The Worldwatch Institute's vision of the Earth re-imagines it as a rational responsive household in which economic action commodifies everything, utilizes anything, wastes nothing, blending the natural and the social into a single but vast set of household accounts whose performativities must constantly weigh consumption against production at every level of analysis from suburbia to the stratosphere in balancing the terrestrial budgets of ecological modernization. As Baudrillard observes, 'it implies practical computation and conceptualization on the basis of a total abstraction, the notion of a world no longer given but instead produced – mastered, manipulated, inventoried, controlled: a world, in short, that has to be constructed' (Baudrillard 1996: 28–9).

The actions of the Worldwatch Institute are frameworks within which new environmentalized social relations of production and consumption can come alive by guarding habitat as the perfect captive source of *habitus* supplies. As Baudrillard observes,

> the great signified, the great referent Nature is dead, replaced by environment, which simultaneously designates and designs its death and the restoration of nature as simulation model. . . . [W]e enter a social

environment of synthesis in which a total abstract communication and an immanent manipulation no longer leave any point exterior to the system.

(Baudrillard 1981: 202).

Rendering wildlife, air, water, habitat, or Nature down into complex new systems of rare goods in the name of environmental protection, and then regulating the social consumption of them through ecological activism shows how green geopolitics are serving as systems of social control over political economy to reintegrate the intractable equations of Nature's wise use along ecological lines of rational consumption.

To re-imagine Mackinder's quest for relative efficiency in green geopolitical terms, one can refer to the Brundtland Report's opening line, 'in the middle of the twentieth century, we saw our planet from space for the first time', which ironically has become a self-fulfilling prophecy by exerting 'a greater impact on thought than did the Copernican revolution of the sixteenth century' (World Commission 1987:1). By seeing the Earth for the first time from space, its 'natural ecosystems' and 'environmental resource base' can be seen and studied in the geopolitical quest to optimize the economic processes of surviving and thriving. The Preamble to Agenda 21 reverberates the tenor of these thoughts for the Brundtland Report's future historians:

> Humanity stands at a defining moment in history. We are confronted with a perpetuation of disparities between and within nations, a worsening of poverty, hunger, ill health and illiteracy, and the continuing deterioration of the ecosystems on which we depend for our wellbeing. However, integration of environment and development concerns and greater attention to them will lead to the fulfilment of basic needs, improved living standards for all, better protected and managed ecosystems and a safer, more prosperous future. No nation can achieve this on its own; but together we can – in a global partnership for sustainable development.
>
> (Grubb *et al.* 1993: 83)

The Preamble to Agenda 21 could as easily be seen as the new charter for a green geopolitics inasmuch as its basic sentiments capture 'humanity's' managerial imperatives in the Earth's geopolitical administration, integrating environmental and developmental systems in a 'global partnership' to better protect all ecosystems and improve living standards for all through technoscientic terraforming.

Once ecology becomes a science of statist administration, its statistical attitudes diffuse through the numerical surveillance of Nature, or Earth and its non-human inhabitants, as well as the study of Culture, or society and its human members, giving us a geopolitics written in the Worldwatchers' green political codes. Government, and now, most importantly, corporate and/or statist ecology, preoccupies itself with 'the conduct of conduct', particularly in consumerism's 'buying of buying' or 'purchasing of purchasing'. *Habitus* is habitat, as any good product semanticist or psycho-demographer working for big businesses knows all too well. The ethical concerns of family, community and nation previously might have guided how conduct was to be conducted; yet, at this juncture, 'the environment' serves increasingly as the most decisive ground for normalizing each individual's behaviour (Odum 1975).

Ecology becomes in the work of a Worldwatch Institute one more formalized disciplinary mode of paying systematic 'attention to the processes of life . . . to invest life through and through' (Foucault 1980: 139) and thereby transforming all living things into biological inventories to develop transnational commerce. The tremendous explosion of global economic prosperity, albeit in highly skewed spatial distributions, after the 1970s energy crises would not have been as possible without the ecology of a Worldwatch Institute to guide ecologically 'the controlled insertion of bodies into the machinery of production and the adjustment of the phenomena of population to economic processes' (Foucault 1980: 141). An anantamo-politics for all of Earth's plants and animals now surfaces from within green geopolitics as strategic plans for a new kind of global administration in which environmentalizing resource managerialists acquire 'the methods of power capable of optimizing forces, aptitudes, and life in general without at the same time making them more difficult to govern' (Foucault 1980: 141).

In the end, one cannot adequately understand the mobilization of green discourses as geo-political policies in present-day regimes, like the United States of America, without seeing how their tactics and institutions assume these environmentalized modes of operation as part and parcel of ordinary practices of governance. Creating a continuous worldwatch over the Earth, despite Vice-President Gore's recent satellite-building directives to NASA, already is a standard operating procedure. Conservationist ethics, resource managerialism, and ecological rhetorics, then, all congeal in green geopolitics as an unusually cohesive power/knowledge formation, whose environmental discourses of relative efficiency in transnational policy organizations, like the Worldwatch Institute, become an integral element of this new geopolitical regime.

Bibliography

Baudrillard, J. (1981) *For A Critique of the Political Economy of the Sign*, St. Louis: Telos Press.
—— (1996) *The System of Objects*, New York: Verso.
Berry, T. (1988) *The Dream of Nature*, San Francisco: Sierra Club Books.
Bourdieu, P. (1984) *Distinction: A Social Critique of the Judgement of Taste*, Cambridge, Mass.: Harvard University Press.

Brown, L. (1981) *Building a Sustainable Society*, New York: Norton.

Brown, L. *et al.* (1995) *State of the World*, New York: Norton.

Brown, L., Flavin, C. and. Postel, S (1991) *Saving the Planet*, New York: Norton.

—— (1996) *State of the World*, New York: Norton.

—— (1997) *State of the World*, New York: Norton.

—— (1998) *State of the World*, New York: Norton.

—— (1999) *State of the World*, New York: Norton.

Christopher, W. (1996) 'Leadership for the next American century', *US Department of State Dispatch* 7, no. 4 22 January 1996: 12.

Clinton, B. (1995) 'Address at Freedom House, 6 October 1995', *Foreign Policy Bulletin* (November/December).

Commoner, B. (1990) *Making Peace with the Planet*, New York: Pantheon.

Cosgrove, D. E. (1994) 'Contested global visions: one-world, whole-earth, and the Apollo Space photographs', *Annals of the Association of American Geographers* 84 (2): 270–94.

Foucault, M. (1980) *The History of Sexuality*, vol. I: *An Introduction*, New York: Vintage.

Fukuyama, F. (1992) *The End of History and the Last Man*, New York: Free Press.

Gore, A. (1992) *Earth in the Balance: Ecology and the Human Spirit*, Boston: Houghton Mifflin.

Grubb, M. *et al.* (1993) *The Earth Summit Agreements: A Guide and Assessment*, London: EarthScan Publications.

Haraway, D. J. (1991) *Simians, Cyborgs, and Women: The Reinvention of Nature*, New York: Routledge.

Jameson, F. (1991) *Postmodernism, or the Cultural Logic of Late Capitalism*, Durham: Duke University Press.

Kaplan, R. (1996) *The Ends of the Earth*, New York: Random House.

Kearns G. (1993) '*Fin de siècle* geopolitics: Mackinder, Hobson and theories of global closure', 9–30 in P. J. Taylor (ed.) *The Political Geography of the Twentieth Century: A Global Analysis*, London: Belhaven.

Knickerbocker, B. (1997) 'Jane Lubchenco', *Christian Science Monitor*, 15 August: 1.

Luke T. W. (1993) 'Green consumerism: ecology and the ruse of recycling', 90–117 in J. Bennett and W. Chaloupka (eds) *In the Nature of Things: Language, Politics and the Environment*, Minneapolis: University of Minnesota Press.

—— (1995) 'On environmentality: geo-power and eco-knowledge in the discourses of contemporary environmentalism', *Cultural Critique* 31 (Fall 1995): 57–81.

—— (1996) 'Liberal society and cyborg subjectivity: the politics of environments, bodies, and nature', *Alternatives* 21: 1–30.

—— (1997) *Ecocritique: Contesting the Politics of Nature, Economy and Culture*, Minneapolis: University of Minnesota Press.

McKibben, B. (1989) *The End of Nature*, New York: Random House.

Mackinder, H. (1904) 'The geographical pivot of history', *Geographical Journal*, 23: 422.

Ó Tuathail, G. (1992) 'Putting Mackinder in his place: material transformation and myth', *Political Geography* 11(1): 100–18.

—— (1996) *Critical Geopolitics: The Politics of Writing Global Space*, Minneapolis: University of Minnesota Press.

Odum, E. (1975) *Ecology: The Link Between the Natural and Social Sciences*, second ed. New York: Holt, Rinehart and Winston.

Ryan, J. (1994) 'Visualizing imperialism: Halford Mackinder and the Colonial Office Visual Instruction Committee', *Ecumene* 1: 157–76.

Sale, K. (1985) *Dwellers in the Land: A Bioregional Vision*, San Francisco: Sierra Club Book.

Sawyer, K. (1998) 'The world turning in a click', *The Washington Post*, 13 March: A1, A13.

Simons, J. (1998) 'How a Vice President fills a cyber-cabinet: with Gore-Techs', *The Wall Street Journal*, 13 March: A1, A6.

Spykman, N. (1938) 'Geography and Foreign Policy II', *American Political Science Review* XXVII (April): 237.

World Commission on Environment and Development (1987) *Our Common Future*, Oxford: Oxford University Press.

WorldWatch Institute (1999) '*Mission Statement*': (http://www.worldwatch.org/).

EPILOGUE

Futures and possibilities

15

GEOPOLITICS, POLITICAL GEOGRAPHY AND SOCIAL SCIENCE

Peter J. Taylor

Let me begin by attempting to put geopolitics into some institutional perspective. As a subdiscipline, it is minuscule, albeit variable in size across countries and over time. For instance,compared to the similar discipline of International Relations (IR) with its university departments and numerous journals, geopolitics is indeed a very poor relation. In fact we can view geopolitics as the periphery of a periphery of a periphery: it has always had an uneasy relation with political geography, which in turn has been located at the edge of human geography, which in turn has never established itself within the core of social science.No doubt this is a rather cruel and crude analogy, but writers on geopolitics do not seem to have been intimidated by this intellectual location. They have discovered geopolitics in everything from high politics to popular culture so that the magnitude of their subject is out of all proportion to the size of their subdiscipline. This is the fascinating, unusual and perhaps unique condition of the intellectual project to which this book contributes.

There are two related reasons why geopolitics has rested uneasily within social science. First, the social sciences as a collective have developed in the twentieth century through what may be termed embedded statism (Taylor 1996b). By this I mean that states have provided the essential setting for analysis, whether as society, economy or political system. However, this setting has been taken for granted. Thus, until spurred by the Marxist critique after 1968, theoretical concern for understanding the state had been muted at best. Second, this statist perspective has had a strong propensity to focus upon one selected state per study thus neglecting the connections and linkages between states. By this I mean that the multiplicity of states as reflected in their inter-relations has been under-researched (Taylor 1995). This is rather an odd outcome in a century of world wars and much other inter-state violence, no wonder Michael Mann (1988: viii) has condemned social science as 'absurdly pacific'. Even in political science, International Relations has been the 'Cinderella' subdiscipline. Given this dual circumstance, if even IR's social science credentials have been in doubt, social science has hardly been fertile ground for the growth of geopolitical research.

But we have to be careful of generalizing about social science. The massive growth of this division of knowledge in the twentieth century has been an American-led phenomenon with sizeable British support. Although drawing on a nineteenth-century European heritage, the social pessimism of these 'founding fathers' of social theory was transmuted into an American optimism by mid-century (Taylor 1996a), one consequence of which was the loss of interest in the state as an abstract frame of reference. However this intellectual vacuum has a longer historical pedigree in first the English and then the US political tradition (Dyson 1980), which has no doubt been an important factor in the nature of the dominant Anglo-American practice of social science. In contrast, the various European traditions of studying social change have not banished the state into the intellectual wilderness, even as the domination of social science has spread beyond its heartland. Geopolitics has reflected this geographical and intellectual tension in the production of knowledge. Its two 'founding fathers' come from each tradition – the Englishman Mackinder contributes especially to the 'geo' which has largely been interpreted as 'global', and the German Ratzel contributes especially to the 'politics' which has largely been interpreted as the power politics of states. Thus, although this combination has not prospered in the world of social science, it has found more space to develop where the state has remained a cogent object of concern outside Anglo-America, as some of the chapters above have illustrated.

This has been the intellectual context within which geopolitics has remained minuscule and marginalized, tending to flower as little more than notorious war-mongering when 'pacific' social science can appear to be severely wanting, for instance German geopolitics leading up the Second World War and, to a much lesser extent, American geopolitics leading up to the Second Cold War (Dalby, 1990). Hence our expectations should be modest when we look at the achievements of geopolitics in understanding social change. We are told that geopolitics sprang from the phobias of the last *fin de siècle* with the implicit violent prescriptions of the subdiscipline being a product of this angst (see chapter two). With the benefit of hindsight we know now that the problems and concerns facing late nineteenth-century world society were solved, at least partially, in the twentieth century by the emergence of both Americanization and socialisms. However, we will look in vain for any consistent and sustained research effort in geopolitics, especially in Anglo-America, on the prime processes and relations of these 'solutions'. Topics such as the development of world consumerism, the spread of welfare states, the Cold War, or even the rise and fall of communism are largely neglected. The only theme where there has been any semblance of continual effort is in the narrow geostrategy area but even here the critical global issue of the nuclear threat to humanity has not figured prominently (the work of William Bunge (1973, 1982, 1989) is an honourable exception). Given this track record, is it worth pursuing the 'futures and possibilities' of geopolitical traditions in the current *fin de siècle* which is my brief here?

The answer is yes for three related reasons. First, geopolitics remains relatively minuscule but, with the massive growth in social science research in recent decades, geopolitics has become larger in absolute terms such that it has reached what might be considered a critical mass for meaningful reproduction. That is, there are enough scholars now working with the term to ensure that it remains a matter of serious intellectual concern in the foreseeable future. Second, geopolitics is moving into the core concerns of contemporary social science. This is symbolized by two publications in particular, John Agnew and Stuart Corbridge's *Mastering Space* (1995) and Gearóid Ó Tuathail's *Critical Geopolitics* (1996), with their political economy and Post-modern/post-structural focuses respectively. Each demonstrates the ways that geopolitical frameworks and analysis can examine contemporary global structures, and each opens up many further avenues to research. Third, the contemporary forms of social change provide a particularly conducive moment for geopolitics to move in from the intellectual periphery. Hence, the fact that recent writings address core social science questions is not a simple matter of subdisciplinary growth. Rather, states have become central to global restructuring in both their political *and* economic roles, a fact that has largely removed the opaque nature of their presence in social science. With states 'coming back in' within the social sciences, geopolitics can contribute something of substance and insight beyond dangerous war-mongering.

Evaluating the futures and possibilities for geopolitics depends fundamentally on how contemporary social change is interpreted. One popular post-Cold War interpretation argues that with the great political conflict concluded, old-style geopolitics is now being replaced by a new 'geo-economics' (Luttwak, 1990). This is, I think, quite an ahistorical view of recent social change. Economic competition between states is, of course, neither new nor novel in terms of its position at the centre of inter-state relations. The wave of protectionism in the late nineteenth century might well be recognized as 'geo-economics'; it is in fact often referred to as 'the new mercantilism' indicating another earlier 'geo-economic era', the original mercantilism of the seventeenth and eighteenth centuries. From this perspective, the current severe economic competition which is running parallel with a politically dominant 'lone superpower' (USA) is analogous to the new mercantilism of a century ago which coincided with a politically dominant global empire (UK). The two political giants were still leading economic powers, but had each lost their earlier hegemonic edges. Consequently, it is not geo-economics which makes the present conditions of social change particularly distinctive, but they are very distinctive in other ways.

According to Wallerstein (1996), there are three critical contemporary changes which are not explicable in cyclical terms, either hegemonic or Kondratieff. First, after several centuries or increasing state power and centralization, in recent decades states have been challenged as never before, both internally, for instance the neo-right 'rolling back the state', and externally, for instance the trans-state processes at the heart of globalization. It is this which has projected the state explicitly to centre stage in social science, with one consequence being that geopolitics looks much less peripheral than heretofore. Rather than being taken for granted, states are seen as part of the power constellations through which social change operates. No longer the implicit sole locus of power, states form an important mosaic of formal powers interacting with spaces of flows such as the world city network. There is nothing new about cross-cutting spaces of territories by spaces of flows (Arrighi 1994) but the fact that it can be viewed this way (Castells 1996) indicates just how embedded states have been in how we view the world. Now liberated from this mask, geopolitics, and political geography more broadly, have possibilities for contributing to understanding emerging space-power relations at all geographical scales.

The second distinctive contemporary change is the erosion of secularism. Related to the changing role of the states – secularism arose in part in state-church conflicts – the rise of new religious fundamentalisms is a surprise of the first order to modernists who had despatched religion to history as pre-modern superstition. There is no need to go down the Huntington (1993) route to appreciate the importance of this trend. It is not a matter of simple boundaries between 'civilizations', for the inter-mingling of peoples and their well-being is far too advanced for that. Rather we have another condition of cross-cutting spaces, of flows of information and images impinging on cultural mosaics at many different geographical scales. This is not new as some of the contributions above demonstrate. However, it is questions of globalization, by bringing different traditions, including modernity, into direct contact and conflict which has, in part, created violent fundamentalisms out of what had previously been simply different and discrete traditions (Giddens 1994: 100). In other words it is the inter-mingling, not the civilization blocks, which represent the new condition. The possibilities for geopolitics and political geographies of differences would seem to be almost endless in this area.

Closely related to the two previous distinctive changes, another surprise is the erosion of faith in science. For more than two centuries, modernity has been premised on beliefs in progress that rely upon technologies grounded in modern science. Contemporary concerns for where this is leading, especially ecologically and genetically, have meant that scientists have been converted from modern heroes to dangerous Frankensteins. Obviously science is not going to disappear but in its more contested situation it may well be that geopolitics has the eclectic tradition to manoeuvre this tricky intellectual minefield. Geopolitical traditions have typically confused epistemological positions: both Mackinder and Ratzel, for instance, have extolled scientific materialism as a means towards romantic nationalist ends. Interestingly, geo-economics practitioners seem to reverse this instrumentalism, using romantic nationalism to promote good materialist profits. The point is, however, that twenty-first century social change is probably not best studied through fundamentalist philosophical positions from the nineteenth century. Geopolitics' crude eclecticism, often rightly derided in the past, might just be a tradition to be cherished in the future.

Bibliography

Agnew, J. and Corbridge, S. (1995) *Mastering Space. Hegemony, Territory and International Political Economy*, London and New York: Routledge.

Arrighi, G. (1994) *The Long Twentieth Century*, London: Verso.

Bunge, W. (1973) 'The geography of human survival', *Annals, Association of American Geographers* 63: 275–95.

—— (1982) *The Nuclear War Atlas*, Victoriaville, Quebec: Society for Human Exploration.

——. (1989) 'Epilogue: our planet is big enough for peace but too small for war', 355–7 in R. J. Johnston and P. J. Taylor (eds) *World in Crisis*, Oxford: Blackwell.

Castells, M. (1996) *The Rise of the Network Society*, Oxford: Blackwell.

Dalby, S. (1990) *Creating the Second Cold War*, London: Pinter.

Dyson, K. H. F. (1980) *The State Tradition in Western Europe*, Oxford: Robertson.

Giddens, A. (1994) 'Living in a post-traditional society', in U. Beck, A. Giddens and S. Lash *Reflexive Modernization*, London: Polity.

Huntington, S. (1993) 'The clash of civilizations', *Foreign Affairs* 72: 22–49.

Luttwak, E. (1990) 'From geopolitics to geo-economics', *National Interest* 20: 17–24.

Mann, M. (1988) *States, War and Capitalism*, Oxford: Blackwell.

Ó Tuathail, G. (1996) *Critical Geopolitics*, Minneapolis: University of Minnesota Press.

Taylor, P. J. (1995) 'Beyond containers: internationally, interstateness, interterritoriality', *Progress in Human Geography* 19: 1–15.

——. (1996a) *The Way the Modern World Works: World Hegemony to World Impasse*, Chichester: Wiley.

—— (1996b) 'Embedded statism and the social sciences: opening up to new spaces', *Environment and Planning* a 28: 1917–28.

Wallerstein, I. (1996) 'The global picture, 1945–2025', 209–25 in T. K. Hopkins and I. Wallerstein (eds) *The Age of Transition. The Trajectory of the World-System, 1945–2025*, London: Zed.

16

IT'S THE LITTLE THINGS

Nigel Thrift

Nowadays, geopolitics tends to be constructed as a discourse which can be understood discursively; thus, 'although a twentieth century concept, geopolitics is best understood as a new field of discourse within the long-established domain of geopower, defined as the entwined historical development of geographical knowledge with state power and its imperatives of governmentality' (Ó Tuathail 1997: 39). In this afterword, I want to suggest that those working in geopolitics have, perhaps, taken this definition a little too literally, producing the world as discursive construction in a way which has problematic consequences for under-standing *how* (and therefore why) geopower is actually practised. In particular, I want to suggest that this exercise in literal transcription leaves out a lot of the 'little things' – 'mundane' objects like files, 'mundane' people like clerks and mundane words like 'the' – which are crucial to how the geopolitical is translated into being.

Now 'discourse' is, of course, a notoriously tricky term and (thankfully) I do not have the space to go into all the twists and turns of the debate. But one thing is certain, the assertion that discourse constitutes or inscribes its object, and that there is no outside to language, is now a commonplace. Even so, when it comes to it, surprisingly large numbers of discourse theorists do still work, explicitly or implicitly, with a linguistic model of sign and referent (or expression and content), and so tend to argue that the world cannot be considered to be a purely discursive construction. For example, that apparent Derridean, Drucilla Cornell (1992: 1) argues that 'very simply, reality is not interpretation all the way down'. Even the current doyenne of discourse theorists, Judith Butler (1993: 68), argues that 'it is not the case that everything including materiality, is always already language'. (Though Butler's position is more nuanced than this statement suggests; Kirby (1997: 126) argues that Butler still works with a model of the split – however performative – between sign and referent, thereby intervening only on to 'the surface of the surface because she assumes that the differentiation of contour-ing is given by/in signification'.)

In other words, in many works on discourse *representation* is continually smuggled back in and all kinds of displaced ghosts and phantoms therefore still haunt this work. Perhaps, then, it may be time to go back to Derrida who, in a sense, tried to mix up sign and referent through a notion of writing which moved beyond the semiological so that 'nature scribbles or flesh reads',

> some sense that word and flesh are utterly implicated, not because 'flesh' is actually a word that moderates the fact of what is being referred to, but because the entity of a word, the identity of a sign, the system of language, and the domain of culture – none of these are autonomously enclosed upon themselves. Rather they are all emergent within a force field of differentiations that has no exteriority in any final sense.
>
> (Kirby 1997: 126–7)

Or, again, though they differ very sharply in certain ways, when put this way one sees the same sense of impatience with the distinction between sign and referent and allegiance to the values of emergence in the work of Deleuze, the same yearning after a sense of ceaseless redistribution, of trajectories and becomings, of ceaseless process.[1]

But, too much of critical geopolitics, it seems to me, is closer to the first sense of discourse than the second. Discourse is a force of inscription on to a world, however subtly that inscription and its relation with the world is theorized. I want to argue that it is this representational relationship which is at the heart of the absences I want to highlight in critical geopolitics as currently practised, a relationship which makes it very difficult to take seriously in all the little things which contribute to geopolitical power, seriously, or, correspondingly the complete range of territorializations that are in play at any time. I think it is this representationalism which begins to explain, for example, the mesmerized attention to texts and images in critical geopolitics, and critical geography more generally, at the expense of other mobiles. I think it is this representationalism which explains some of the apparent attraction to theorists like Virilio with their rampant humanisms and technological determinisms.[2] I think it is this representationalism which explains some of the tendency to millennial either/or exaggerations in the over-dramatized style of writers like Bauman (Osborne 1998).

Let me try and suggest some of the ways in which this discursive model produces difficulties in understanding how geopower is deployed. I will consider three different arenas. The first of these is the object world. Producing geopower involves the construction and distribution of objects at a distance, objects which must stay stable if they are to be projected (Wise 1997). At a simple level, this might require the construction and distribution of subjects, material and weapons in a particular place at a particular time. At a more complex level, this must require the construction of numerous bureaucratic/logistical practices (which primarily shuffle paper), as opposed to theories about bureaucracy. This kind of work has been stimulated of late by interventions by Latour and others into the world of bureaucratic procedure. According to Latour (1993: 28), it is the production of 'facts' – 'mute but endowed or entrusted with meaning' through devices like text, drawings, numbers and symbols of many different kinds affiliated to deracinated particulars, making legible at a distance what was far away and indistinct, though being able to travel without altering their character or profile – which is crucial. Latour calls these travelling inscriptions 'immutable mobiles'. They are able to be combined with similarly constituted items through devices like lists, tables, charts, maps, statistics, the prose of reports, and so on to produce models of instances of times and spaces in other times and spaces (Poovey 1998). At these other sites, the accumulated information can be summed up through the application of further devices (files, indexes, catalogues, bibliographies, and so on) and then stored or put to work. In turn, the growth of an archive at these 'centres of calculation' produces a set of bureaucratic imaginaries with a voracious appetite for further information, a kind of grand fiction with undoubted 'reality' effects. For example, in the case of the British Empire, Richards (1993) shows that these sorts of effects were maintained by an imperial imaginary that envisaged the archive as an interface between knowledge and the state. Understood as such an interface, Richards (1993: 14) argues that 'the archive was less a specific institution than an entire epistemological complex for representing a comprehensive knowledge within the domain of Empire'. In turn, such archives produce their own imaginaries, full of

> hallucination produced by the archive itself, a kind of self-haunting that was generated at the interface between knowledge and the state. In this sense the phantoms of the archive that Richards points to are no more than a product of the archive state's mania for total knowledge – an obsession which served, in turn, to reconfigure the relationship between the realities of empire and the fantasies of global administration, and to propel it and its border disputes through the twentieth century.[3]
>
> (Hevia 1998: 256)

The resonances with the study of twentieth century geopolitics are clear here, I hope.[4] What we see is not some extraordinary, hermetically sealed apparatus of representation but an ordinary, hesitant set of *practices* shot through with doubts and phantoms, and the attention has therefore been shifted 'from the correspondence between representations and real objects and from the moral ambivalence of the colonial encounter to the material practices – census taking, map-making, ethnographic and natural history description, and the collecting and inventorying of these various inscriptional media' (Hevia 1998: 239–40). Practices make correspondences, but the correspondences are never perfect.

The second arena is the human body. I have argued elsewhere that the body eludes discursive inscription through special qualities of embodiment which fashion semblances and conjure social worlds (Thrift 1996). Yet these qualities have been generally lost in the study of critical geopolitics. This is even though embodiments must be extraordinarily important to critical geopolitics, not just as vectors of force, but as sites of performance in their own right.[5] The lack of attention to embodiment, except as inscribed representation, perhaps explains why critical geopolitics has produced so little work of an ethnographic bent. Then again, it may suggest why critical geopolitics has had such difficulty writing about certain aspects of gender. For example, women are written out of critical geopolitics though they are so often there (for example in the 'clerical labour force that has made the complex communications, money transfers and arms shipments possible' (Enloe 1989: 9)) because they do not figure in documents and texts. This case is well put in Christine Sylvester's recent (1998) remarkable paper 'Handmaids tales of Washington power'. Sylvester shows that women are crucial to the deployment of geopolitical power but are rarely represented as its handmaidens: therefore there 'has got to be a body/sex/gender power analysis of [geopolitical] modes of decision-making' (Sylvester 1998: 52). For example, in the White House during the Cuban missile crisis:

> When at the 'blank white spaces at the edge of print', one stumbles on Kennedy's purposive dismissal of 'women' he dined with at the Alsops, one realizes the potential abyss of body pre-programming into which one stares. When one actually sees a handmaid cited as ahead of her 'man' in entrapping Washington's currents, one realizes that there may be many counter-memories of events and decisions submerged in citations to 'men' in their limited and somewhat inflexible organizations . . . do not handmaids in the organization of the Cuban missile crisis warrant investigation on the grounds that they seep about in many different ways – in the historical organization of power, discourse and bodies? Without them, how can [models of decision-making] be decided?
>
> (Sylvester 1998: 59)

In other words, women's bodies have citational force even though they may not be cited and it follows that critical geopolitics needs to be 'repopulated' by the women who turn out to run such large parts of its apparatuses.

Let us finally come to one more arena: the arena of words. After all, here we might be thought to have the clearest example of representation at work, the word. Yet, what we do not get from critical geopolitics is a clear enough sense of how words function to bring about geopolitical change and it is not possible to do so as long as geopolitical forces continue to be framed as 'big' and 'commanding' (with all the masculine overtones). Some of the most potent geopolitical forces are, I suspect, lurking in the 'little' 'details' of people's lives, what is '"carried" in the specific variabilities of their activities' (Shotter and Billig 1998: 23), in the context of utterances. And these variabilities have immediate consequences. Thus,

> As Bakhtin notes, and as is confirmed by the work in conversational analysis, 'we sensitively catch the smallest shift in intonation, the slightest interruption of voices in anything of importance to us in another person's practical everyday discourse. All those verbal sideward glances, reservations, loopholes, hints, thrusts do not slip past our ear, are not foreign to our own lips' (Bakhtin 1984: 201). And we in turn show our stance to what they do or say also in fleeting bodily reactions, facial expressions, sounds of approval or disapproval, etc. Indeed, even in the continuously responsive unfolding of non-linguistic activities between ourselves and others – in a dance, in a handshake, or even a mere chance collision on the street – we are actively aware of whether the other's motives are, so to speak, 'in tune' or 'at odds' with ours. And in our sense of their attunement or lack of it, we can sense their attitude to us as intimate or distant, friendly or hostile, deferential or arrogant, and so on.
>
> (Shotter and Billig 1998: 23)

Thus, very effective work has been done in disciplines like anthropology and discursive psychology (Billig 1995, 1997) which attempts to provide a sense of how national identity and an accompanying geopolitical stance are inscribed through the smallest of details. Thus, for example, national identity is not accomplished in grand displays which incite the citizen to wave the flag in a fit of patriotic fervour. Instead, it goes on in more mundane citations:

> it is done unobtrusively on the margins of conscious awareness by little words such as 'the' and 'we'. Each day we read or hear phrases such as 'the prime minister', 'the nation', or the 'weather'. The definite article assumes deictically the national borders. It points to the homeland: but while we, the readers or listeners, understand the pointing, we do not follow it with our consciousness – it is a 'seen but unnoticed' feature of our everyday discourse.[6]
>
> (Shotter and Billig 1998: 20)

Such work goes some way towards understanding the deep, often unconscious aggressions which lurk behind so much geopolitical 'reasoning', which through small details build a sense of 'us' as not like 'them', and from which political programmes then flow as infractions are identified and made legible.[7]

In these few brief comments, I hoped to have outlined a parallel agenda for critical geopolitics, one still based on discourse, but on discourse understood in a broader way, and one which is less taken in by representation and more attuned to actual practices. In turn, such an agenda leads us away from interpretation of hyperbolic written and drawn rhetorics (which, I suspect, are often read by only a few and taken in by even fewer) towards the (I hesitate to say 'real') work of discourse, the constant hum of practices and their attendant territorializations within which geopower ferments and sometimes boils over.

Notes

1 Derrida and Deleuze's projects differ very sharply in certain respects. As Deleuze (1973, cited in Smith 1997: xv–xvi) once remarked:

> As for the method of deconstruction of texts, I see clearly what it is, I admire it a lot, but it has nothing to do with my own method. I do not present myself as a commentator on texts. For me, a text is merely a small cog in an extra-textual practice. It is not a question of commentating on the text by a method of deconstruction, or by a method of textual practice, or by other methods; it is a question of seeing what *use* it has in the extra-textual practice that belongs to the text.

The position is, given the two authors' affinities, closer to that of Foucault.

> In *The Archaeology of Knowledge*, repeated reference is made to typing the letters 'AZERT' on a piece of paper as an illustration of a statement – the basic unit of a discursive function – that simultaneously becomes such in the relation to something else. The typed letters correspond to the layout of keys on a standard French typewriter, but, for Foucault, it is not in their simple duplication that they become 'active' in a discursive sense. Nor is the relationship that of signification, where the typed characters correspond to material signifiers for the signified of the French typing series. It is instead in the way that the documentary fragment formed by the typed sheet of paper exists at an intersection of a secondary field of elements constituted by the typewriter keys, casing and fingers of the typist. The ordering of this non-discursive field supports and makes possible that of the discursive, which in its turn exerts a kind of holding power back on the elements.
> The further example of a plot on a graph makes this apparent: the graph of, say, intelligence quotient (IQ) itself refers to nothing. It marks the organization of a field of knowledge which quantifies any given population of industry according to a set of principles. It also numbers the intersection of a technology of psychological testing applied to a particular group of people in a specific place (school, work place, interview procedure). In a later reworking, Foucault goes on to show how the ordering of a filing cabinet with the proportion of files compiling information on individual cases

comprises the non-discursive conditions by which the 'contagion-discipline' strategy is enacted, involved with a very specific kind of judicial knowledge of criminology and the delinquent.

(Brown 1997: 71).

See also the comments on Latour that follow.

2 Egged on by breathless modernist notions of the new which ignore the fact that not everything can possibly be new and extraordinary in any way of life at any time.

3 These echo of these imaginaries and their urge to total knowledge has perhaps been best analysed by Sterling and Gibson (1995) and Stephenson (1996).

4 Thus, many phantoms are often mundane dis-locations. For example, the case of the nineteenth-century British Empire does not seem in its travails so different from many current forms:

> at the centre, the policy making apparatus of empire found it difficult to keep up with the activities of its subordinate parts: the work of summariz-ing and indexing information lagged behind the production of new knowledge by field-agents, while sub-centres of calculation might employ their own methods for indexing and storing information, independent of their related parts. As a result, new information was often literally misplaced or forgotten at cavernous collection sites and collateral storage facilities. Temporal lag and inconsistent filing meant that the centre occasionally found itself working at cross-purposes with its agents on the frontiers of empire. This disorganization led at times to moments of panic-stricken disavowal or conflict between field agents and policy makers in London.
>
> The archive state was also affected by a fundamental contradiction in its effort to create an epistemological empire. As decoding and recoding net-works spread, technological change accelerated processes of circulation and collection. Thus even as the archive state strove to produce total knowl-edge, even as more 'reality' captured by the coding networks was being reduced to two dimensional surfaces, knowledge itself appeared to be tran-sitory . . . This not only raised the spectre of epistemological relations, but suggested a kind of constant deferral of the ultimate goal of the project, making knowledge producers and indexers little more than worker bees in a seemingly endless process of collecting and filing.
>
> (Hevia 1998: 248)

5 A reference back to Butler is appropriate here, but note Thrift (in press).

6 And in simple little phrases such as 'the sun never sets on the British Empire'.

7 In other words, I want critical geopolitics to become something much closer to the kind of analysis of the new world order offered by Wise (1997) or even Linde-Laursen (1995).

Bibliography

Bakhtin, M. M. (1984) *Problems of Dostoevsky's Poetics*, Minneapolis: University of Minnesota Press.

Billig, M. (1995) *Banal Nationalism*, London: Sage.

—— (1997) 'Keeping the white queen in play', 149–57 in M. Fine *et al.* (eds) *Off-White*, London: Routledge.

Brown, S. D. (1997) 'In the wake of disaster: stress, hysteria and the event', 64–90 in K. Hetherington and R. Munro (eds) *Ideas of Difference*, Oxford: Blackwell.

Butler, J. (1993) *Bodies that Matter. On the Discursive Limits of 'Sex'*, New York: Routledge.

Cornell, D. (1992) *The Philosophy of the Limit*, New York: Routledge.

Enloe, C. (1989) *Bananas, Beaches and Babes: Making Feminist Sense of International Politics*, Berkeley: University of California Press.

Hevia, J. L. (1998) 'The archive state and the fear of pollution: from the Opium Wars to Fu Manchu', *Cultural Studies* 12, 234–69.

Kirby, V. (1997) *Telling Flesh: The Substance of the Corporeal*, London: Routledge.

Latour, B (1993) *We Have Never Been Modern*, Hassocks: Harvester.

Linde-Laursen, A. (1995) 'SMall differences, large issues. The making and re-making of a national border', *South Atlantic Quarterly* 94: 1123–43.

Ó Tuathail, G. (1997) 'At the end of geopolitics? Reflections on a plural problematic at the century's end', *Alternatives* 22: 35–55.

Osborne, T. (1998) *Aspects of Enlightenment. Social Theory and the Ethics of Truth*, London: UCL Press.

Poovey, M. (1998) *A History of the Modern Fact. Problems of Knowledge in the Sciences of Wealth and Society*, Chicago: Chicago University Press.

Richards, T. (1993) *The Imperial Archive*, London: Verso.

Shotter, J. and Billig, M. (1998) 'A Bakhtinian psychology: from out of the heads of individuals and into the dialogues between them', 13–29 in M. M. Bell and M. Gardiner (eds) *Bahktin and the Human Sciences. No Last Words*, London: Sage.

Smith, D. W. (1997) 'A life of pure immanence: Deleuze's critique et clinique project', xi–ivi in *Gilles Deleuze. Essays, Critical and Clinical*, Minneapolis: University of Minnesota Press.

Stephenson, N. (1996) *The Diamond Age*, Harmondsworth, Penguin.

Sterling, B. and Gibson, W. (1995) *The Difference Engine,* London: Orion.

Sylvester, C. (1998) 'Handmaids tales of Washington power: the abject and the real Kennedy White House', *Body and Society* 4: 39–66.

Thrift, N. J. (1996) *Spatial Formations*, London: Sage.

—— (in press) 'Afterwords', *Environment and Planning D. Society and Space* 18, in press.

Wise, J. M. (1997) *Exploring Technology and Social Space*, London: Sage.

INDEX

9 780415 172493